Birkhäuser

Advances in Mathematical Fluid Mechanics

Lecture Notes in Mathematical Fluid Mechanics

Complementing the *Advances in Mathematical Fluid Mechanics* series, the *Lecture Notes* subseries showcase topics of current interest in the field of mathematical fluid mechanics. Titles may present new results or novel perspectives on existing areas of research, or serve as extended surveys of certain topics. Traditional lecture notes covering core concepts for graduate/doctoral students are also included. All manuscripts are peer-reviewed to meet the highest standards of scientific literature.

Tomáš Bodnár • Giovanni P. Galdi •
Šárka Nečasová
Editors

Stochastics in Fluids

The 2023 Prague-Sum Workshop Lectures

Birkhäuser

Editors
Tomáš Bodnár
Department of Technical Mathematics
Czech Technical University in Prague
Prague 2, Czech Republic

Giovanni P. Galdi
Dept. of Mechanical Engineering
University of Pittsburgh
Pittsburgh, PA, USA

Šárka Nečasová
Institute of Mathematics
Czech Academy of Sciences
Prague 1, Czech Republic

ISSN 2297-0320 ISSN 2297-0339 (electronic)
Advances in Mathematical Fluid Mechanics
ISSN 2510-1374 ISSN 2510-1382 (electronic)
Lecture Notes in Mathematical Fluid Mechanics
ISBN 978-3-032-02298-1 ISBN 978-3-032-02299-8 (eBook)
https://doi.org/10.1007/978-3-032-02299-8

This book is published under the imprint Birkhäuser, www.birkhauser-science.com by the registered company Springer Nature Switzerland AG
The registered company address is: Gewerbestrasse 11, 6330 Cham, Switzerland

This book is dedicated to our dear colleague, passionate mathematician, and great friend, who will be missed in our workshops and our lives.

Petr Kučera
(1960–2025)

Preface

This book is based on lectures being presented at a workshop associated to the thematic summer school named "Stochastics in Fluids," which was held in Prague in August 2023 in the series "Prague-Sum" events started back in 2011.

The aim of this collection of contributions is to present the full versions of talks given during the workshop. The topics cover various aspects of processes and methods in contemporary physics as well as in mathematical modeling and analysis of fluids flows problems. All the contributions are written by experts in the corresponding fields, which makes this volume an excellent presentation of the state-of-the-art knowledge in this area.

The summer school series Prague-Sum is generously co-sponsored by the European Research Community on Flow, Turbulence and Combustion (ERCOFTAC) and by the RSJ Foundation. In addition, the "Stochastics in Fluids" thematic school was supported by Praemium Academiæ of Š. Nečasová and the Institute of Mathematics of the Czech Academy of Sciences by RVO:67985840.

Prague, Czech Republic — Tomáš Bodnár
Pittsburgh, PA, USA — Giovanni P. Galdi
Prague, Czech Republic — Šárka Nečasová
May 2025

Contents

Contributors

Tomáš Bodnár Institute of Mathematics, Czech Academy of Sciences, Prague, Czech Republic

Alexander Drobny Institute of Mathematics, Kassel University, Kassel, Germany

Elfriede Friedmann Institute of Mathematics, Kassel University, Kassel, Germany

Ondřej Kreml Institute of Mathematics, Czech Academy of Sciences, Prague, Czech Republic

Anna Lancmanová Faculty of Mechanical Engineering, Czech Technical University in Prague, Prague, Czech Republic

Pavel Ludvík Faculty of Science, Palacký University Olomouc, Olomouc, Czech Republic

Václav Mácha Institute of Mathematics of the Czech Academy of Sciences, Praha, Czech Republic

Šárka Nečasová Institute of Mathematics, Czech Academy of Sciences, Prague, Czech Republic

Andrzej F. Nowakowski School of Mechanical, Aerospace and Civil Engineering, University of Sheffield, Sheffield, UK

Ana Radoševič University of Zagreb, Faculty of Economics and Business, Zagreb, Croatia

Gianmarco Sperone Facultad de Matemáticas, Pontificia Universidad Católica de Chile, Santiago, Chile

Tong Tang Yang Zhou University, , Yangzhou, China

On Wild Solutions to the Compressible Euler System

Ondřej Kreml

Abstract In this chapter, we survey recent results concerning the existence and properties of so-called wild solutions to the compressible Euler system in two spatial dimensions. These solutions are constructed using the convex integration method developed in this context by De Lellis and Székelyhidi (Ann Math 170:1417–1436;2009, Arch Rational Mech Anal 195:225–260;2010). We introduce the Riemann problem for the compressible Euler system, classify its one-dimensional self-similar solutions, and summarize results related to the uniqueness and nonuniqueness of these solutions within the class of admissible weak solutions. We also discuss the existence of wild solutions for regular initial data and the failure of criteria based on maximal dissipation of energy to select physically relevant solutions.

1 The Compressible Euler System

The isentropic Euler system of gas dynamics is one of the typical examples of a hyperbolic system of conservation laws. In the absence of external forces, it reads as

$$\partial_t \rho + \operatorname{div}_x(\rho \mathbf{v}) = 0 \tag{1}$$

$$\partial_t(\rho \mathbf{v}) + \operatorname{div}_x(\rho \mathbf{v} \otimes \mathbf{v}) + \nabla_x p(\rho) = 0, \tag{2}$$

where the unknowns are the density $\rho(t, x)$ and the velocity $\mathbf{v}(t, x)$. The pressure $p(\rho)$ is a given function of the density. We assume that the pressure law satisfies $p' > 0$, which ensures that the systems (1)–(2) is hyperbolic on the regions where $\rho > 0$.

O. Kreml (✉)
Institute of Mathematics, Czech Academy of Sciences, Prague, Czech Republic
e-mail: kreml@math.cas.cz

T. Bodnár et al. (eds.), *Stochastics in Fluids*, Lecture Notes in Mathematical Fluid Mechanics, https://doi.org/10.1007/978-3-032-02299-8_1

Unless specified otherwise, we will consider the systems (1)–(2) in $Q_T := (0, T) \times \mathbb{R}^2$ and accompanied with the initial conditions

$$(\rho, \mathbf{v})(0, \cdot) = (\rho^0, \mathbf{v}^0)(\cdot) \qquad \text{in } \mathbb{R}^2. \tag{3}$$

Mathematical theory of systems of hyperbolic conservation laws remains far from fully understood in more than one spatial dimension. Even in the simplest cases, such as the one-dimensional Burgers equation, solutions to these systems exhibit a lack of regularity. Singularities, such as shock waves or contact discontinuities, can form in finite time—even when the initial data are smooth. This breakdown of regularity presents significant analytical challenges as it is unreasonable to work with strong solutions unless sufficiently short time interval is prescribed. However, in the absence of additional admissibility criteria, weak solutions to these systems are generally nonunique. In one space dimension, the notion of admissibility in the form of entropy inequalities was used to restore uniqueness; for more details about this nowadays classical theory, we refer, for example, to the monograph [11]. Recent results in multiple space dimensions demonstrate that similar approach there faces significant difficulties. As a result, many fundamental questions about existence, uniqueness, and stability of solutions remain open.

1.1 *Weak Solutions: Admissibility*

We recall the usual definition of a weak solution to (1)–(3).

Definition 1.1 (Weak Solution) A pair $(\rho, \mathbf{v}) \in L^\infty(Q_T)$ is a weak solution to (1)–(3) if the following identities hold for every test functions $\psi \in C_c^\infty([0, T) \times \mathbb{R}^2)$, $\boldsymbol{\varphi} \in C_c^\infty([0, T) \times \mathbb{R}^2)$:

$$\int_0^T \int_{\mathbb{R}^2} [\rho \partial_t \psi + \rho \mathbf{v} \cdot \nabla_x \psi] \, \mathrm{d}x \mathrm{d}t + \int_{\mathbb{R}^2} \rho^0(x) \psi(x, 0) \, \mathrm{d}x = 0 \tag{4}$$

$$\begin{aligned} &\int_0^T \int_{\mathbb{R}^2} [\rho \mathbf{v} \cdot \partial_t \boldsymbol{\varphi} + \rho (\mathbf{v} \otimes \mathbf{v}) : \nabla_x \boldsymbol{\varphi} + p(\rho) \mathrm{div}_x \boldsymbol{\varphi}] \, \mathrm{d}x \mathrm{d}t \\ &\quad + \int_{\mathbb{R}^2} \rho^0(x) \mathbf{v}^0(x) \cdot \boldsymbol{\varphi}(x, 0) \, \mathrm{d}x = 0 \end{aligned} \tag{5}$$

While there is abundance of entropies (in the sense of the theory of hyperbolic conservation laws) for the one-dimensional Euler system, in multiple space dimension, there is only one entropy which is the total energy of the system. Introducing the internal energy $\varepsilon : \mathbb{R}^+ \to \mathbb{R}$ through the law

$$p(\rho) = \rho^2 \varepsilon'(\rho), \tag{6}$$

we define the total energy as

$$\eta(\rho, \mathbf{v}) = \rho\varepsilon(\rho) + \frac{1}{2}\rho|\mathbf{v}|^2. \tag{7}$$

This motivates the definition of admissible weak solutions.

Definition 1.2 (Admissible Weak Solution) A bounded weak solution $(\rho, \mathbf{v})$ of (1)–(3) is admissible if it satisfies the following inequality for every nonnegative test function $\varphi \in C_c^\infty([0, T) \times \mathbb{R}^2)$:

$$\int_0^T \int_{\mathbb{R}^2} \left[\left(\rho\varepsilon(\rho) + \rho\frac{|\mathbf{v}|^2}{2} \right) \partial_t\varphi + \left(\rho\varepsilon(\rho) + \rho\frac{|\mathbf{v}|^2}{2} + p(\rho) \right) \mathbf{v} \cdot \nabla_x\varphi \right] \mathrm{d}x\mathrm{d}t$$
$$+ \int_{\mathbb{R}^2} \left(\rho^0(x)\varepsilon(\rho^0(x)) + \rho^0(x)\frac{|\mathbf{v}^0(x)|^2}{2} \right) \varphi(x, 0)\, \mathrm{d}x \geq 0 \tag{8}$$

Note that the admissibility in the form of (mathematical) entropy inequality (8) is nothing else than the natural (physical) energy inequality associated with the studied system.

In the whole chapter, we will use the following notation. Vectors are denoted by bold letters and their components are given by

$$\mathbf{a} = (a_1, a_2).$$

The space variable, despite being also a vector, is denoted simply as $x = (x_1, x_2)$. We use notation $B_r(x_0)$ for an open ball of radius r and center $x_0 \in \mathbb{R}^2$. Matrices are denoted by blackboard bold letters, and $\mathbb{I}$ denotes the identity matrix. Moreover, $\mathcal{S}_0^{2\times 2}$ denotes the set of symmetric traceless 2×2 matrices. The inequality $\mathbb{A} < \mathbb{B}$ between two 2×2 matrices is understood to mean that $\mathbb{B} - \mathbb{A}$ is a positive definite matrix. We use standard notation for Lebesgue spaces and spaces of continuous and smooth functions. If needed, we emphasize that the functions take values in X by using notation $C(\Omega; X)$, $L^p(\Omega; X)$, etc.

1.2 1D Riemann Problem

The one-dimensional Riemann problem plays a central role in the theory and numerical solution of hyperbolic conservation laws. It is the initial value problem where the data consists of two constant states separated by a single discontinuity. In the context of the studied two-dimensional problem (1)–(3), the Riemann initial data take the form

$$(\rho^0(x), \mathbf{v}^0(x)) := \begin{cases} (\rho_-, \mathbf{v}_-) \text{ if } x_2 < 0 \\ \\ (\rho_+, \mathbf{v}_+) \text{ if } x_2 > 0, \end{cases} \tag{9}$$

where $\rho_\pm$ and $\mathbf{v}_\pm$ are constants. Despite its apparent simplicity, the Riemann problem encapsulates the fundamental wave interactions that arise in hyperbolic systems, such as shocks, rarefactions, and contact discontinuities.

The 1D theory for the Riemann problem for systems (1)–(3) with initial data (9) is nowadays well understood; we refer to [11] and [8, Section 2] for more details. Rewriting the system using the momentum $\mathbf{m} = \rho\mathbf{v}$ instead of the velocity field and assuming that the solution does not depend on x_1, one obtains

$$\begin{aligned} &\partial_t \rho + \partial_{x_2} m_2 = 0 \\ &\partial_t m_1 + \partial_{x_2} \tfrac{m_1 m_2}{\rho} = 0 \\ &\partial_t m_2 + \partial_{x_2} \left(\tfrac{m_2^2}{\rho} + p(\rho) \right) = 0 \end{aligned} \tag{10}$$

with initial data

$$(\rho^0(x), \mathbf{m}^0(x)) := \begin{cases} (\rho_-, \mathbf{m}_-) = (\rho_-, \rho_- \mathbf{v}_-) \text{ if } x_2 < 0 \\ \\ (\rho_+, \mathbf{m}_+) = (\rho_+, \rho_+ \mathbf{v}_+) \text{ if } x_2 > 0. \end{cases} \tag{11}$$

Introducing the state vector $\mathbf{U} := (\rho, m_1, m_2)$, we observe that the system (10) takes the form

$$\partial_t \mathbf{U} + \partial_{x_2} F(\mathbf{U}) = 0.$$

Self-similar solutions to this system of equations are functions of a single variable $\frac{x_2}{t}$, i.e., $\mathbf{U}(t, x_2) = \tilde{\mathbf{U}}(\frac{x_2}{t})$.

The system (10) is strictly hyperbolic on the part of the state space where $\rho > 0$ (see [11]) since the Jacobian matrix $DF(\mathbf{U})$ has three real distinct eigenvalues

$$\lambda_1 = \frac{m_2}{\rho} - \sqrt{p'(\rho)}, \qquad \lambda_2 = \frac{m_2}{\rho}, \qquad \lambda_3 = \frac{m_2}{\rho} + \sqrt{p'(\rho)}, \tag{12}$$

where we recall that we assume $p' > 0$. The theory yields the existence of three Riemann invariants of the system (10) (for definitions, see [11]). More precisely, the functions

$$w_3 = \frac{m_2}{\rho} + \int_0^\rho \frac{\sqrt{p'(\tau)}}{\tau} d\tau, \qquad w_2 = \frac{m_1}{\rho}, \qquad w_1 = \frac{m_2}{\rho} - \int_0^\rho \frac{\sqrt{p'(\tau)}}{\tau} d\tau \tag{13}$$

are, respectively, (1- and 2-), (1- and 3-), and (2- and 3-) Riemann invariants.

It is not difficult to observe that the 2-characteristic family of the system (10) is linearly degenerate, i.e., $D\lambda_2 \cdot R_2 = 0$, where $R_2 = (0, 1, 0)^T$ is the associated right eigenvector to λ_2. In fact, the state variable m_1 appears only in the second equation of (10), and thus the system can be decoupled. It can be shown that if the initial data (11) of the Riemann problem satisfy $v_{-1} = v_{+1}$, i.e., $\frac{m_{-1}}{\rho_-} = \frac{m_{+1}}{\rho_+}$, then the first component of the velocity of the self-similar solution has to be equal to this constant, $v_1 = \frac{m_1}{\rho} = v_{-1} = v_{+1}$. If $v_{-1} \neq v_{+1}$, then the self-similar solution contains a 2−contact discontinuity on which the solution exhibits a jump from v_{-1} on the left to v_{+1} on the right while the density and second component of the velocity are constant across the contact discontinuity. The speed of this contact discontinuity is given by $v_{-2} = v_{+2}$, and the admissibility condition (8) across the discontinuity is satisfied as an equality.

The properties of the 1- and 3-characteristic families depend on the choice of the pressure law. If the pressure law satisfies condition

$$2p'(\rho) + \rho p''(\rho) > 0 \tag{14}$$

for all $\rho > 0$, then the 1− and 3−characteristic families are genuinely nonlinear. Typical example of pressure law satisfying (14) is the polytropic law

$$p(\rho) = K\rho^{\gamma} \tag{15}$$

with $K > 0$ and $\gamma \geq 1$.

However, there are pressure laws which violate (14). One of such examples, which we will work with later in this chapter, is the Chaplygin gas pressure law

$$p(\rho) = -\frac{1}{\rho}, \tag{16}$$

for which the 1- and 3-characteristic families are linearly degenerate.

1.2.1 Genuinely Nonlinear Pressure

Genuinely nonlinear characteristic families give rise to self-similar solutions in the form of admissible shocks or rarefaction waves, while linearly degenerate characteristic families give rise to self-similar solutions in the form of contact discontinuities.

Shocks need to satisfy the Rankine-Hugoniot jump conditions

$$s(r_R - r_L) = r_R u_R - r_L u_L \tag{17}$$

$$s(r_R u_R - r_L u_L) = r_R u_R^2 - r_L u_L^2 + p(r_R) - p(r_L), \tag{18}$$

where $s \in \mathbb{R}$ is the shock speed and $(r_L, \mathbf{u}_L = (a, u_L))$ and $(r_R, \mathbf{u}_R = (a, u_R))$ are the constant states on the left, resp. right side of the shock with $a \in \mathbb{R}$ being an arbitrary first component of the velocity which remains constant across the shock. The admissibility condition (8) on the shock translates to the inequality

$$(u_R - u_L)\left(p(r_L) + p(r_R) - 2r_L r_R \frac{\varepsilon(r_L) - \varepsilon(r_R)}{r_L - r_R}\right) \leq 0. \tag{19}$$

Lemma 1.1 *Let $p(\rho)$ satisfy* (14). *Then, it holds*

$$p(r_L) + p(r_R) - 2r_L r_R \frac{\varepsilon(r_R) - \varepsilon(r_L)}{r_R - r_L} > 0 \tag{20}$$

for any $r_L \neq r_R$, $r_L, r_R > 0$.

Proof Without loss of generality, we assume $r_R > r_L$, and we introduce variable $z = \frac{r_R}{r_L} > 1$. Using (6), we rewrite the expression on the left-hand side of (20) multiplied by $\frac{z-1}{r_L} > 0$ as

$$f(z) := (z-1)r_L\varepsilon'(r_L) + z^2(z-1)r_L\varepsilon'(r_L z) - 2z(\varepsilon(r_L z) - \varepsilon(r_L)).$$

We calculate first and second derivative of f to obtain

$$f'(z) = r_L\varepsilon'(r_L) + (3z^2 - 4z)r_L\varepsilon'(r_L z) + z^2(z-1)r_L^2\varepsilon''(r_L z) - 2(\varepsilon(r_L z) - \varepsilon(r_L))$$

and

$$\begin{aligned} f''(z) &= (6z-6)r_L\varepsilon'(r_L z) + (6z^2 - 6z)r_L^2\varepsilon''(r_L z) + z^2(z-1)r_L^3\varepsilon'''(r_L z) \\ &= \frac{z-1}{z}\left(6r_L z\varepsilon'(r_L z) + 6r_L^2 z^2\varepsilon''(r_L z) + r_L^3 z^3\varepsilon'''(r_L z)\right). \end{aligned}$$

However, (14) can be rewritten using (6) as

$$6\rho\varepsilon'(\rho) + 6\rho^2\varepsilon''(\rho) + \rho^3\varepsilon'''(\rho) > 0; \tag{21}$$

therefore, we conclude that $f''(z) > 0$ for all $z > 1$. Since $f(1) = f'(1) = 0$, we end up with $f(z) > 0$ for all $z > 1$, and the proof is finished. □

As a consequence of (19) and Lemma 1.1, one obtains that for pressure laws satisfying (14) admissible shocks need to satisfy $u_L \geq u_R$.

The characterization of rarefaction waves is provided by the statement that every i-Riemann invariant is constant along any i-rarefaction wave; see [11, Theorem 7.6.6].

With this, we can state the following Lemma providing the full wave analysis of (10)–(11) for pressure laws satisfying (14) and therefore characterization of types of self-similar solutions. In most generality, the self-similar solution consists of four constant states connected by three waves. On the left, the constant state coincides with the left part of the Riemann initial data, i.e., $(\rho_-, (\rho_- v_{-1}, \rho_- v_{-2}))$. This state is connected to constant state $(\rho_m, (\rho_m v_{-1}, \rho_m v_m))$ with a 1−wave, which is either a shock or a rarefaction wave. Then, a jump in the first component of the velocity occurs across the 2−contact discontinuity resulting in constant state $(\rho_m, (\rho_m v_{+1}, \rho_m v_m))$. Finally, a 3−wave, which is either a shock or a rarefaction wave, connects this state to $(\rho_+, (\rho_+ v_{+1}, \rho_+ v_{+2}))$.

Lemma 1.2 *Let $(\rho_\pm, \mathbf{v}_\pm)$ be given constant states such that $\rho_\pm > 0$, $\mathbf{v}_\pm = (v_{\pm 1}, v_{\pm 2})$, and let $p(\rho)$ satisfy* (14)*:*

(1) If

$$v_{+2} - v_{-2} \geq \int_0^{\rho_-} \frac{\sqrt{p'(\tau)}}{\tau} \mathrm{d}\tau + \int_0^{\rho_+} \frac{\sqrt{p'(\tau)}}{\tau} \mathrm{d}\tau, \tag{22}$$

then there exists a unique self-similar solution to the Riemann problem (10)–(11) *consisting of a 1−rarefaction wave, a 2−contact discontinuity, and a 3−rarefaction wave. The intermediate state is vacuum, i.e., $\rho_m = 0$.*

(2) If

$$\left| \int_{\rho_-}^{\rho_+} \frac{\sqrt{p'(\tau)}}{\tau} \mathrm{d}\tau \right| < v_{+2} - v_{-2} < \int_0^{\rho_-} \frac{\sqrt{p'(\tau)}}{\tau} \mathrm{d}\tau + \int_0^{\rho_+} \frac{\sqrt{p'(\tau)}}{\tau} \mathrm{d}\tau, \tag{23}$$

then there exists a unique self-similar solution to the Riemann problem (10)–(11) *consisting of a one-rarefaction wave, a 2-contact discontinuity, and a 3-rarefaction wave. The density of the intermediate state ρ_m and its second component of the velocity v_m are given as a unique solution of the system of equations*

$$v_{+2} - v_{-2} = \int_{\rho_m}^{\rho_-} \frac{\sqrt{p'(\tau)}}{\tau} \mathrm{d}\tau + \int_{\rho_m}^{\rho_+} \frac{\sqrt{p'(\tau)}}{\tau} \mathrm{d}\tau \tag{24}$$

$$v_m = v_{-2} + \int_{\rho_m}^{\rho_-} \frac{\sqrt{p'(\tau)}}{\tau} \mathrm{d}\tau. \tag{25}$$

(3) If $\rho_- > \rho_+$ and

$$-\sqrt{\frac{(\rho_- - \rho_+)(p(\rho_-) - p(\rho_+))}{\rho_- \rho_+}} < v_{+2} - v_{-2} < \int_{\rho_+}^{\rho_-} \frac{\sqrt{p'(\tau)}}{\tau} \mathrm{d}\tau, \tag{26}$$

then there exists a unique self-similar solution to the Riemann problem (10)–(11) *consisting of a 1-rarefaction wave, a 2-contact discontinuity, and an admissible 3-shock. The density of the intermediate state ρ_m and its second component of the velocity v_m are given as a unique solution of the system of equations*

$$v_{+2} - v_{-2} = \int_{\rho_m}^{\rho_-} \frac{\sqrt{p'(\tau)}}{\tau} \mathrm{d}\tau - \sqrt{\frac{(\rho_m - \rho_+)(p(\rho_m) - p(\rho_+))}{\rho_m \rho_+}} \tag{27}$$

$$v_m = v_{-2} + \int_{\rho_m}^{\rho_-} \frac{\sqrt{p'(\tau)}}{\tau} \mathrm{d}\tau. \tag{28}$$

(4) *If* $\rho_- < \rho_+$ *and*

$$-\sqrt{\frac{(\rho_+ - \rho_-)(p(\rho_+) - p(\rho_-))}{\rho_+\rho_-}} < v_{+2} - v_{-2} < \int_{\rho_-}^{\rho_+} \frac{\sqrt{p'(\tau)}}{\tau}\mathrm{d}\tau, \tag{29}$$

then there exists a unique self-similar solution to the Riemann problem (10)–(11) *consisting of an admissible* 1*-shock, a* 2*-contact discontinuity, and a* 3*-rarefaction wave. The density of the intermediate state* ρ_m *and its second component of the velocity* v_m *are given as a unique solution of the system of equations*

$$v_{+2} - v_{-2} = \int_{\rho_m}^{\rho_+} \frac{\sqrt{p'(\tau)}}{\tau}\mathrm{d}\tau - \sqrt{\frac{(\rho_m - \rho_-)(p(\rho_m) - p(\rho_-))}{\rho_m\rho_-}} \tag{30}$$

$$v_m = v_{-2} - \sqrt{\frac{(\rho_m - \rho_-)(p(\rho_m) - p(\rho_-))}{\rho_m\rho_-}}. \tag{31}$$

(5) *If*

$$v_{+2} - v_{-2} < -\sqrt{\frac{(\rho_+ - \rho_-)(p(\rho_+) - p(\rho_-))}{\rho_+\rho_-}}, \tag{32}$$

then there exists a unique self-similar solution to the Riemann problem (10)–(11) *consisting of an admissible* 1*-shock, a* 2*-contact discontinuity, and an admissible* 3*-shock. The density of the intermediate state* ρ_m *and its second component of the velocity* v_m *are given as a unique solution of the system of equations*

$$\begin{aligned} v_{+2} - v_{-2} = &-\sqrt{\frac{(\rho_m - \rho_-)(p(\rho_m) - p(\rho_-))}{\rho_m\rho_-}} \\ &-\sqrt{\frac{(\rho_m - \rho_+)(p(\rho_m) - p(\rho_+))}{\rho_m\rho_+}} \end{aligned} \tag{33}$$

$$v_m = v_{-2} - \sqrt{\frac{(\rho_m - \rho_-)(p(\rho_m) - p(\rho_-))}{\rho_m\rho_-}}. \tag{34}$$

(6) *If*

$$v_{+2} - v_{-2} = \left|\int_{\rho_-}^{\rho_+} \frac{\sqrt{p'(\tau)}}{\tau}\mathrm{d}\tau\right| \tag{35}$$

and $\rho_- \neq \rho_+$, then there exists a unique self-similar solution to the Riemann problem (10)–(11) *consisting of a rarefaction wave and a 2-contact discontinuity. If $\rho_- < \rho_+$, then the rarefaction wave is a 3−rarefaction wave, while for $\rho_- > \rho_+$ the rarefaction wave is a 1-rarefaction wave.*

(7) If

$$v_{+2} - v_{-2} = -\sqrt{\frac{(\rho_+ - \rho_-)(p(\rho_+) - p(\rho_-))}{\rho_+\rho_-}} \tag{36}$$

and $\rho_- \neq \rho_+$, then there exists a unique self-similar solution to the Riemann problem (10)–(11) *consisting of an admissible shock and a 2-contact discontinuity. If $\rho_- < \rho_+$, then the shock is a 1−shock, while for $\rho_- > \rho_+$ the shock is a 3-shock.*

(8) If $\rho_- = \rho_+$ and $v_{-2} = v_{+2}$, then there exists a unique self-similar solution to the Riemann problem (10)–(11) *consisting of a 2-contact discontinuity.*

In all cases, if $v_{-1} = v_{+1}$, then the 2−contact discontinuity does not appear in the solution.

Proof This is a consequence of the classical theory for hyperbolic conservation laws, and we refer to [11, Chapter 9] for general theory and to [8, Section 2] for application of this theory to the Euler system with power law pressure which generates genuinely nonlinear 1- and 3-characteristic families. □

1.2.2 Chaplygin Gas Pressure

We will not study the case when $2p'(\rho) + \rho p''(\rho) < 0$ in this chapter. While the 1- and 3-characteristic families of (10) are still genuinely nonlinear in this case, the rarefaction wave curves in the state space become concave which yields nonstandard behavior of simple waves.

We will however demonstrate some of this behavior on the case of the Chaplygin gas pressure law (16) which has the property $2p'(\rho) + \rho p''(\rho) = 0$. In this case, all three characteristic families are linearly degenerate, and therefore all simple waves are contact discontinuities. These discontinuities need to satisfy the Rankine-Hugoniot conditions (17)–(18). It is straightforward to deduce that the speeds of the contact discontinuities equal to

$$s_1 = \lambda_1 = \frac{m_2}{\rho} - \frac{1}{\rho}, \qquad s_2 = \lambda_2 = \frac{m_2}{\rho}, \qquad s_3 = \lambda_3 = \frac{m_2}{\rho} + \frac{1}{\rho}. \tag{37}$$

The energy is conserved across the contact discontinuities as is demonstrated by the fact that

$$p(r_L) + p(r_R) - 2r_L r_R \frac{\varepsilon(r_R) - \varepsilon(r_L)}{r_R - r_L} = 0 \tag{38}$$

for pressure law (16); see (19).

If there exists a bounded self-similar solution, then it in general consists of four constant states connected by three contact discontinuities. On the left, the constant state coincides with the left part of the Riemann initial data, i.e., $(\rho_-, (\rho_- v_{-1}, \rho_- v_{-2}))$. This state is connected to constant state $(\rho_m, (\rho_m v_{-1}, \rho_m v_m))$ with a 1−contact discontinuity. Then a jump in the first component of the velocity occurs across the 2−contact discontinuity resulting in constant state $(\rho_m, (\rho_m v_{+1}, \rho_m v_m))$. Finally, a 3−contact discontinuity connects this state to $(\rho_+, (\rho_+ v_{+1}, \rho_+ v_{+2}))$.

Lemma 1.3 *Let $(\rho_\pm, \mathbf{v}_\pm)$ be given constant states such that $\rho_\pm > 0$, $\mathbf{v}_\pm = (v_{\pm 1}, v_{\pm 2})$, and let $p(\rho)$ be given by* (16)*:*

(1) If

$$v_{+2} - v_{-2} > -\frac{1}{\rho_+} - \frac{1}{\rho_-} \quad \text{and} \quad v_{+2} - v_{-2} \neq \pm\left(\frac{1}{\rho_+} - \frac{1}{\rho_-}\right), \tag{39}$$

then there exists a unique self-similar solution to the Riemann problem (10)–(11) *consisting of a three contact discontinuities. The density of the intermediate state ρ_m and its second component of the velocity v_m are given as*

$$\frac{1}{\rho_m} = \frac{1}{2}(v_{+2} - v_{-2}) + \frac{1}{2}\left(\frac{1}{\rho_+} + \frac{1}{\rho_-}\right) \tag{40}$$

$$v_m = \frac{1}{2}(v_{+2} + v_{-2}) + \frac{1}{2}\left(\frac{1}{\rho_+} - \frac{1}{\rho_-}\right). \tag{41}$$

(2) If

$$v_{+2} - v_{-2} = \frac{1}{\rho_+} - \frac{1}{\rho_-}, \tag{42}$$

then there exists a unique self-similar solution to the Riemann problem (10)–(11) *consisting of a 1-contact discontinuity and a 2-contact discontinuity.*

(3) If

$$v_{+2} - v_{-2} = \frac{1}{\rho_-} - \frac{1}{\rho_+}, \tag{43}$$

then there exists a unique self-similar solution to the Riemann problem (10)–(11) *consisting of a 2-contact discontinuity and a 3-contact discontinuity.*

(4) If $\rho_- = \rho_+$ and $v_{-2} = v_{+2}$, then there exists a unique self-similar solution to the Riemann problem (10)–(11) *consisting of a 2-contact discontinuity.*

In all cases, if $v_{-1} = v_{+1}$*, then the* 2*-contact discontinuity does not appear in the solution. If*

$$v_{+2} - v_{-2} \le -\frac{1}{\rho_+} - \frac{1}{\rho_-}, \tag{44}$$

then the Riemann problem (10)–(11) *does not possess a self-similar solution belonging to the class of BV functions.*

Proof This is a consequence of standard wave analysis of the Riemann problem. See [2, Section 2] for more details. □

In the case (44), there exists a generalized measure-valued solution called a δ-shock. In order to describe it, we introduce a solution in the form

$$(\rho, (\rho v_1, \rho v_2))(t, x_2) = \begin{cases} (\rho_-, (\rho_- v_{-1}, \rho_- v_{-2})) & \text{for } x_2 < \sigma t \\ (\omega(t)\delta_{x_2 - \sigma t}, \xi, \sigma) & \text{for } x_2 = \sigma t \\ (\rho_+, (\rho_+ v_{+1}, \rho_+ v_{+2})) & \text{for } x_2 > \sigma t, \end{cases} \tag{45}$$

where the δ-shock speed σ is also the value of the second component of the velocity on the δ-shock, ξ is the value of the first component of the velocity on the δ-shock, and together with the density weight $\omega(t)$ they satisfy the generalized Rankine-Hugoniot conditions

$$\frac{\mathrm{d}\omega(t)}{\mathrm{d}t} = \sigma(\rho_+ - \rho_-) - (\rho_+ v_{+2} - \rho_- v_{-2}), \tag{46}$$

$$\frac{\mathrm{d}\omega(t)\xi}{\mathrm{d}t} = \sigma(\rho_+ v_{+1} - \rho_- v_{-1}) - (\rho_+ v_{+1} v_{+2} - \rho_- v_{-1} v_{-2}), \tag{47}$$

$$\frac{\mathrm{d}\omega(t)\sigma}{\mathrm{d}t} = \sigma(\rho_+ v_{+2} - \rho_- v_{-2}) - \left(\rho_+ v_{+2}^2 - \rho_- v_{-2}^2 - \frac{1}{\rho_+} + \frac{1}{\rho_-}\right). \tag{48}$$

We set the initial data $\omega(0) = 0$ and we define $\frac{1}{\rho} = 0$ for $x_2 = \sigma t$. System (46)–(48) can be solved directly, and its solution is

$$\omega(t) = (\rho_- v_{-2} - \rho_+ v_{+2})t, \tag{49}$$

$$\sigma = \frac{1}{2}(v_{+2} + v_{-2}), \tag{50}$$

$$\xi = \frac{1}{2}(v_{+1} + v_{-1}) \tag{51}$$

in the case $\rho_+ = \rho_-$ and

$$\omega(t) = At, \tag{52}$$

$$\sigma = \frac{\rho_+ v_{+2} - \rho_- v_{-2} + A}{\rho_+ - \rho_-}, \tag{53}$$

$$\xi = \frac{(\rho_+ v_{+1} - \rho_- v_{-1})A + \rho_- \rho_+ (v_{+2} - v_{-2})(v_{+1} - v_{-1})}{(\rho_+ - \rho_-)A} \tag{54}$$

with

$$A = \sqrt{\rho_+ \rho_- \left((v_{+2} - v_{-2})^2 - \left(\frac{\rho_- - \rho_+}{\rho_- \rho_+} \right)^2 \right)}$$

in the case $\rho_+ \neq \rho_-$.

The introduction of δ-shocks requires also refinement of definition of solutions. We refer to [2, Section 2] for details about the sense in which such solution satisfies the Eqs. (1)–(2) and in what sense the energy inequality (8) is satisfied.

2 The Convex Integration Scheme

The roots of convex integration trace back to the work of Nash in the 1950s, particularly his groundbreaking results on isometric embeddings of Riemannian manifolds [30]. Together with Kuiper [26], the authors showed that flexibility—rather than rigidity—can dominate in low regularity, allowing for wildly counterintuitive constructions. The method hinged on a geometric insight: if a differential relation is underdetermined and satisfies a certain convexity condition, then "flexible" solutions can be built by iteratively correcting approximate solutions with oscillatory perturbations.

This flexibility idea was then abstracted and generalized by Gromov [19] in the 1970s through his theory of h-principles, which formalized convex integration as a general method for solving differential relations.

The convex integration technique was introduced into the fluid dynamics context by De Lellis and Székelyhidi [13, 14] and then developed by others in various directions, the main one being the proof of the famous Onsager's conjecture for incompressible Euler equations [3, 21].

2.1 Main Ideas

The idea to apply the convex integration scheme deduced for the incompressible Euler system to achieve results for the compressible Euler system (1)–(2) was first used in [14] and then successfully applied in many other works. This technique provides solutions which contain oscillations in the velocity field while the density field remains intact; in other words, the density part of the solution is the same as the density part of the subsolution; for more details, see Sect. 2.2.

Markfelder [28] was the first to develop a fully compressible convex integration scheme to handle the system (1)–(2); the same author revisited this approach in [29] and developed a scheme able to generate also the energy and energy flux. Debiec, Skipper, and Wiedemann [15] while working on a related scheme observed that a fully compressible convex integration approach does not yield a larger Λ-convex hull, and therefore adding oscillations to the density field as well would not lead to more solutions than one obtains using the "incompressible" version of the convex integration technique. Therefore, the "incompressible" version, which we introduce here, is fully sufficient to obtain all possible convex integration results as long as we are interested in bounded solutions and not continuous or more regular solutions. For construction of Hölder continuous solutions to system (1)–(2), we refer to [18].

The core idea of the whole L^∞ construction is the following Lemma:

Lemma 2.1 *Let* $(\tilde{\mathbf{v}}, \tilde{\mathbb{U}}) \in \mathbb{R}^2 \times \mathcal{S}_0^{2\times 2}$ *and* $C > 0$ *be such that* $\tilde{\mathbf{v}} \otimes \tilde{\mathbf{v}} - \tilde{\mathbb{U}} < \frac{C}{2}\mathbb{I}$*. For any open set* $\Omega \subset \mathbb{R} \times \mathbb{R}^2$*, there are infinitely many maps* $(\underline{\mathbf{v}}, \underline{\mathbb{U}}) \in L^\infty(\mathbb{R} \times \mathbb{R}^2)$ *taking values in* $\mathbb{R}^2 \times \mathcal{S}_0^{2\times 2}$ *with the following properties:*

- $\underline{\mathbf{v}}$ *and* $\underline{\mathbb{U}}$ *vanish identically outside* Ω*.*
- $\mathrm{div}_x \underline{\mathbf{v}} = 0$ *and* $\partial_t \underline{\mathbf{v}} + \mathrm{div}_x \underline{\mathbb{U}} = 0$*.*
- $(\tilde{\mathbf{v}} + \underline{\mathbf{v}}) \otimes (\tilde{\mathbf{v}} + \underline{\mathbf{v}}) - (\tilde{\mathbb{U}} + \underline{\mathbb{U}}) = \frac{C}{2}\mathbb{I}$ *a.e. on* Ω*.*

Lemma 2.1 can be viewed as a special case of the following more general statement which will be useful in Sect. 4.2:

Lemma 2.2 *Let* $\Omega \subset \mathbb{R} \times \mathbb{R}^2$ *be an open set. Let* $(\tilde{\rho}, \tilde{\mathbf{m}}, \tilde{\mathbb{U}}, \tilde{C}) \in C(\Omega; \mathbb{R}^+ \times \mathbb{R}^2 \times \mathcal{S}_0^{2\times 2} \times \mathbb{R}^+)$ *such that*

$$\frac{\tilde{\mathbf{m}} \otimes \tilde{\mathbf{m}}}{\tilde{\rho}} - \tilde{\mathbb{U}} < \frac{\tilde{C}}{2}\mathbb{I} \tag{55}$$

pointwise in Ω*. Let us assume furthermore that* $\tilde{C}$ *is a bounded function on* Ω*. Then, there are infinitely many maps* $(\underline{\mathbf{m}}, \underline{\mathbb{U}}) \in L^\infty(\mathbb{R} \times \mathbb{R}^2, \mathbb{R}^2 \times \mathcal{S}_0^{2\times 2})$ *such that*

- $\underline{\mathbf{m}}$ *and* $\underline{\mathbb{U}}$ *vanish identically outside* Ω*.*
-

$$\mathrm{div}_x \underline{\mathbf{m}} = 0 \tag{56}$$

$$\partial_t \underline{\mathbf{m}} + \mathrm{div}_x \underline{\mathbb{U}} = 0. \tag{57}$$

- *It holds*

$$\frac{(\tilde{\mathbf{m}} + \underline{\mathbf{m}}) \otimes (\tilde{\mathbf{m}} + \underline{\mathbf{m}})}{\tilde{\rho}} - (\tilde{\mathbb{U}} + \underline{\mathbb{U}}) = \frac{\tilde{C}}{2}\mathbb{I} \tag{58}$$

a.e. in Ω*.*

The proof of Lemma 2.1 is a minor variant of the proofs of core Lemmas of De Lellis and Székelyhidi in [13, 14], and it was adapted to this formulation in [7]. Lemma 2.2 follows the same steps, uses some ideas introduced by Chiodaroli in [6], and was proved in [10]. Here, we just briefly mention key steps in the proof of Lemma 2.2.

Proof We define X_0 to be the space of $(\underline{\mathbf{m}}, \underline{\mathbb{U}}) \in C_c^\infty(\Omega; \mathbb{R}^2 \times \mathcal{S}_0^{2\times 2})$ which satisfy (56)–(57) and the pointwise inequality $\frac{(\tilde{\mathbf{m}}+\underline{\mathbf{m}})\otimes(\tilde{\mathbf{m}}+\underline{\mathbf{m}})}{\tilde{\rho}} - (\tilde{\mathbb{U}}+\underline{\mathbb{U}}) < \frac{\tilde{C}}{2}\mathbb{I}$. We then take the closure X of X_0 in the L^∞ weak* topology. Since X is a bounded (weakly*) closed subset of L^∞, we obtain that the L^∞ weak* topology is metrizable on X, hence producing a complete metric space (X, d). It suffices to show that on a residual set (in the sense of Baire category) (58) holds.

For any $N \in \mathbb{N}$, we define the map I_N as follows: to $(\underline{\mathbf{m}}, \underline{\mathbb{U}})$, we associate the corresponding restrictions of these maps to $(-N, N) \times B_N(0)$. We then consider I_N as a map from (X, d) to Y, where Y is the space $L^\infty((-N, N) \times B_N(0); \mathbb{R}^2 \times \mathcal{S}_0^{2\times 2})$ endowed with the strong L^2 topology. Then, as in [13, Lemma 4.5], I_N is a Baire-1 map, and hence, from a classical theorem in Baire category, its points of continuity form a residual set in X.

The proof is finished once we show that if $(\underline{\mathbf{m}}, \underline{\mathbb{U}})$ is a point of continuity of I_N, then (58) holds a.e. on $(-N, N) \times B_N(0)$, because the set of those maps at which all I_N are continuous is also a residual set.

The desired property follows from the following statement, which is up to a minor differences proved in [6, Lemma 4.5]:

If $(\underline{\mathbf{m}}, \underline{\mathbb{U}}) \in X_0$, then there is a sequence $(\mathbf{m}_k, \mathbb{U}_k) \subset X_0$ converging weakly* to $(\underline{\mathbf{m}}, \underline{\mathbb{U}})$ for which

$$\liminf_k \|\tilde{\mathbf{m}} + \mathbf{m}_k\|_{L^2(\Gamma)} \geq \|\tilde{\mathbf{m}} + \underline{\mathbf{m}}\|^2_{L^2(\Gamma)} + \beta \left(\int_\Gamma \tilde{C}\tilde{\rho}\, \mathrm{d}x\mathrm{d}t - \|\tilde{\mathbf{m}} + \underline{\mathbf{m}}\|^2_{L^2(\Gamma)} \right)^2,$$

where $\Gamma = (-N, N) \times B_N(0)$ and β depends only on Γ.

We usc this statement as follows: We fix a point $(\underline{\mathbf{m}}, \underline{\mathbb{U}}) \in X$ where I_N is continuous and assume by contradiction that (58) does not hold on Γ. Then there is a sequence $(\underline{\mathbf{m}}_k, \underline{\mathbb{U}}_k) \subset X_0$ converging weakly* to $(\underline{\mathbf{m}}, \underline{\mathbb{U}})$. Since the latter is a point of continuity for I_N, we then have that $\underline{\mathbf{m}}_k \to \underline{\mathbf{m}}$ strongly in $L^2(\Gamma)$. To each $(\underline{\mathbf{m}}_k, \underline{\mathbb{U}}_k)$, we find a sequence $\{(\mathbf{m}_{k,j}, \mathbb{U}_{k,j})\}$ such that

$$\liminf_j \|\tilde{\mathbf{m}} + \mathbf{m}_{k,j}\|_{L^2(\Gamma)} \geq \|\tilde{\mathbf{m}} + \underline{\mathbf{m}}_k\|^2_{L^2(\Gamma)} + \beta \left(\int_\Gamma \tilde{C}\tilde{\rho}\, \mathrm{d}x\mathrm{d}t - \|\tilde{\mathbf{m}} + \underline{\mathbf{m}}_k\|^2_{L^2(\Gamma)} \right)^2$$

and $(\mathbf{m}_{k,j}, \mathbb{U}_{k,j}) \rightharpoonup^\star (\underline{\mathbf{m}}_k, \underline{\mathbb{U}}_k)$. Then, a diagonal argument allows to conclude the existence of a sequence $(\mathbf{m}_{k,j(k)}, \mathbb{U}_{k,j(k)})$ which converges weakly* to $(\underline{\mathbf{m}}, \underline{\mathbb{U}})$ and such that

$$\liminf_k \|\tilde{\mathbf{m}} + \mathbf{m}_{k,j(k)}\|_{L^2(\Gamma)} \geq \|\tilde{\mathbf{m}} + \underline{\mathbf{m}}\|^2_{L^2(\Gamma)} + \beta \left(\int_\Gamma \tilde{C}\tilde{\rho}\, \mathrm{d}x\mathrm{d}t - \|\tilde{\mathbf{m}} + \underline{\mathbf{m}}\|^2_{L^2(\Gamma)} \right)^2$$

$$> \|\tilde{\mathbf{m}} + \underline{\mathbf{m}}\|^2_{L^2(\Gamma)}.$$

This contradicts the assumption that $(\underline{\mathbf{m}}, \underline{\mathbb{U}})$ is a point of continuity for I_N.

□

2.2 *Subsolutions*

We start with a definition of fan partition, which was introduced in [7] and its generalization from [10].

Definition 2.1 (Fan Partition) A fan partition of $(0, \infty) \times \mathbb{R}^2$ consists of finitely many open sets $P_-, P_1, \ldots, P_N, P_+$ of the following form,

$$P_- = \{(t, x) : t > 0 \quad \text{and} \quad x_2 < \nu_- t\} \tag{59}$$

$$P_+ = \{(t, x) : t > 0 \quad \text{and} \quad x_2 > \nu_+ t\} \tag{60}$$

$$P_i = \{(t, x) : t > 0 \quad \text{and} \quad \nu_{i-1} t < x_2 < \nu_i t\}, \tag{61}$$

where $\nu_- = \nu_0 < \nu_1 < \ldots < \nu_N = \nu_+$ is an arbitrary collection of real numbers.

Definition 2.2 (Generalized Fan Partition) A generalized fan partition of $(0, \infty) \times \mathbb{R}^2$ consists of finitely many sets $P_-, P_1, \ldots, P_N, P_+$ such that

$$P_- = \{(t, x) : t > 0 \quad \text{and} \quad x_2 < \widetilde{\nu}_-(t)\} \tag{62}$$

$$P_+ = \{(t, x) : t > 0 \quad \text{and} \quad \widetilde{\nu}_+(t) < x_2\} \tag{63}$$

$$P_i = \{(t, x) : t > 0 \quad \text{and} \quad \widetilde{\nu}_{i-1}(t) < x_2 < \widetilde{\nu}_i(t)\}, \tag{64}$$

where $\widetilde{\nu}_i(t)$ is a collection of continuous functions satisfying $\widetilde{\nu}_-(t) = \widetilde{\nu}_0(t) < \widetilde{\nu}_1(t) < \ldots < \widetilde{\nu}_N(t) = \widetilde{\nu}_+(t)$ for all $t > 0$ and $\widetilde{\nu}_i(0) = 0$ for all $i = 0, \ldots, N$.

Now, we can define admissible subsolutions. The first definition is from [7] while the second one from [10].

Definition 2.3 (Admissible Fan Subsolutions) An *admissible fan subsolution* to the compressible Euler system (1)–(3) with initial data (9) is a triple $(\overline{\rho}, \overline{\mathbf{v}}, \overline{\mathbb{U}})$:

$(0,\infty)\times\mathbb{R}^2 \to (\mathbb{R}^+,\mathbb{R}^2,\mathcal{S}_0^{2\times 2})$ of piecewise constant functions satisfying the following requirements:

- There is a fan partition $P_-, P_1, \ldots, P_N, P_+$ of $(0,\infty)\times\mathbb{R}^2$ such that

$$(\overline{\rho},\overline{\mathbf{v}},\overline{\mathbb{U}}) = \sum_{i=1}^{N}(\rho_i,\mathbf{v}_i,\mathbb{U}_i)\chi_{P_i} + (\rho_-,\mathbf{v}_-,\mathbb{U}_-)\chi_{P_-} + (\rho_+,\mathbf{v}_+,\mathbb{U}_+)\chi_{P_+}$$

  where $\rho_i, \mathbf{v}_i,$ and $\mathbb{U}_i$ are constants with $\rho_i > 0$, $\mathbb{U}_\pm = \mathbf{v}_\pm\otimes\mathbf{v}_\pm - \frac{1}{2}|\mathbf{v}_\pm|^2\mathbb{I}$.
- For every $i \in \{1,\ldots,N\}$, there exists a positive constant C_i such that

$$\mathbf{v}_i\otimes\mathbf{v}_i - \mathbb{U}_i < \frac{C_i}{2}\mathbb{I}. \tag{65}$$

- The triple $(\overline{\rho},\overline{\mathbf{v}},\overline{\mathbb{U}})$ solves the following system in the sense of distributions:

$$\partial_t\overline{\rho} + \operatorname{div}_x(\overline{\rho}\,\overline{\mathbf{v}}) = 0 \tag{66}$$

$$\partial_t(\overline{\rho}\,\overline{\mathbf{v}}) + \operatorname{div}_x\left(\overline{\rho}\,\overline{\mathbb{U}}\right) + \nabla_x\left(p(\overline{\rho}) + \frac{1}{2}\left(\sum_i C_i\rho_i\chi_{P_i} + \overline{\rho}|\overline{\mathbf{v}}|^2\chi_{P_+\cup P_-}\right)\right) = 0 \tag{67}$$

- The triple $(\overline{\rho},\overline{\mathbf{v}},\overline{\mathbb{U}})$ satisfies the following inequality in the sense of distributions:

$$\partial_t\left(\overline{\rho}\varepsilon(\overline{\rho})\right) + \operatorname{div}_x\left[\left(\overline{\rho}\varepsilon(\overline{\rho}) + p(\overline{\rho})\right)\overline{\mathbf{v}}\right] + \partial_t\left(\overline{\rho}\frac{|\overline{\mathbf{v}}|^2}{2}\chi_{P_+\cup P_-}\right) \tag{68}$$
$$+ \operatorname{div}_x\left(\overline{\rho}\frac{|\overline{\mathbf{v}}|^2}{2}\overline{\mathbf{v}}\chi_{P_+\cup P_-}\right) + \sum_{i=1}^{N}\left[\partial_t\left(\rho_i\frac{C_i}{2}\chi_{P_i}\right) + \operatorname{div}_x\left(\rho_i\,\overline{\mathbf{v}}\frac{C_i}{2}\chi_{P_i}\right)\right] \le 0$$

In Sect. 4.2, we will study the system (1)–(2) with initial data

$$(\rho^0,\mathbf{v}^0) = \begin{cases} (\rho_-^0,\mathbf{v}_-^0) \text{ if } x_2 < 0 \\ (\rho_+^0,\mathbf{v}_+^0) \text{ if } x_2 > 0, \end{cases} \tag{69}$$

where $(\rho_\pm^0,\mathbf{v}_\pm^0)$ are C^1 functions on $(-\infty,0)$ and $(0,\infty)$, respectively. For this initial data, we define generalized admissible fan subsolutions as follows:

Definition 2.4 (Generalized Admissible Fan Subsolution) A generalized admissible fan subsolution to the compressible Euler system (1)–(3) with initial data (69) is a quadruple of piecewise continuous functions $(\overline{\rho}, \overline{\mathbf{m}}, \overline{\mathbb{U}}, \overline{C}) : (0, \infty) \times \mathbb{R}^2 \to (\mathbb{R}^+, \mathbb{R}^2, \mathcal{S}_0^{2\times 2}, \mathbb{R}^+)$ satisfying the following properties:

- There exists a generalized fan partition P_-, P_1, P_+ of $(0, \infty) \times \mathbb{R}^2$ such that

$$(\overline{\rho}, \overline{\mathbf{m}}, \overline{\mathbb{U}}, \overline{C}) = \sum_{i=-,1,+} (\rho_i, \mathbf{m}_i, \mathbb{U}_i, C_i)\chi_{P_i}, \tag{70}$$

  where $(\rho_i, \mathbf{m}_i, \mathbb{U}_i, C_i)$ are continuous functions on P_i, $\mathbb{U}_\pm = \frac{\mathbf{m}_\pm \otimes \mathbf{m}_\pm}{\rho_\pm} - \frac{1}{2}\frac{|\mathbf{m}_\pm|^2}{\rho_\pm}\mathbb{I}$, $C_\pm = \frac{|\mathbf{m}_\pm|^2}{\rho_\pm}$ and

$$\lim_{t\to 0^+} (\rho_\pm, \mathbf{m}_\pm)(t, x) = (\rho_\pm^0, \rho_\pm^0 \mathbf{v}_\pm^0)(x) \qquad \text{for } \pm x_2 > 0. \tag{71}$$

- It holds

$$\frac{\mathbf{m}_1 \otimes \mathbf{m}_1}{\rho_1} - \mathbb{U}_1 < \frac{C_1}{2}\mathbb{I} \tag{72}$$

  pointwise in P_1.
- The quadruple $(\overline{\rho}, \overline{\mathbf{m}}, \overline{\mathbb{U}}, \overline{C})$ solves the following system of partial differential equations in the sense of distributions

$$\partial_t \overline{\rho} + \mathrm{div}_x \overline{\mathbf{m}} = 0 \tag{73}$$

$$\partial_t \overline{\mathbf{m}} + \mathrm{div}_x \overline{\mathbb{U}} + \nabla_x \left(p(\overline{\rho}) + \frac{\overline{C}}{2} \right) = 0. \tag{74}$$

- There exists $K(t) : (0, \infty) \to \mathbb{R}^+$ such that for all $(x, t) \in P_1$ it holds

$$\left(\varepsilon(\rho_1) + \frac{p(\rho_1)}{\rho_1} + \frac{C_1}{2\rho_1} \right)(x, t) = K(t), \tag{75}$$

  and the following inequality is satisfied in the sense of distributions

$$\partial_t \left(\overline{\rho}\varepsilon(\overline{\rho}) + \frac{\overline{C}}{2} \right) + \mathrm{div}_x \left(\left(\overline{\rho}\varepsilon(\overline{\rho}) + p(\overline{\rho}) + \frac{\overline{C}}{2} \right) \frac{\overline{\mathbf{m}}}{\overline{\rho}} \right) \le 0. \tag{76}$$

The problem of finding infinitely many admissible weak solution reduces to finding one admissible subsolution as is demonstrated by the following Propositions:

Proposition 2.1 *Let p be any C^1 function and $(\rho_\pm, \mathbf{v}_\pm)$ be such that there exists at least one admissible fan subsolution $(\overline{\rho}, \overline{\mathbf{v}}, \overline{\mathbb{U}})$ of* (1)–(3) *with initial data* (9). *Then there are infinitely many bounded admissible solutions $(\rho, \mathbf{v})$ to* (1)–(3) *with initial data* (9) *such that $\rho = \overline{\rho}$.*

Proposition 2.2 *Let p be any C^1 function and $(\rho^0, \mathbf{v}^0)$ be such that there exists at least one admissible generalized fan subsolution $(\overline{\rho}, \overline{\mathbf{m}}, \overline{\mathbb{U}}, \overline{C})$ to the Euler equations* (1)–(3) *with initial data* (69). *Then there exist infinitely many bounded admissible weak solutions $(\rho, \mathbf{v})$ to* (1)–(3) *such that $\rho = \overline{\rho}$.*

Both Propositions 2.1 and 2.2 are proved using Lemma 2.1 (Lemma 2.2, resp.) in domains P_i $(i = 1, \dots, N-1)$ (or P_1, resp.) of the associated fan partition. It is a matter of direct computation to check that $(\overline{\rho}, \overline{\mathbf{v}} + \underline{\mathbf{v}})$ is an admissible weak solution in case of Proposition 2.1. In case of Proposition 2.2, $(\overline{\rho}, \frac{\overline{\mathbf{m}}+\underline{\mathbf{m}}}{\overline{\rho}})$ is a weak solution, and we use (75) in order to conclude that the admissibility condition holds, as

$$
\begin{aligned}
&\operatorname{div}_x \left(\left(\rho\varepsilon(\rho) + p(\rho) + \frac{C}{2} \right) \frac{\underline{\underline{\mathbf{m}}}}{\rho} \right) \\
&= \operatorname{div}_x \left(\left(\rho_1\varepsilon(\rho_1) + p(\rho_1) + \frac{C_1}{2} \right) \frac{\underline{\underline{\mathbf{m}}}}{\rho_1} \right) = K(t)\operatorname{div}_x \underline{\underline{\mathbf{m}}} = 0.
\end{aligned}
$$

For more details, see [7, Section 3] and [10, Section 3]. In the remainder of this chapter, we will call solutions obtained by the convex integration technique as a result of Proposition 2.1 or 2.2 wild solutions.

3 Wild Solutions for Riemann Initial Data

The question of uniqueness or nonuniqueness of the 1D self-similar solution to the Riemann problem among all admissible weak solutions to the multidimensional problem has been studied extensively recently. The issue can be split into three categories.

3.1 Rarefaction Waves

First results of this type concerned uniqueness of rarefaction waves; these results were formulated by Chen and Chen [4] and later by Feireisl and Kreml [17]. Both these results use the relative energy method in order to prove the following type of theorem:

Theorem 3.1 *Let $p(\rho)$ be an increasing convex function. Let $(\tilde{\rho}, \tilde{\mathbf{v}})$ be a self-similar solution to* (1)–(3) *with initial data* (9) *consisting only of rarefaction waves. Let $(\rho, \mathbf{v})$ be an admissible weak solution to the same problem emanating from the same initial data. Then*

$$\rho = \tilde{\rho} \quad \textit{and} \quad \mathbf{v} = \tilde{\mathbf{v}}.$$

The relative energy method is based on the quantity

$$\mathcal{E}(\rho, \mathbf{v}|r, \mathbf{u}) = \frac{1}{2}\rho|\mathbf{v} - \mathbf{u}|^2 + \rho(\varepsilon(\rho) - \varepsilon(r)) - r\varepsilon'(r)(\rho - r)$$
$$= \frac{1}{2}\rho|\mathbf{v} - \mathbf{u}|^2 + (H(\rho) - H(r) - H'(r)(\rho - r)),$$

where $H(r) = r\varepsilon(r)$, which is used to measure a distance between a weak solution $(\rho, \mathbf{v})$ and a couple of sufficiently smooth test functions $(r, \mathbf{u})$. Then an inequality for $\int_{\mathbb{R}^2} \mathcal{E}(\rho, \mathbf{v}|r, \mathbf{u})(\tau, x)\,\mathrm{d}x$ is deduced by means of $\int_{\mathbb{R}^2} \mathcal{E}(\rho, \mathbf{v}|r, \mathbf{u})(0, x)\,\mathrm{d}x$ and a remainder term $\int_0^\tau \int_{\mathbb{R}^2} R(\rho, r, \mathbf{v}, \mathbf{u})\,\mathrm{d}x\mathrm{d}t$.

Using the rarefaction wave solution $(\tilde{\rho}, \tilde{\mathbf{v}})$ as the test function $(r, \mathbf{u})$ and estimating the remainder term yield the conclusion by means of the Gronwall lemma.

Interestingly enough, both [4] and [17] use the assumption that the pressure is convex. Indeed, the remainder term produces on the right-hand side of the relative energy inequality a term of the type

$$-\int_0^\tau \int_{\mathbb{R}^2} \big(p(\rho) - p'(\tilde{\rho})(\rho - \tilde{\rho}) - p(\tilde{\rho})\big)\partial_{x_2}\tilde{v}_2\,\mathrm{d}x\mathrm{d}t.$$

The convexity of the pressure together with the fact that the velocity component of the rarefaction wave solution is nondecreasing makes this term nonpositive, and the argument works.

This however leaves a gap between the pressures for which uniqueness of rarefaction waves is proved and all genuinely nonlinear pressures satisfying (14) allowing for standard behavior of simple waves. One example of a genuinely nonlinear pressure law for which the method of [4, 17] does not provide uniqueness of rarefaction waves is

$$p(\rho) = \ln(1 + \rho).$$

3.2 Shocks

The study of nonuniqueness of admissible weak solutions for Riemann data allowing the formation of a shock was started in [7]. There however the Riemann problem was used simply as a tool to construct nonuniqueness for Lipschitz initial data, more on this topic in Sect. 4.1. This result however motivated further research in this direction, and Chiodaroli and Kreml [8] showed that for pressure laws $p(\rho) = \rho^\gamma$, $\gamma \geq 1$ and for any Riemann data (9) satisfying $v_{-1} = v_{+1}$ which give rise to the self-similar solution consisting of two shocks (see point (5) in Lemma 1.2) there exist infinitely many admissible weak solutions to the Euler system (1)–(2).

Based on Proposition 2.1, it is enough to show that for this Riemann initial data there exists a single admissible fan subsolution. It turns out that it is enough to work with a fan partition consisting of three sets P_-, P_1, and P_+. Since the fan subsolution is piecewise constant, Eqs. (66)–(67) together with inequalities (65) and (68) reduce to a system of six algebraic equations and four inequalities for eight unknowns $\nu_- < \nu_+, C > 0, \rho_1 > 0, \mathbf{v}_1 = (\alpha, \beta)$, and $\mathbb{U}_1$ which has two independent components $(\mathbb{U}_1)_{11} = \gamma_1, (\mathbb{U}_1)_{12} = \gamma_2$; see [8, Proposition 4.1]. The arising system of equations and inequalities consists of the following:

- Rankine-Hugoniot conditions on the left interface:

$$\nu_-(\rho_- - \rho_1) = \rho_- v_{-2} - \rho_1 \beta \tag{77}$$

$$\nu_-(\rho_- v_{-1} - \rho_1 \alpha) = \rho_- v_{-1} v_{-2} - \rho_1 \gamma_2 \tag{78}$$

$$\nu_-(\rho_- v_{-2} - \rho_1 \beta) = \rho_- v_{-2}^2 + \rho_1 \gamma_1 + p(\rho_-) - p(\rho_1) - \rho_1 \frac{C}{2} \tag{79}$$

- Rankine-Hugoniot conditions on the right interface:

$$\nu_+(\rho_1 - \rho_+) = \rho_1 \beta - \rho_+ v_{+2} \tag{80}$$

$$\nu_+(\rho_1 \alpha - \rho_+ v_{+1}) = \rho_1 \gamma_2 - \rho_+ v_{+1} v_{+2} \tag{81}$$

$$\nu_+(\rho_1 \beta - \rho_+ v_{+2}) = -\rho_1 \gamma_1 - \rho_+ v_{+2}^2 + p(\rho_1) - p(\rho_+) + \rho_1 \frac{C}{2} \tag{82}$$

- Subsolution condition:

$$\alpha^2 + \beta^2 < C \tag{83}$$

$$\left(\frac{C}{2} - \alpha^2 + \gamma_1\right)\left(\frac{C}{2} - \beta^2 - \gamma_1\right) - (\gamma_2 - \alpha\beta)^2 > 0 \tag{84}$$

- Admissibility condition on the left interface:

$$\nu_-(\rho_- \varepsilon(\rho_-) - \rho_1 \varepsilon(\rho_1)) + \nu_- \left(\rho_- \frac{|v_-|^2}{2} - \rho_1 \frac{C}{2}\right)$$
$$\leq [(\rho_- \varepsilon(\rho_-) + p(\rho_-)) v_{-2} - (\rho_1 \varepsilon(\rho_1) + p(\rho_1))\beta] + \left(\rho_- v_{-2} \frac{|v_-|^2}{2} - \rho_1 \beta \frac{C}{2}\right); \tag{85}$$

- Admissibility condition on the right interface:

$$
\begin{aligned}
&\nu_+(\rho_1\varepsilon(\rho_1) - \rho_+\varepsilon(\rho_+)) + \nu_+\left(\rho_1\frac{C}{2} - \rho_+\frac{|v_+|^2}{2}\right) \\
&\leq [(\rho_1\varepsilon(\rho_1) + p(\rho_1))\beta - (\rho_+\varepsilon(\rho_+) + p(\rho_+))v_{+2}] + \left(\rho_1\beta\frac{C}{2} - \rho_+ v_{+2}\frac{|v_+|^2}{2}\right)
\end{aligned}
\tag{86}
$$

Careful analysis of the arising system of algebraic equations and inequalities yields the desired conclusion.

This line of research was continued by the same authors in [9] and independently by Klingenberg and Markfelder [23]. In both works, the authors expanded the results of [8] to the case when the Riemann data produce a self-similar solution in the form of a shock and a rarefaction wave; however, the value $v_{+2} - v_{-2}$ had to be sufficiently close to the value $-\sqrt{\frac{(\rho_+-\rho_-)(p(\rho_+)-p(\rho_-))}{\rho_+\rho_-}}$; see Lemma 1.2, points (3) and (4). The method of the proof remains the same as in [8] and relies on the analysis of the arising system of algebraic identities and inequalities.

Klingenberg and Markfelder moreover complemented this analysis with a clever idea that the infinitely many solutions produced by Proposition 2.1 can be patched with a standard shock or a rarefaction wave if one ensures that the shock or rarefaction wave lies outside of the region P_1. This way, they were able to prove that whenever the Riemann data is such that $v_{-1} = v_{+1}$ and the self-similar solution consists of a shock or a shock and a rarefaction wave, problems (1)–(3) with such initial data admit infinitely many admissible weak solutions.

This line of research was concluded with the work of Březina et al. [1], who showed how to add the contact discontinuity created by assuming $v_{-1} \neq v_{-2}$. This is resolved by considering fan partition consisting of four sets P_-, P_1, P_2, and P_+, where the speed of the middle interface ν_1 is set to be equal to the second component of the velocity field of the subsolution on the left and on the right of this interface, which are set to coincide, i.e., $\nu_1 = \beta_1 = \beta_2$, and, moreover, the equality of densities of the subsolution on the left- and right-hand sides of the subsolution is prescribed, i.e., $\rho_1 = \rho_2$. It is not difficult to show that with these choices the proof reduces to previous already known cases.

All works mentioned above can be summarized into the following theorem:

Theorem 3.2 *Let $p(\rho) = K\rho^\gamma$ with $K > 0$ and $\gamma \geq 1$. Let $(\rho_-, \mathbf{v}_-)$ and $(\rho_+, \mathbf{v}_+)$ be constant states such that the self-similar solution to* (1)–(3) *with initial data* (9) *contains at least one shock. Then, there exist infinitely many admissible weak solutions to* (1)–(2) *with this initial data.*

3.3 Contact Discontinuities

With results of previous two subsections at hand, the only unresolved case for power law pressure is the case of Riemann data giving rise to a contact discontinuity or in more general to rarefaction waves and a contact discontinuity. This case remains open to this day; however, there are results for other pressure laws hinting at the fact that even contact discontinuities may be nonunique.

The first result of this type is due to Březina, Kreml, and Mácha [2] who studied the system (1)–(2) with the Chaplygin gas pressure law (16). Using the same approach as in the case of shocks, they were able to prove the following theorem:

Theorem 3.3 *Let $p(\rho)$ satisfy* (16). *Let $(\rho_-, \mathbf{v}_-)$ and $(\rho_+, \mathbf{v}_+)$ be constant states such that*

$$v_{+2} - v_{-2} < -\max\left\{\frac{1}{\rho_-}, \frac{1}{\rho_+}\right\}. \tag{87}$$

Then, there exist infinitely many admissible weak solutions to (1)–(3) *with initial data* (9).

Comparing condition (87) with Lemma 1.3, we observe that Theorem 3.3 provides two pieces of information: First, for any Riemann initial data which give rise to the solution in a form of a δ-shock, there exist infinitely many admissible weak solutions. This means that while in one dimension there is no bounded self-similar weak solution for this initial data, considering the same problem in multiple space dimensions allows for existence of bounded weak solutions.

Second, for

$$-\frac{1}{\rho_-} - \frac{1}{\rho_+} < v_{+2} - v_{-2} < -\max\left\{\frac{1}{\rho_-}, \frac{1}{\rho_+}\right\},$$

there is a self-similar solution to the 1D problem which consists of two or three contact discontinuities (depending on whether $v_{-1} = v_{+1}$ or not), and by Theorem 3.3, the 2D problem with the same initial data gives rise to infinitely many admissible weak solutions. Therefore, this provides an example that a self-similar solution consisting only of contact discontinuities is nonunique in the class of admissible weak solutions to the 2D problem.

Another interesting statement was recently proved by Krupa and Székelyhidi [25], who considered the case when the self-similar solution contains only the contact discontinuity created by different first components of the velocities. They proved nonuniqueness of this solution in the class of bounded admissible weak solutions to the 2D problem while treating the pressure as an additional degree of freedom. Their theorem reads as follows:

Theorem 3.4 *There exists a smooth pressure function $p(\rho) : (0, \infty) \to \mathbb{R}$ such that $p, p' > 0$, and the system* (1)–(3) *with this pressure law admits initial data*

(9) *with* $\rho_- = \rho_+$, $v_{-2} = v_{+2}$, *and* $v_{-1} \neq v_{+1}$ *which gives rise to infinitely many admissible weak solutions with density uniformly bounded away from 0.*

The proof of Theorem 3.4 is computer-assisted; however, the main idea of looking for fan subsolution is preserved. The subsolution here contains three regions where the convex integration Lemma 2.1 is applied. The fact that the pressure law is an additional degree of freedom is used as follows: The values $\varepsilon(\rho_i)$ for $i = -, 1, 2, 3, +$ are represented by constants $\hat{\varepsilon}_i$, and the values $p(\rho_i)$ are represented by another set of constants $\rho_i^2 \hat{\varepsilon}_i'$. Then, the following set of inequalities

$$\hat{\varepsilon}_j - \hat{\varepsilon}_i - \hat{\varepsilon}_i'(\rho_j - \rho_i) > 0 \quad \text{for } i, j = -, 1, 2, 3, \quad i \neq j \tag{88}$$

is added to the set of algebraic identities and inequalities related to the existence of a fan subsolution. Inequalities (88) ensure that there is a function $p(\rho)$ with $p' > 0$ such that $\rho_i^2 \hat{\varepsilon}_i' = p(\rho_i)$ and $\hat{\varepsilon}_i = \varepsilon(\rho_i)$ with $p(\rho), \varepsilon(\rho)$ interrelated through (6); see [25, Lemma 3.2].

The idea to treat the pressure law as a degree of freedom is in a sense quite strong. In fact, the authors make a conjecture, that for any Riemann initial data there exists a pressure law satisfying $p' > 0$ for which the systems (1)–(2) with this Riemann initial data have infinitely many bounded admissible weak solutions.

While not directly related to the existence of wild solutions, Krupa and Székelyhidi proved in [25] another interesting theorem:

Theorem 3.5 *Let* $p(\rho) : (0, \infty) \to \mathbb{R}$ *such that* $p, p' > 0$. *On* $(0, \infty) \times \mathbb{R}^2$, *consider a bounded admissible weak solution* $(\rho, \mathbf{v})$ *to* (1)–(3) *with initial data* (9) *with* $\rho_- = \rho_+$, $v_{-2} = v_{+2}$, *and* $v_{-1} \neq v_{+1}$, *such that* $(\rho, \mathbf{v})$ *is independent of* x_1. *Then,* $(\rho, \mathbf{v})$ *is the self-similar solution consisting of a single contact discontinuity.*

In other words, any one-dimensional admissible weak solution to the Riemann problem which gives rise to a single contact discontinuity has to coincide with this contact discontinuity solution. The authors use the Lagrangian formulation of the problem, relative energy method, and a mollification argument to prove this theorem.

We conclude this subsection by mentioning another interesting result related to contact discontinuities in hyperbolic systems. Kang et al. [22] studied the full Euler system, where temperature is an additional unknown. This system allows for self-similar solutions to the Riemann problem which contain a contact discontinuity not related to shear (i.e., different values of the first component of the velocity). More precisely, the Riemann problem has a self-similar solution consisting of a single contact discontinuity if the velocity and the pressure of the left and right states of the Riemann data coincide. The authors in [22] showed that this solution in 3D is unique and stable in the class of vanishing dissipation limits of solutions to the associated Navier-Stokes-Fourier system. The proof of this statement is based on the theory of a-contraction up to a shift, which was developed by the first two authors in their previous works. The idea is the construction of a weight function which is bounded and bounded from below, such that discontinuous solutions enjoy a contraction property for the corresponding weighted relative entropy up to a shift.

4 Wild Solutions for Regular Initial Data

While the uniqueness or the lack of uniqueness of admissible solutions to the Riemann problem is of interest itself, originally, it was used to study a different question: Does smoothness of initial data help to prevent the emergence of wild solutions? The answer to this question is negative, as was shown in a series of works. The investigation started with the original paper of De Lellis and Székelyhidi [14], who first applied the convex integration theory developed for incompressible Euler system to the study of (1)–(2). Their result states that for any pressure law $p(\rho)$ there exists bounded initial data $(\rho^0, \mathbf{v}^0)$ for which there are infinitely many bounded admissible weak solutions to (1)–(2) with initial data (3).

This result was improved by Chiodaroli [6], who showed that for any initial density $\rho^0 \in C^1$ there exists bounded initial velocity $\mathbf{v}^0$ such that locally in time there are infinitely many bounded admissible weak solutions to (1)–(2) with initial data (3). Feireisl [16] then proved a similar property globally in time under the condition that $\sup|\nabla_x \rho^0|$ is sufficiently small, and Luo, Xie and Xin [27] considered possibly discontinuous piecewise C^1 initial densities.

All these results implied that the existence of wild solutions was related to a suitably chosen initial velocity field, which in all these results was just a bounded function with no further properties.

4.1 Lipschitz Initial Data

A significant step forward in this line of research was achieved by Chiodaroli et al. [7] who proved the following:

Theorem 4.1 *Let $p(\rho) = \rho^2$. There are Lipschitz initial data $(\rho^0, \mathbf{v}^0)$ for which there are infinitely many bounded admissible solutions $(\rho, \mathbf{v})$ of* (1)–(3) *on* $[0, \infty) \times \mathbb{R}^2$ *with* $\inf \rho > 0$. *These solutions are all locally Lipschitz on a finite interval on which they all coincide with the unique classical solution.*

The proof is based on the study of an auxiliary Riemann problem and a suitable choice of Riemann initial data. Indeed, the well-known observation is that if Riemann initial data (9) give rise to a self-similar solution consisting of a rarefaction wave, then switching the left and right states implies that the rarefaction wave evolves backward in time; in other words, such Riemann initial data are generated by a collapsing compression wave.

The authors were able to find an example of such Riemann data, which are generated by a collapsing compression wave $(\rho^c(t, x), \mathbf{v}^c(t, x))$ defined for $t < 0$

and for which there exists a fan admissible subsolution and therefore infinitely many bounded admissible weak solutions. More specifically, the Riemann data used in the proof of Theorem 4.1 are $\rho_- = 1, \mathbf{v}_- = (-\frac{1}{4}, 2\sqrt{2})$, $\rho_+ = 4, \mathbf{v}_+ = (-\frac{1}{4}, 0)$. The initial data from Theorem 4.1 are then chosen as, for example, $(\rho^0(x), \mathbf{v}^0(x)) = (\rho^c(-1, x), \mathbf{v}^c(-1, x))$.

As the authors in [7] commented, the choice of the pressure law $p(\rho) = \rho^2$ was merely to make the calculations easier. With Theorem 3.2 at hand, Theorem 4.1 can be easily generalized to any power law pressure $p(\rho) = K\rho^\gamma$ with $K > 0$ and $\gamma \geq 1$, as it is not difficult to observe that any collapsing compression wave (i.e., backward rarefaction wave) gives rise to Riemann data for which the self-similar solution contains at least one shock.

The fact that all admissible weak solutions have to coincide with the Lipschitz solution, i.e., with the compression wave until it collapses to a Riemann state, is a consequence of the weak-strong uniqueness property stated, for example, in [11, Theorem 5.3.1].

4.2 Smooth Initial Data

The final piece of information showing that existence of wild solutions is not really related to the lack of regularity of the initial data was provided by Chiodaroli, Kreml, Mácha, and Schwarzacher in [10], who proved the following Theorem:

Theorem 4.2 *Let $p(\rho) = \rho^2$ and let $T > 0$. There are initial data $(\rho^0, v^0) \in C^\infty(\mathbb{R}^2)$ and $\delta_0 > 0$, such that there exist infinitely many bounded admissible weak solutions to the Euler system* (1)–(3) *on the time interval* $(0, T+\delta_0)$. *These solutions all coincide with the unique C^∞ solution to the Euler system* (1)–(3) *on the time interval* $(0, T)$.

The idea of the authors is to mollify the sharp corners of the compression wave to make it smooth instead of merely Lipschitz. When the compression wave collapses, it gives rise to generalized Riemann data of the type (69) instead of a couple of constant states. They showed that the behavior of ρ and v_2 near the arising discontinuity can be described by means of function

$$f_0(x_2) = 1 - \frac{2}{\pi} \arctan(\log|\log(x_2)|); \tag{89}$$

for more details, see [10, Section 2].

In order to construct a generalized admissible fan subsolution, several steps need to be done which are trivial in the case of piecewise constant subsolution but require close attention here.

First, the subsolution equals to the self-similar solution in regions P_- and P_+, and, therefore, it is defined by the method of characteristics there. It needs to be ensured that the curves $\tilde{\nu}_\pm(t)$ defining the generalized fan partition are such that the characteristics do not intersect in P_- and P_+. Second, Rankine-Hugoniot conditions of a similar type as (77)–(82) as well as admissibility conditions similar to (85)–(86) need to be satisfied on curves $x_2 = \tilde{\nu}_\pm(t)$ with $\nu_\pm(t) = \tilde{\nu}'_\pm(t)$. Finally, in the region P_1, the subsolution conditions similar to (83)–(84) need to be satisfied together with Eqs. (73)–(74), the condition (75), and the admissibility inequality (76).

The subsolution is constructed to be independent of x_1 with the following properties:

- ρ_1 is constant in the whole P_1.
- K is constant in the whole P_1.
- β, which here stands for the second component of the momentum $\mathbf{m}_1$ of the subsolution in the region P_1, depends only on t and is independent of x_2.

Then, most of conditions arising from (73) to (76) in region P_1 are satisfied trivially except for one, which arises from the momentum equation (74) and is rewritten as

$$\partial_t \beta - \partial_{x_2} \varepsilon_2 = 0, \tag{90}$$

where $\varepsilon_2 = \frac{C_1}{2} + \gamma_1$. It turns out that ε_2 needs to be linear in x_2 which moreover simplifies (90) to

$$\partial_t \beta = \frac{\varepsilon_{2R} - \varepsilon_{2L}}{l} =: -\frac{\varepsilon_\Delta(t)}{l}, \tag{91}$$

where

$$l = l(t) = \tilde{\nu}_+(t) - \tilde{\nu}_-(t) > 0 \tag{92}$$

and $\varepsilon_{2L} = \varepsilon_2(t, \tilde{\nu}_-(t))$ and $\varepsilon_{2R} = \varepsilon_2(t, \tilde{\nu}_+(t))$.

The values of $\nu_\pm(t)$, therefore also $\tilde{\nu}_\pm(t)$, and consequently of l depend on ε_Δ. The same holds also for $\beta(t)$, so the simple-looking equation (91) is in fact rewritten more precisely as

$$\partial_\varepsilon \beta(t, \varepsilon_\Delta(t)) \frac{\mathrm{d}}{\mathrm{d}t} \varepsilon_\Delta(t) = -\frac{\varepsilon_\Delta(t)}{l(t, \varepsilon_\Delta(t))} - \partial_t \beta(t, \varepsilon_\Delta(t)). \tag{93}$$

The existence of a solution to this equation accompanied with natural initial condition $\varepsilon_\Delta(0) = 0$ cannot be achieved directly from the Picard-Lindelöf theorem; instead, the authors show the existence by means of the Banach fixed point argument.

The whole construction then works for sufficiently small times, for which the values of the constructed subsolution do not differ too much from constant values given by Riemann data related to Lipschitz compression wave used in Theorem 4.1.

4.3 Wild Initial Data Are Dense

While the results of Theorems 4.1 and 4.2 show examples of initial data for which there are infinitely many admissible weak solutions, a natural follow-up question arises: How rare are such initial data?

An interesting result in this direction was provided by Chen et al. [5]. The authors work on a 2D or 3D torus and consider a weaker type of solution, namely, solutions which satisfy (8) only for test functions $\varphi = \chi_{[0,\tau]}(t)$ for all $t \in (0, T)$, i.e., solutions satisfying

$$\int_{\mathbb{T}^n} \left(\rho\varepsilon(\rho) + \rho\frac{|\mathbf{v}|^2}{2} \right) (t, \cdot)\,\mathrm{d}x \leq \int_{\mathbb{T}^n} \left(\rho^0\varepsilon(\rho^0) + \rho^0\frac{|\mathbf{v}^0|^2}{2} \right) \mathrm{d}x \tag{94}$$

for all $t \in (0, T)$.

The authors prove the following theorem:

Theorem 4.3 *Assume that $p(\rho)$ is given by* (15) *with $1 < \gamma \leq 1 + \frac{2}{n}$, $n = 2, 3$. Then, for any $\varepsilon > 0$ and any $(\rho^0, \mathbf{v}^0)$ such that $\rho^0\varepsilon(\rho^0) + \rho^0\frac{|\mathbf{v}^0|^2}{2} \in L^1(\mathbb{T}^n)$, there exist infinitely many $(r^0, \mathbf{u}^0)$ satisfying*

$$r^0 > 0, \qquad r^0\varepsilon(r^0) + r^0\frac{|\mathbf{u}^0|^2}{2} \in L^1(\mathbb{T}^n)$$

and

$$\|\rho^0 - r^0\|_{L^\gamma}^\gamma + \left\| \frac{\mathbf{v}^0}{\sqrt{\rho^0}} - \frac{\mathbf{u}^0}{\sqrt{r^0}} \right\|_{L^2}^2 < \varepsilon$$

such that for each of such initial values $(r^0, \mathbf{u}^0)$ there exist infinitely many weak solutions $(\rho, \mathbf{v})$ to the systems (1)–(2) *satisfying* (94).

In other words, in every neighborhood of every finite energy initial datum, there exists a wild initial datum, i.e., another finite energy initial datum which moreover allows for the existence of infinitely many weak solutions satisfying (94). This result is also based on the convex integration method; more precisely, the authors use this argument twice: First, a time-localized convex integration is performed to construct a subsolution with its wild initial datum, and then a second convex integration is used to pass from this subsolution to infinitely many solutions.

We conclude with a remark that while Theorem 4.3 provides an abundance of wild initial data the question of existence of an admissible weak solution to (1)–(3) for any finite energy initial data is still an open problem.

5 Failure of Energy Dissipation Criteria

Most results in Sects. 3 and 4 suggest that the energy inequality (8) is not strong enough to single out a unique weak solution. This gives rise to further natural question: What is a good selection principle which could work in favor of selecting a unique physical solution? This question is open to this day, in part because some other candidates for selection criteria fail to accomplish expected results.

5.1 Energy-Conserving Solutions

Klingenberg and Markfelder [24] observed that replacing the inequality (8) with an equation does not help to get rid of wild solutions as they found a fan subsolution satisfying (68) with an equality sign, thus giving rise to infinitely many solutions satisfying the energy equation. Moreover, they used the same Riemann data which were used in [7], therefore giving rise to the following:

Theorem 5.1 *Let* $p(\rho) = \rho^2$. *There exist Lipschitz initial data* $(\rho^0, \mathbf{v}^0)$ *such that there are infinitely many energy-conserving weak solutions to* (1)–(3).

In order to prove Theorem 5.1, the authors examined further the Riemann data and the fan subsolution used in [7] and figured out that it is possible to adjust some components of the piecewise constant subsolution to achieve equations instead of inequalities in (85)–(86).

5.2 Global Maximal Dissipation Rate Criterion

Further energy-related criteria which may look suitable to eliminate wild solutions and select physical ones use the idea that the admissible solution dissipates maximal possible amount of energy.

We define the total energy of an admissible weak solution as

$$E[\rho, \mathbf{v}](t) = \int_{\mathbb{R}^2} \left(\rho\varepsilon(\rho) + \rho \frac{|\mathbf{v}|^2}{2} \right) \mathrm{d}x. \tag{95}$$

The right derivative of $E[\rho, \mathbf{v}](t)$ defines the energy dissipation rate of $(\rho, \mathbf{v})$ at time t:

$$D[\rho, \mathbf{v}](t) = \frac{\mathrm{d}_+ E[\rho, \mathbf{v}](t)}{\mathrm{d}t} \tag{96}$$

Since we work with solutions with positive densities at infinity, we restrict the integration domains in order to work with finite values; hence, we introduce

$$E_L[\rho, \mathbf{v}](t) = \int_{(-L,L)^2} \left(\rho\varepsilon(\rho) + \rho \frac{|\mathbf{v}|^2}{2} \right) \mathrm{d}x \tag{97}$$

$$D_L[\rho, \mathbf{v}](t) = \frac{\mathrm{d}_+ E_L[\rho, \mathbf{v}](t)}{\mathrm{d}t}. \tag{98}$$

Now, we can define solutions satisfying the global maximal dissipation rate criterion.

Definition 5.1 (Global Maximal Dissipation Rate Criterion) A weak solution $(\rho, \mathbf{v})$ of (1)–(3) satisfies the global maximal dissipation rate criterion if there exists $L^* > 0$ such that there is no other weak solution $(\overline{\rho}, \overline{\mathbf{v}})$ with the property that for some $\tau \geq 0$ $(\overline{\rho}, \overline{\mathbf{v}})(x, t) = (\rho, \mathbf{v})(x, t)$ on $\mathbb{R}^2 \times [0, \tau]$ and $D_L[\overline{\rho}, \overline{\mathbf{v}}](\tau) < D_L[\rho, \mathbf{v}](\tau)$ for all $L \geq L^*$.

A similar form of maximal dissipation criterion was proposed by Dafermos [11] who investigated the equivalence of this criterion to other admissibility criteria for hyperbolic conservation laws in the one-dimensional case. He proved that for a single equation this criterion is equivalent to the viscosity criterion in the class of piecewise smooth solutions. Hsiao [20] proved, in the class of piecewise smooth solutions, the equivalence of the entropy rate criterion and the viscosity criterion for the one-dimensional system of equations of non-isentropic gas dynamics in Lagrangian formulation with pressure laws $p(\rho) = \rho^\gamma$ for $\gamma \geq \frac{5}{3}$ while the same equivalence was disproved for $\gamma < \frac{5}{3}$. See also [12] for further analysis on the relation between this kind of criterion and admissibility of solutions for a more general class of evolutionary equations.

In [16], Feireisl proved global existence of infinitely many admissible weak solutions for certain initial data and showed that none of these solutions satisfy the global maximal dissipation rate criterion. While this suggests possible effectiveness of this criterion to rule out wild solutions constructed by the convex integration method, there is an interesting result in the opposite direction proved by Chiodaroli and Kreml [8].

Theorem 5.2 *Let* $p(\rho) = K\rho^\gamma$, $K > 0$, $1 \leq \gamma < 3$. *There exist Riemann data* (9) *for which the self-similar solution to* (1)–(3) *emanating from these data does not satisfy the global maximal dissipation rate criterion.*

In other words, there exist Riemann data (9) for which some of the infinitely many wild solutions constructed in Theorem 3.2 dissipate more energy than the self-similar solution, suggesting in particular that the global maximal dissipation rate criterion would not pick the self-similar solution as the admissible (or physical) one.

The solutions constructed by application of the convex integration method to a given piecewise constant fan subsolution all dissipate the same amount of energy, and the dissipation is concentrated only on the interfaces given by $x_2 = \nu_{\pm}t$. Since the 1D self-similar solution dissipates energy only on admissible shocks, the proof in [8] is based on the analysis of the properties of the fan subsolution and comparison with the self-similar solution.

5.3 Local Maximal Dissipation Rate Criterion

Since the global maximal dissipation rate criterion fails to select the self-similar 1D solution, Feireisl proposed to study a local version of this criterion. To this end, we introduce the energy dissipation measure

$$\partial_t \left(\rho\varepsilon(\rho) + \rho\frac{|\mathbf{v}|^2}{2} \right) + \mathrm{div}_x \left(\left(\rho\varepsilon(\rho) + \rho\frac{|\mathbf{v}|^2}{2} + p(\rho) \right) \mathbf{v} \right) =: -\mu[\rho, \mathbf{v}]. \tag{99}$$

It is clear that for any admissible weak solution $\mu[\rho, \mathbf{v}]$ is a nonnegative measure.

Definition 5.2 (Local Maximal Dissipation Rate Criterion) A weak solution $(\rho, \mathbf{v})$ of (1)–(2) satisfies the local maximal dissipation rate criterion if there is no other weak solution $(\overline{\rho}, \overline{\mathbf{v}})$ with the property that for some $\tau \geq 0$, $(\overline{\rho}, \overline{\mathbf{v}})(x, t) = (\rho, \mathbf{v})(x, t)$ on $\mathbb{R}^2 \times [0, \tau]$ and $\mu[\overline{\rho}, \overline{\mathbf{v}}] > \mu[\rho, \mathbf{v}]$ on $(\tau, T) \times \mathbb{R}^2$.

Unlike the global maximal dissipation rate criterion, which allowed to compare any two solutions starting from the same initial data and to identify solution dissipating more energy, the order defined by the local maximal dissipation rate criterion is not complete, i.e., there are solutions $(\rho, \mathbf{v})$ and $(\overline{\rho}, \overline{\mathbf{v}})$ starting from the same initial data for which none of the inequalities $\mu[\overline{\rho}$ and $\overline{\mathbf{v}}] \geq \mu[\rho, \mathbf{v}]$, $\mu[\overline{\rho}, \overline{\mathbf{v}}] \leq \mu[\rho, \mathbf{v}]$ hold. Typically, this happens to solutions constructed in [8] which are used to prove Theorem 5.2.

Even though this criterion requires a solution to dissipate maximal amount of energy at every point in space, there are situations where even this criterion does not select the self-similar solution to the Riemann problem. This was first demonstrated in [2], where the authors used the fact that self-similar solutions to the problem with Chaplygin gas pressure law do not dissipate energy, since they consist only of contact discontinuities. Any wild solution which satisfies (8) as a strict inequality therefore beats the self-similar solution in the eyes of the local maximal dissipation rate criterion.

Theorem 5.3 *Let $p(\rho) = -\rho^{-1}$. Assume that the Riemann initial data* (9) *satisfy*

$$\max\left\{\frac{1}{\rho_-}, \frac{1}{\rho_+}\right\} < v_{-2} - v_{+2} < \frac{1}{\rho_-} + \frac{1}{\rho_+}. \tag{100}$$

Then, the self-similar solution to (1)–(3) *consisting of two (for* $v_{-1} = v_{+1}$*) or three (for* $v_{-1} \neq v_{+1}$*) contact discontinuities does not satisfy the local maximal dissipation rate criterion.*

Recently, Markfelder [29] examined this criterion as well in the context of power law pressure with the help of a new convex integration approach he developed. He was able to take Riemann data giving rise to a self-similar solution in the form of just one admissible shock, and he proved the existence of a fan subsolution consisting of five regions P_-, P_1, P_2, P_3, and P_+ with the speed of the second interface ν_1 being equal to the speed of the shock of the self-similar solution. The author then observed that it is possible to construct this subsolution in such a way that the dissipation of the energy across this interface is larger than the dissipation of the energy on the shock for the self-similar solution. Hence, he obtained the following theorem:

Theorem 5.4 *Let* $p(\rho) = \rho^2$. *There exist Riemann initial data* (9) *for which the self-similar solution to* (1)–(3) *does not satisfy the local maximal dissipation rate criterion.*

6 Conclusion

The convex integration technique has profoundly impacted the study of the compressible Euler equations in multiple space dimensions. This method exploits the flexibility of the differential inclusions associated with the Euler system, allowing for the construction of solutions that exhibit intricate oscillatory behavior. Convex integration demonstrates that the set of weak solutions can be surprisingly vast, even in cases where energy conservation, energy dissipation, or other usual constraints may be expected to select a unique or physically meaningful solution.

To this day, a major unresolved challenge lies in identifying suitable selection criteria that distinguish physically relevant solutions from the broader set of mathematically admissible ones. It becomes clear that additional principles or mechanisms must be sought to capture the correct physical solutions.

One promising approach is based on the vanishing viscosity method, which considers solutions of the compressible Navier-Stokes equations as the viscosity tends to zero. The hope is that these viscous approximations will converge to physically meaningful solutions of the inviscid Euler equations. Of course, in reality, all physical fluids possess some nonzero viscosity, no matter how small, meaning that the Euler system should be viewed always as an approximation rather than an exact description of observed phenomena. However, the full scope and limitations of this approach, especially in multiple space dimensions and in the presence of shocks and other singularities, are still not completely understood. Nevertheless, the vanishing viscosity method remains a key candidate in the ongoing search for a rigorous selection criterion in compressible inviscid fluid dynamics.

Acknowledgments The work of Ondřej Kreml was supported by the Institute of Mathematics, Czech Academy of Sciences (RVO 67985840), and by the Praemium Academiae of Š. Nečasová.

References

1. Březina, J., Chiodaroli, E., Kreml, O.: Contact discontinuities in multi-dimensional isentropic Euler equations. Electron. J. Differential Equations **2018**(94), 1–11 (2018)
2. Březina, J., Kreml, O., Mácha, V.: Non-uniqueness of delta shocks and contact discontinuities in the multi-dimensional model of Chaplygin gas. NoDEA Nonlinear Differential Equations Appl. **28**, Paper No. 13, 24 pp. (2021)
3. Buckmaster, T., De Lellis, C., Székelyhidi, L., Vicol, V.: Onsager's conjecture for admissible weak solutions. Comm. Pure Appl. Math. **72**, 229–274 (2019)
4. Chen, G.-Q., Chen, J.: Stability of rarefaction waves and vacuum states for the multidimensional Euler equations. J. Hyperbolic Differ. Equ. **4**, 105–122 (2007)
5. Chen, R. M., Vasseur, A. F., Yu, C.: Global ill-posedness for a dense set of initial data to the isentropic system of gas dynamics. Adv. Math. **393**, Paper No. 108057, 46 pp. (2021)
6. Chiodaroli, E.: A counterexample to well–posedness of entropy solutions to the compressible Euler system, J. Hyperbolic Differ. Equ. **11**, 493–519 (2014)
7. Chiodaroli, E., De Lellis, C., Kreml, O.: Global ill-posedness of the isentropic system of gas dynamics. Comm. Pure Appl. Math. **68**, 1157–1190 (2015)
8. Chiodaroli, E., Kreml, O.: On the energy dissipation rate of solutions to the compressible isentropic Euler system. Arch. Rational Mech. Anal. **214**, 1019–1049 (2014)
9. Chiodaroli, E., Kreml, O.: Non-uniqueness of admissible weak solutions to the Riemann problem for the isentropic Euler equations. Nonlinearity **31**, 1441–1460 (2018)
10. Chiodaroli, E., Kreml, O., Mácha, V., Schwarzacher, S.: Non-uniqueness of admissible weak solutions to the compressible Euler equations with smooth initial data. Trans. Amer. Math. Soc. **374**, 2269–2295 (2021)
11. Dafermos, C. M.: Hyperbolic conservation laws in continuum physics. Third edition. Springer-Verlag, Berlin (2010)
12. Dafermos, C. M.: Maximal dissipation in equations of evolution. J. Differ. Equ. **252**, 567–587 (2012)
13. De Lellis, C., Székelyhidi, L.: The Euler equations as a differential inclusion, Ann. of Math. **170**, 1417–1436 (2009)
14. De Lellis, C., Székelyhidi, L.: On admissibility criteria for weak solutions of the Euler equations, Arch. Rational Mech. Anal. **195**, 225–260 (2010)
15. Debiec, T., Skipper, J., Wiedemann, E.: A general convex integration scheme for the isentropic compressible Euler equations. J. Hyperbolic Differ. Equ. **20**, 95–117 (2023)
16. Feireisl, E.: Maximal dissipation and well-posedness for the compressible Euler system, J. Math. Fluid Mech. **16**, 447–461 (2014)
17. Feireisl, E., Kreml, O.: Uniqueness of rarefaction waves in multidimensional compressible Euler system, J. Hyperbolic Differ. Equ. **12**, 489–499 (2015)
18. Giri, V., Kwon, H.: On non-uniqueness of continuous entropy solutions to the isentropic compressible Euler equations. Arch. Ration. Mech. Anal. **245**, 1213–1283 (2022)
19. Gromov, M.: Partial differential relations. Ergeb. Math. Grenzgeb. (3), Springer-Verlag, Berlin (1986)
20. Hsiao, L.: The entropy rate admissibility criterion in gas dynamics. J. Differ. Equ. **38**, 226–238 (1980)
21. Isett, P.: A proof of Onsager's conjecture, Ann. of Math. **188**, 871–963 (2018)
22. Kang, M.-J., Vasseur, A. F., Wang, Y.: Uniqueness of a planar contact discontinuity for 3D compressible Euler system in a class of zero dissipation limits from Navier-Stokes-Fourier system. Comm. Math. Phys. **384**, 1751–1782 (2021)

23. Klingenberg, C., Markfelder, S.: The Riemann problem for the multidimensional isentropic system of gas dynamics is ill-posed if it contains a shock. Arch. Rational Mech. Anal. **227**, 967–994 (2018)
24. Klingenberg, C., Markfelder, S.: Non-uniqueness of energy-conservative solutions to the isentropic compressible two-dimensional Euler equations. J. Hyperbolic Differ. Equ. **15**, 721–730 (2018)
25. Krupa, S. G., Székelyhidi, L.: Contact Discontinuities for 2-D Isentropic Euler are Unique in 1-D but Wildly Non-unique Otherwise. Comm. Math. Phys. **406**, Paper No. 109 (2025)
26. Kuiper, N. H.: On C^1-isometric imbeddings. I, II, Indag. Math. **17**, 545–556, 683–689 (1955)
27. Luo, T., Xie, C., Xin, Z.: Non-uniqueness of admissible weak solutions to compressible Euler systems with source terms. Adv. Math. **291**, 542–583 (2016)
28. Markfelder, S.: Convex integration applied to the multi-dimensional compressible Euler equations. Springer, Cham, (2021)
29. Markfelder, S.: A new convex integration approach for the compressible Euler equations and failure of the local maximal dissipation criterion. Nonlinearity **37**, Paper No. 115022, 60 pp (2024)
30. Nash, J.: The imbedding problem for Riemannian manifolds, Ann. of Math. **63**, 20–63 (1956)

Comparative Study of 2D-2D FEM-FSI and 2D-1D Reduced FD Models for Respiratory Flow Simulation

Anna Lancmanová, Alexander Drobny, Elfriede Friedmann, and Tomáš Bodnár

Dedicated to the memory of Goldi (Brigitte Schirle), a cherished soul of 1974

Abstract This contribution presents a computational comparison and cross-validation of a coupled fluid-structure interaction (FSI) model considering an oscillatory (direction reversing) flow of an incompressible fluid in a partially elastic channel. The solution of a full 2D FSI model solved by a finite-element method (FEM) is compared with a simplified coupled 2D-1D model solved by finite difference method (FDM). Besides providing a direct comparison of instantaneous velocity profiles with the Womersley analytical solution in the rigid part of the channel, the data from the elastic part considering fluid-structure interaction are compared as well. The results show a good mutual agreement of both models, with obvious limitations in the elastic region, where the computational setups for both models can't be exactly identical.

A. Lancmanová
Faculty of Mechanical Engineering, Czech Technical University in Prague, Prague, Czech Republic
e-mail: anna.lancmanova@fs.cvut.cz

A. Drobny · E. Friedmann
Institute of Mathematics, Kassel University, Kassel, Germany
e-mail: adrobny@mathematik.uni-kassel.de; friedmann@mathematik.uni-kassel.de

T. Bodnár (✉)
Institute of Mathematics, Czech Academy of Sciences, Prague, Czech Republic
e-mail: bodnar@math.cas.cz

T. Bodnár et al. (eds.), *Stochastics in Fluids*, Lecture Notes in Mathematical Fluid Mechanics, https://doi.org/10.1007/978-3-032-02299-8_2

1 Introduction

The main motivation for this work comes from the modeling of respiratory flows. In addition to being an important biological and medical problem, the mathematical modeling and numerical simulations of respiratory flows are a highly challenging problem, both theoretically and computationally. The airflow in the respiratory system exhibits several characteristic properties that make the problem particularly difficult to handle [1]:

The first challenge, not addressed in this paper, is the geometrical complexity of the human airways. They consist of a system of branching channels with a treelike (or fractal-like) structure, which makes the simulations quite complex and computationally expensive [2].

The second difficulty is associated with the naturally unsteady character of respiration, where the inhaling-exhaling cycle periodically repeats (under normal breathing conditions), leading to changes in both the magnitude and the orientation of the fluid flow. This distinct oscillatory character has further important consequences both at the physical and at the mathematical level. From a physical perspective, the periodic in time transition from laminar to turbulent flow regime is quite unique and rarely observed in other applications. It is also very different from the pulsatile flow of blood as it is known in hemodynamics [3, 4], where the flow velocity magnitude changes periodically but preserves its orientation (mean flow direction). From the mathematical (and numerical) point of view, one of the major difficulties associated with the oscillatory (direction reversing) flow behavior is that it necessarily affects the character of the boundary conditions considered and imposed on the inlet/outlet boundaries of the simulated part of the respiratory tract [5, 6].

The third challenge associated with the realistic simulations of the respiratory problems is connected to the fluid-structure interaction (FSI) between the airflow on one side and the airways and lung tissues on the other side. The structural and mechanical properties of airways differ substantially starting from the upper airways, passing through the successive branching channels up to the deep lungs; see [7] or [8–10]. While the upper airways may appear rigid, the lower airways and alveolar regions deform easily under internal or external pressure. These FSI effects are not only nonnegligible but also essential for the whole pulmonary and respiratory mechanics.

The aforementioned characteristic challenges of respiratory flows necessitate certain modeling simplifications to reduce the complexity and computational cost. One such technique is dimensionality reduction of the problem, which leads to approximate models with lower dimensions. These typically one-dimensional models keep the temporal characteristics of the original problem, while their spatial resolution is reduced to one dimension only (by cross-sectional averaging, considering distributed parameters) [11–14]. These simplified models can be used to describe the flow and other features in the periphery of the modeled system, while the main airways remain fully discretized in 2D (or 3D) to properly capture the flow domain complexity. The full- and lower- dimensional models are coupled at a common interface to provide the appropriate feedback and boundary conditions for both partial models [15]. Further details on the reduced-dimensional models and their coupling with the full-dimensional counterparts can be found in the review article [1].

In the work presented in this paper, the investigation is focusing on a coupled problem, where the fully discretized model considers the flow in the rigid upper airways, while the reduced-dimensional simplified model describes the airflow in the compliant (elastic) lower part of the respiratory tract. In this way, the geometrical (3D/2D) complexity can be captured in the rather rigid upper airways, while the airway elasticity and associated FSI effects are only considered in the coupled 1D model of the deep lungs.

The aim of this paper is to analyze and quantify the impact of the dimensional reduction in the coupled FSI problem. For simplicity, the oscillatory flow in a straight pipe (represented here by a 2D channel only) is considered, where one part of the channel has rigid walls, while in the remaining part the walls are elastic, allowing FSI effects to be observed. In the full original (multidimensional) setting, the 2D problem is considered, discretized, and solved in both the rigid and the elastic part of the channel. The results are compared with a much simpler approach, where the 2D problem is solved only in the rigid part of the channel while in the elastic part of the channel the full model is replaced by a reduced 1D FSI model.

In this work, we investigate a coupled problem in which the fully discretized model captures the flow in the rigid upper airways, while a reduced-dimensional simplified model describes the airflow in the compliant (elastic) lower part of the respiratory tract. This approach allows the geometric (3D/2D) complexity to be captured in the rather rigid upper airways, while airway elasticity and the associated FSI effects are accounted for only in the coupled 1D model of the deep lungs.

2 Problem Setting

As mentioned in the Introduction, the problem addressed in this work is motivated by a reciprocating airflow in a partially compliant tube. Two alternative approaches are selected and mutually compared.

The first approach, a fully 2D model with both a rigid and an elastic segment, is solved numerically using the finite element method (FEM) in an Arbitrary Lagrangian-Eulerian (ALE) framework.

The second, simplified approach is formulated as a coupled 2D-1D problem, where the flow in the rigid part of the tube is modeled as a two-dimensional channel flow, which is subsequently coupled to a simplified (dimensionally reduced) one-dimensional FSI model representing the elastic part of the tube. In this simplified 2D-1D coupled model, both parts are discretized by the finite difference method (FDM).

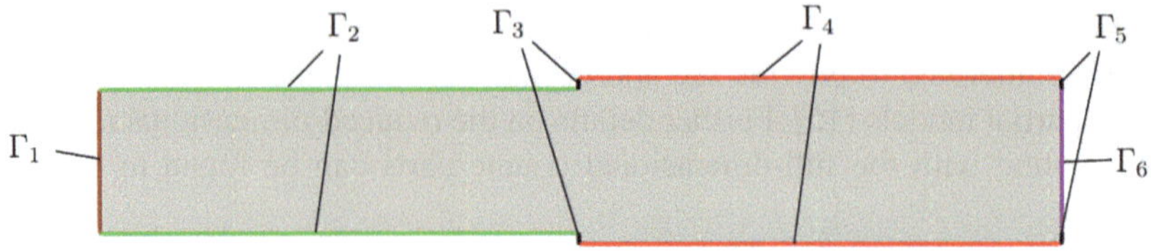

Fig. 1 The domain used in FEM, with annotations indicating the different boundaries

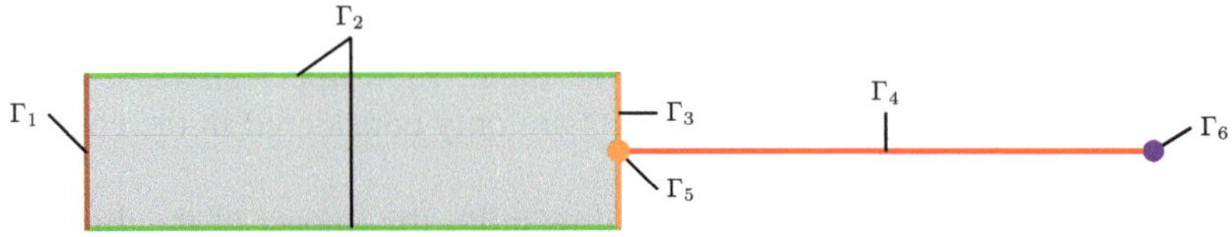

Fig. 2 The domain used in FDM with different boundaries

This basic setting naturally involves two distinct problems that need to be formulated, along with separate computational frameworks to assure the mutual compatibility and comparability of both cases. The two reference domains shown in Figs. 1 and 2 correspond to the considered numerical approaches used in the simulations. Figure 1 illustrates the schematic of the domain used in the FEM-ALE framework where $\Gamma_1, \Gamma_2, and \Gamma_6$ are boundaries of the fluid domain and $\Gamma_3, \Gamma_4, and \Gamma_5$ are boundaries of the structure domain. Figure 2 illustrates the corresponding reduced domain for the FDM solution, in which the two-dimensional rigid (fluid) domain is shown in gray while the red segment represents the coupled one-dimensional model extension.

The general problem definition remains identical for these two approaches, ensuring comparability between them. In the problem formulation, a rigid tube is coupled to a flexible tube, which requires different treatments in the FEM and FDM frameworks; however, in both approaches, the airflow is driven by a pressure difference between Γ_1 and Γ_6. On Γ_1, the inlet pressure varies periodically in time at an angular frequency $\omega = 2\pi f = \frac{4Wo^2\nu}{D^2}$, f referring to the ordinary frequency and amplitude A_ω set to one, while the outlet pressure (at Γ_6) is fixed as

$$\begin{aligned} p &= l A_\omega \cos(\omega t) \quad && \text{on } \Gamma_1 \, , \\ p &= 0 && \text{on } \Gamma_6 \, . \end{aligned} \tag{1}$$

The structural boundaries (channel walls) are defined in such a way that Γ_2 represents the rigid wall (where the displacement $u = 0$) while Γ_4 is left free to move. The no-slip condition $v = 0$ for velocity is imposed on Γ_2. For the airflow simulations presented in this study, the values of physical parameters were chosen to approximate those of the human trachea. The airflow simulations presented in this study use physical parameters chosen to approximate those of the human trachea. Unless stated otherwise, the parameters summarized in Table 1 are used throughout

Table 1 Physical parameters used in the simulations

Parameter	Description	Value
ρ	Air density	$1.024\,\mathrm{kg}\cdot\mathrm{m}^{-3}$
ν	Kinematic viscosity	$1.958\times10^{-5}\,\mathrm{m}^2\,\mathrm{s}^{-1}$
Wo	Womersley number	2.41
A_ω	Amplitude	$1.0\,\mathrm{Pa\,m}^{-1}$
E	Young modulus	2.9 MPa
κ_E	Poisson ratio	0.48
D	Diameter	0.018 m
h_0	Tube wall thickness	0.0014 m
l	Total length	0.12 m
l_1	Length of rigid part	$l/2$ (test case 2), l (test case 1)
l_2	Length of elastic part	$l/2$ (test case 2), 0 (test case 1)

the study. For validation purposes in Sect. 5.1, the problem is simplified to consider only the rigid part of the domain. The setup remains the same, except that the original elastic tube is replaced by a rigid one, resulting in the parameter setting $l_1 = l$ and $l_2 = 0$.

3 Finite Element Solver

The finite element method (FEM) is a general approach for numerically solving differential or integral equations. Other methods are finite difference and finite volume methods. FEM allows for the analysis of complex geometries, general boundary conditions, and variable or nonlinear material properties. Its clear structure facilitates the development of FEM software for various applications, and numerous open-source codes are already available. FEM is supported by a solid mathematical foundation, enabling precise control over computations, assessment of solution accuracy, and estimation of errors.

3.1 Coupled 2D-2D Model with Rigid-Elastic Walls

There are many approaches for dealing with the FSI problem numerically. Since in our application there are no topology changes or too large deformations which could lead to possible mesh degeneration, the Arbitrary Lagrangian-Eulerian (ALE) method [16] is the natural choice. To this end, we derive a monolithic variational formulation on the reference domain which will then be discretized in time and in space. The nonlinear system is linearized using Newton's method. Alternative approaches include partitioned methods, which are mostly used for weakly coupled problems (see, for example, [17]), and Eulerian methods, which are commonly chosen in the presence of topology changes (see, for example, [18]).

FEM allows for the analysis of structural mechanics (solids) and fluid dynamics (fluids), as well as their interaction. A key challenge in the coupled 2D-2D problem is ensuring a stable and accurate fluid-structure coupling. While structural equations are typically formulated in a Lagrangian framework and fluid equations in a Eulerian framework, the Arbitrary Lagrangian-Eulerian (ALE) method [16] is used to bridge this gap. This approach is one of the most commonly employed methods in the literature; see, for example, [19, 20].

In the following, we denote the fluid domain at time $t \in I$, where $I = [0, T]$, $T > 0$ is the time interval, by $\Omega_f := \Omega_f(t) \subset \mathbb{R}^2$. The current solid domain is denoted by $\Omega_s := \Omega_s(t) \subset \mathbb{R}^2$. The whole domain is denoted by $\Omega := \Omega_f \cup \Gamma_{\text{interface}} \cup \Omega_s$ with the common interface $\Gamma_{\text{interface}} := \Gamma_{\text{interface}}(t) = \partial\Omega_f \cap \partial\Omega_s$. The corresponding reference domains and interface are denoted by $\hat{\Omega}$, $\hat{\Omega}_f$, $\hat{\Omega}_s$, and $\hat{\Gamma}_{\text{interface}}$, respectively. The boundary is divided as follows,

$$\partial\hat{\Omega} = \hat{\Gamma}_1 \cup \hat{\Gamma}_2 \cup \hat{\Gamma}_3 \cup \hat{\Gamma}_4 \cup \hat{\Gamma}_5 \cup \hat{\Gamma}_6,$$

and similarly for the current boundary $\partial\Omega$. This is illustrated in Fig. 3 (top) where the reference fluid domain is shown in blue while the (elastic) structure domain is highlighted in red. An example of the deformed current domain is shown in Fig. 3.

In the following, we denote points in the reference domain by $\hat{x} \in \hat{\Omega}$ and points in the current domain by $x \in \Omega$. We define the deformation of the structure domain $\hat{\Omega}_s$ as the following mapping

$$\hat{\boldsymbol{T}}_s : I \times \hat{\Omega}_s \to \Omega_s \quad \text{with } x = \hat{\boldsymbol{T}}_s(t, \hat{x})$$

Fig. 3 The reference domain (top) and an exemplary current domain (bottom) used in FEM

and the deformation gradient $\hat{\mathbf{F}} : \hat{\Omega}_s \to \mathbb{R}^{2\times 2}$ by

$$\hat{\mathbf{F}} := \hat{\mathbf{F}}(\hat{\boldsymbol{u}}_s) := \hat{\nabla}\hat{\boldsymbol{T}}_s = \hat{\mathbb{I}} + \hat{\nabla}\hat{\boldsymbol{u}}_s$$

with the identity matrix $\hat{\mathbb{I}}$ and the displacement $\hat{\boldsymbol{u}}_s$.

The ALE formulation is based on an artificial displacement $\hat{\boldsymbol{u}}_f(t,\hat{x})$ in the fluid domain and on the ALE mapping

$$\hat{\boldsymbol{T}}_f : I \times \hat{\Omega}_f \to \Omega_f\,, \quad x = \hat{\boldsymbol{T}}_f(t,\hat{x}) = \hat{x} + \hat{\boldsymbol{u}}_f(t,\hat{x})\,. \tag{2}$$

The deformation gradient is defined as

$$\hat{\mathbf{F}} := \hat{\mathbf{F}}(\hat{\boldsymbol{u}}_f) := \hat{\nabla}\hat{\boldsymbol{T}}_f = \hat{\mathbb{I}} + \hat{\nabla}\hat{\boldsymbol{u}}_f\,.$$

In the following, we assume that $\hat{\boldsymbol{T}}_s$ and $\hat{\boldsymbol{T}}_f$ are C^1-diffeomorphisms.

The fluid-structure interaction problem between an incompressible Newtonian fluid and an elastic structure is formulated as follows,

$$\begin{aligned}
\rho_f \partial_t \boldsymbol{v}_f + \rho_f(\boldsymbol{v}_f \cdot \nabla)\boldsymbol{v}_f - \operatorname{div}\mathbb{T}_f &= \rho_f \boldsymbol{f} && \text{in } \Omega_f\,,\, t \in I\,,\\
\operatorname{div}\boldsymbol{v}_f &= 0 && \text{in } \Omega_f\,,\, t \in I\,,\\
\hat{\rho}_s^0 \partial_t \hat{\boldsymbol{v}}_s - \widehat{\operatorname{div}}(\hat{\Pi}) &= \hat{\rho}_s^0 \hat{\boldsymbol{f}}_s && \text{in } \hat{\Omega}_s\,,\, t \in I\,,\\
\partial_t \hat{\boldsymbol{u}}_s - \hat{\boldsymbol{v}}_s &= 0 && \text{in } \hat{\Omega}_s\,,\, t \in I
\end{aligned} \tag{3}$$

with the Cauchy stress tensor

$$\mathbb{T}_f := -p_f \mathbb{I} + \rho_f \nu_f (\nabla \boldsymbol{v}_f + \nabla \boldsymbol{v}_f^T)\,, \tag{4}$$

where $\mathbb{I}$ is the identity matrix.

For the elasticity model, we use a linear model where the Piola-Kirchhoff stress tensor is given by

$$\hat{\Pi} := 2\mu\hat{\boldsymbol{\varepsilon}} + \lambda \operatorname{tr}(\hat{\boldsymbol{\varepsilon}})\hat{\mathbb{I}} \tag{5}$$

with $\hat{\boldsymbol{\varepsilon}} = \frac{1}{2}(\hat{\nabla}\hat{\boldsymbol{u}}_s + \hat{\nabla}\hat{\boldsymbol{u}}_s^T)$. Here, p_f is the fluid pressure, $\boldsymbol{v}_f$ the fluid velocity, $\hat{\boldsymbol{v}}_s$ the structure velocity, $\hat{\boldsymbol{u}}_s$ the displacement, ν_f the kinematic viscosity, and $\hat{\rho}_s^0$ and ρ_f the densities of the solid and fluid, respectively. Finally, μ and λ are the Lamé coefficients (see Sect. 4.1.4 or [21, 22]). For the structure density, we choose $\hat{\rho}_s^0 = 1000\,\text{kg}\,\text{m}^{-3}$. For the structure equations, we used a mixed formulation with $\hat{\boldsymbol{v}}_s = \partial_t \hat{\boldsymbol{u}}_s$ which will allow us to use the same time discretization method for fluid and solid equations.

The interface conditions are continuity of the velocities, displacements, and normal stresses,

$$\begin{aligned} \boldsymbol{v}_f &= \boldsymbol{v}_s \quad \text{on } \Gamma_{\text{interface}} \, , \\ \hat{\boldsymbol{u}}_f &= \hat{\boldsymbol{u}}_s \quad \text{on } \hat{\Gamma}_{\text{interface}} \, , \\ \hat{J}\hat{\mathbb{T}}_f \hat{\mathbf{F}}^{-T} \hat{\boldsymbol{n}}_f + \hat{\Pi}\hat{\boldsymbol{n}}_s &= 0 \quad \text{on } \hat{\Gamma}_{\text{interface}} \, , \end{aligned} \tag{6}$$

where $\hat{\boldsymbol{n}}_f$ and $\hat{\boldsymbol{n}}_s$ are the outward normal vectors and $\hat{J} = \det \hat{\mathbf{F}}$.

In the following, the formulation will be transformed into a common framework using the ALE transformation (2). Since the fluid deformations in the human airway application are rather small, we use a simple model for computing the displacement $\hat{\boldsymbol{u}}_f$ for the ALE transformation [23, 24],

$$-\widehat{\operatorname{div}}(\alpha_u \hat{\nabla} \hat{\boldsymbol{u}}_f) = 0 \, , \; \hat{\boldsymbol{u}}_f = \hat{\boldsymbol{u}}_s \text{ on } \hat{\Gamma}_{\text{interface}} \, , \; \hat{\boldsymbol{u}}_f = 0 \text{ on } \partial\hat{\Omega}_f \setminus \hat{\Gamma}_{\text{interface}}, \tag{7}$$

with a diffusion parameter $\alpha_u := \alpha_u(\hat{x})$.

3.2 *Variational Formulation of the FSI Problem*

In this section, we present the monolithic variational formulation and outline the key steps for the numerical discretization. The monolithic FSI formulation in ALE coordinates and their discretization is based on [25, 26] and [27].

To derive the monolithic variational formulation, the fluid equations are transformed into the ALE framework and formulated in the reference domain. To this end, we define the following function spaces:

$$\begin{aligned} \hat{V} &:= \{\hat{\boldsymbol{v}}_f \in H^1(\hat{\Omega}_f)^2 : \hat{\boldsymbol{v}}_f = 0 \text{ on } \hat{\Gamma}_2 \, , \hat{\boldsymbol{v}}_f = \hat{\boldsymbol{v}}_s \text{ on } \hat{\Gamma}_{\text{interface}}\} \, , \\ \hat{U} &:= \{\hat{\boldsymbol{u}}_f \in H^1(\hat{\Omega}_f)^2 : \hat{\boldsymbol{u}}_f = 0 \text{ on } \hat{\Gamma}_1 \cup \hat{\Gamma}_2 \cup \hat{\Gamma}_6 \, , \hat{\boldsymbol{u}}_f = \hat{\boldsymbol{u}}_s \text{ on } \hat{\Gamma}_{\text{interface}}\} \, , \\ \hat{V}_0 &:= \{\hat{\boldsymbol{\psi}}_f \in H^1(\hat{\Omega}_f)^2 : \hat{\boldsymbol{\psi}}_f = 0 \text{ on } \hat{\Gamma}_2 \, , \hat{\boldsymbol{\psi}}_f = \hat{\boldsymbol{\psi}}_s \text{ on } \hat{\Gamma}_{\text{interface}}\} \, , \\ \hat{U}_0 &:= \{\hat{\boldsymbol{\psi}}_f \in H^1(\hat{\Omega}_f)^2 : \hat{\boldsymbol{\psi}}_f = 0 \text{ on } \hat{\Gamma}_1 \cup \hat{\Gamma}_2 \cup \hat{\Gamma}_6 \cup \hat{\Gamma}_{\text{interface}}\} \, , \\ \hat{W} &:= \{\hat{\boldsymbol{w}}_s \in H^1(\hat{\Omega}_s)^2 : \hat{\boldsymbol{w}}_s = 0 \text{ on } \hat{\Gamma}_3 \cup \hat{\Gamma}_5\} \, , \\ \hat{L}_f &:= L^2(\hat{\Omega}_f) \, , \\ \hat{L}_s^2 &:= L^2(\hat{\Omega}_s)^2 \end{aligned}$$

Using the simple harmonic equation for the ALE transformation (7) and transforming the integrals to the reference domains, the variational formulation reads as follows:

Find $\{\hat{\boldsymbol{v}}_f, \hat{\boldsymbol{v}}_s, \hat{\boldsymbol{u}}_f, \hat{\boldsymbol{u}}_s, \hat{p}_f\} \in \hat{V} \times \hat{W} \times \hat{U} \times \hat{W} \times \hat{L}_f$ such that $\hat{\boldsymbol{v}}_f(0) = \hat{\boldsymbol{v}}_f^0$, $\hat{\boldsymbol{v}}_s(0) = \hat{\boldsymbol{v}}_s^0$, $\hat{\boldsymbol{u}}_f(0) = \hat{\boldsymbol{u}}_f^0$, $\hat{\boldsymbol{u}}_s(0) = \hat{\boldsymbol{u}}_s^0$ and for almost all time steps $t \in I$ it holds

$$
\begin{aligned}
\left(\hat{J}\hat{\rho}_f \partial_t \hat{\boldsymbol{v}}_f, \hat{\boldsymbol{\psi}}_f^v\right)_{\hat{\Omega}_f} + \left(\hat{\rho}_f \hat{J}(\hat{\mathbf{F}}^{-1}(\hat{\boldsymbol{v}}_f - \partial_t \hat{\boldsymbol{u}}_f)\cdot\hat{\nabla})\hat{\boldsymbol{v}}_f, \hat{\boldsymbol{\psi}}_f^v\right)_{\hat{\Omega}_f}& \\
+ \left(\hat{J}\hat{\mathbb{T}}_f\hat{\mathbf{F}}^{-T}, \hat{\nabla}\hat{\boldsymbol{\psi}}_f^v\right)_{\hat{\Omega}_f} - \left\langle \hat{J}\hat{\boldsymbol{h}}_f\hat{\mathbf{F}}^{-T}\hat{\boldsymbol{n}}_f, \hat{\boldsymbol{\psi}}_f^v\right\rangle_{\hat{\Gamma}_1}& \\
- \left\langle \hat{J}\hat{\boldsymbol{h}}_f\hat{\mathbf{F}}^{-T}\hat{\boldsymbol{n}}_f, \hat{\boldsymbol{\psi}}_f^v\right\rangle_{\hat{\Gamma}_6} &= 0 \;\forall \hat{\boldsymbol{\psi}}_f^v \in \hat{V}_0 \,, \\
\left(\widehat{\mathrm{div}}(\hat{J}\hat{\mathbf{F}}^{-1}\hat{\boldsymbol{v}}_f), \hat{\psi}_f^p\right)_{\hat{\Omega}_f} &= 0 \;\forall \hat{\psi}_f^p \in \hat{L}_f \,, \\
\left(\hat{\rho}_s^0 \partial_t \hat{\boldsymbol{v}}_s, \hat{\boldsymbol{\psi}}_s^v\right)_{\hat{\Omega}_s} + \left(\hat{\Pi}, \hat{\nabla}\hat{\boldsymbol{\psi}}_s^v\right)_{\hat{\Omega}_s} &= 0 \;\forall \hat{\boldsymbol{\psi}}_s^v \in \hat{W}_s \,, \\
\left(\partial_t \hat{\boldsymbol{u}}_s - \hat{\boldsymbol{v}}_s, \hat{\boldsymbol{\psi}}_s^u\right)_{\hat{\Omega}_s} &= 0 \;\forall \hat{\boldsymbol{\psi}}_s^u \in \hat{L}_s^2 \,, \\
\left(\alpha_u \hat{\nabla}\hat{\boldsymbol{u}}_f, \hat{\nabla}\hat{\boldsymbol{\psi}}_f^u\right)_{\hat{\Omega}_f} &= 0 \;\forall \hat{\boldsymbol{\psi}}_f^u \in \hat{U}_0
\end{aligned}
$$

with

$$
\hat{\mathbb{T}}_f = -\hat{p}_f\hat{\mathbb{I}} + \hat{\rho}_f\nu_f(\hat{\nabla}\hat{\boldsymbol{v}}_f\hat{\mathbf{F}}^{-1} + \hat{\mathbf{F}}^{-T}\hat{\nabla}\hat{\boldsymbol{v}}_f^T) \,.
$$

Here, $(\cdot,\cdot)_{\hat{\Omega}_f}$ denotes the usual $L^2(\hat{\Omega}_f)$ product. The first two interface conditions in (6) are both incorporated into the Sobolev spaces $\hat{V}$ and $\hat{U}$, respectively. The continuity of the normal stresses is fulfilled in a variational sense:

$$
\int_{\hat{\Gamma}_{\text{interface}}} \hat{J}\hat{\mathbb{T}}_f\hat{\mathbf{F}}^{-T}\hat{\boldsymbol{n}}_f\hat{\boldsymbol{\psi}}_f^v \,\mathrm{d}\hat{s} + \int_{\hat{\Gamma}_{\text{interface}}} \hat{\Pi}\hat{\boldsymbol{n}}_s\hat{\boldsymbol{\psi}}_s^v \,\mathrm{d}\hat{s} = 0
$$

The pressure condition (1) is included as a Neumann condition in the variational formulation with

$$
\begin{aligned}
\hat{\boldsymbol{h}}_f &= \hat{\rho}_f\nu_f\hat{\mathbf{F}}^{-T}\hat{\nabla}\hat{\boldsymbol{v}}_f^T + lA_\omega\cos(\omega t)\hat{I} && \text{on } \hat{\Gamma}_1 \,, \\
\hat{\boldsymbol{h}}_f &= \hat{\rho}_f\nu_f\hat{\mathbf{F}}^{-T}\hat{\nabla}\hat{\boldsymbol{v}}_f^T && \text{on } \hat{\Gamma}_6 \,.
\end{aligned}
$$

For a more general formulation and more complex models like viscoelastic fluids and nonlinear elasticity, we refer to [28] and [29].

For completeness, we also state the variational formulation for the pure Navier-Stokes case where the domain and boundaries simplify as indicated in Fig. 4. To this

Fig. 4 The domain and different boundaries for the pure Navier-Stokes (rigid walls channel) case

end, we define the function spaces

$$V := \{\boldsymbol{v}_f \in H^1(\Omega_f)^2 : \boldsymbol{v}_f = 0 \text{ on } \Gamma_2\} ,$$
$$L_f := L^2(\Omega_f) .$$

Then, the variational formulation reads as follows:

Find $\{\boldsymbol{v}_f, p_f\} \in V \times L_f$ such that $\boldsymbol{v}_f(0) = \boldsymbol{v}_f^0$ and for almost all time steps $t \in I$ it holds

$$\rho_f(\partial_t \boldsymbol{v}_f, \boldsymbol{\psi}^v)_\Omega + \rho_f((\boldsymbol{v}_f \cdot \nabla)\boldsymbol{v}_f, \boldsymbol{\psi}^v)_{\Omega_f}$$
$$+(\mathbb{T}, \nabla\boldsymbol{\psi}^v)_{\Omega_f} - \langle \boldsymbol{h}_f, \boldsymbol{\psi}^v\rangle_{\Gamma_1} - \langle \boldsymbol{h}_f, \boldsymbol{\psi}^v\rangle_{\Gamma_6} = 0 \; \forall \boldsymbol{\psi}^v \in V ,$$
$$(\text{div}\boldsymbol{v}, \psi^p)_\Omega = 0 \; \forall \psi^p \in L_f$$

with

$$\mathbb{T} = -p_f \mathbb{I} + \rho_f \nu_f (\nabla \boldsymbol{v}_f + \nabla \boldsymbol{v}_f^T)$$

and

$$\boldsymbol{h}_f = \rho_f \nu_f \nabla \boldsymbol{v}_f^T n_f + l A_\omega \cos(\omega t) \boldsymbol{n}_f \quad \text{on } \Gamma_1 ,$$
$$\boldsymbol{h}_f = \rho_f \nu_f \nabla \boldsymbol{v}_f^T \boldsymbol{n}_f \quad \text{on } \Gamma_6 .$$

3.3 *Discretization*

In the following, we briefly recapitulate the steps for the numerical discretization.

For the temporal discretization, we use the shifted Crank-Nicolson scheme (see, for example, [30]). The time derivatives are approximated using a difference quotient. The pressure terms and the equation for the ALE transformation are treated implicitly.

Here, we will only show the discretization for the structure terms and refer to [29] for details.

First, the time interval $I = [0, T]$ is divided into discrete time steps

$$0 = t_0 < t_1 < \cdots < t_N = T$$

with $I_n := (t_{n-1}, t_n]$ and $\Delta t_n := t_n - t_{n-1}$.

The structure terms are split into terms involving time derivatives and remaining terms $\hat{A}^s(\hat{\boldsymbol{U}})(\hat{\boldsymbol{\psi}}) = \hat{A}^s_T(\hat{\boldsymbol{U}})(\hat{\boldsymbol{\psi}}) + \hat{A}^s_R(\hat{\boldsymbol{U}})(\hat{\boldsymbol{\psi}})$ with

$$\hat{A}^s_T(\hat{\boldsymbol{U}})(\hat{\boldsymbol{\psi}}) = \left(\hat{\rho}^0_s \partial_t \hat{\boldsymbol{v}}_s, \hat{\boldsymbol{\psi}}^v_s\right)_{\hat{\Omega}_s} + \left(\partial_t \hat{\boldsymbol{u}}_s, \hat{\boldsymbol{\psi}}^u_s\right)_{\hat{\Omega}_s},$$

$$\hat{A}^s_R(\hat{\boldsymbol{U}})(\hat{\boldsymbol{\psi}}) = \left(\hat{\Pi}, \hat{\nabla}\hat{\boldsymbol{\psi}}^v_s\right)_{\hat{\Omega}_s} - \left(\hat{\boldsymbol{v}}_s, \hat{\boldsymbol{\psi}}^u_s\right)_{\hat{\Omega}_s}.$$

Then, we have

$$\hat{A}^s_T(\hat{\boldsymbol{U}}^{n,\Delta t_n})(\hat{\boldsymbol{\psi}}) \approx \frac{1}{\Delta t_n}\hat{\rho}^0_s \left(\hat{\boldsymbol{v}}^n_s - \hat{\boldsymbol{v}}^{n-1}_s, \hat{\boldsymbol{\psi}}^v_s\right)_{\hat{\Omega}_s} + \frac{1}{\Delta t_n}\left(\hat{\boldsymbol{u}}^n_s - \hat{\boldsymbol{u}}^{n-1}_s, \hat{\boldsymbol{\psi}}^u_s\right)_{\hat{\Omega}_s},$$

and the problem after temporal discretization reads as follows:

For a given previous timestep solution $\hat{\boldsymbol{U}}^{n-1} = \left\{\hat{\boldsymbol{v}}^{n-1}_f, \hat{\boldsymbol{v}}^{n-1}_s, \hat{\boldsymbol{u}}^{n-1}_f, \hat{\boldsymbol{u}}^{n-1}_s, \hat{p}^{n-1}_f\right\}$, find $\hat{\boldsymbol{U}}^n = \left\{\hat{\boldsymbol{v}}^n_f, \hat{\boldsymbol{v}}^n_s, \hat{\boldsymbol{u}}^n_f, \hat{\boldsymbol{u}}^n_s, \hat{p}^n_f\right\}$ such that

$$\begin{aligned}
&\hat{A}^s_T(\hat{\boldsymbol{U}}^{n,\Delta t_n})(\hat{\boldsymbol{\psi}}) + \hat{A}^f_T(\hat{\boldsymbol{U}}^{n,\Delta t_n})(\hat{\boldsymbol{\psi}}) + \hat{A}_I(\hat{\boldsymbol{U}}^n)(\hat{\boldsymbol{\psi}}) + \theta\hat{A}^s_R(\hat{\boldsymbol{U}}^n)(\hat{\boldsymbol{\psi}}) + \theta\hat{A}^f_R(\hat{\boldsymbol{U}}^n)(\hat{\boldsymbol{\psi}}) \\
&= -(1-\theta)\hat{A}^s_R(\hat{\boldsymbol{U}}^{n-1})(\hat{\boldsymbol{\psi}}) - (1-\theta)\hat{A}^f_R(\hat{\boldsymbol{U}}^{n-1})(\hat{\boldsymbol{\psi}})
\end{aligned}$$

with $\theta = 0.5 + O(\Delta t)$, $\Delta t := \max\{\Delta t_n\}$, and $\hat{\boldsymbol{U}}^n = \hat{U}(t_n)$. Here, $\hat{A}_I(\hat{\boldsymbol{U}}^n)(\hat{\boldsymbol{\psi}})$ includes the terms which are treated implicitly, $\hat{A}^f_T(\hat{\boldsymbol{U}}^{n,\Delta t_n})(\hat{\boldsymbol{\psi}})$ includes the terms involving time derivatives, and $\hat{A}^f_R(\hat{\boldsymbol{U}}^n)(\hat{\boldsymbol{\psi}})$ includes the remaining fluid terms [29].

The spatial discretization is done using the finite element method. For the finite elements, we choose quadratic continuous elements for the velocity and the displacement and linear continuous elements for the pressure: $Q^2_2 \times Q^2_2 \times Q_1$. The combination of velocity and pressure space satisfies the inf-sup condition (see, for example, [27]).

Since the equations are nonlinear, we use Newton's method for linearization. For details on the temporal and spatial discretization and the corresponding Gateaux derivatives, we refer to [29]. The resulting linear system is solved using the parallel direct solver Mumps [31].

4 Finite Difference Solver

Due to geometrical simplicity of the considered case, the finite difference method was chosen, as it enables easy and accurate approximation of the governing equations on a structured, orthogonal, equispaced grid. It allows for straightforward implementation and simple modification of numerous schemes and algorithms that were tested for the considered case.

4.1 *Coupled 2D-1D Model with Rigid-Elastic Walls*

4.1.1 Formulation of the 2D Model

As for the fully 2D FEM model, here also the well-known system of Navier-Stokes equations for an incompressible, Newtonian fluid with constant density is used. In this case, however, the conservative (or divergence) form of the governing system was chosen, because it better corresponds to the coupled 1D FSI system written in a similar manner. Again, the governing system consists of the linear momentum balance equation written now in the form

$$\frac{\partial \boldsymbol{v}}{\partial t} + \nabla \cdot (\boldsymbol{v} \otimes \boldsymbol{v}) = -\frac{1}{\rho} \nabla p + \nu \Delta \boldsymbol{v} \,, \tag{8}$$

which is accompanied by the divergence-free incompressibility constraint

$$\nabla \cdot \boldsymbol{v} = 0 \,. \tag{9}$$

To solve this governing system of partial differential equations, a simple explicit finite difference central scheme is used, together with the pressure correction method for velocity-pressure coupling. The Jacobi solver is employed for the discretized Poisson pressure-correction equation. For more detailed information regarding the implementation of this numerical approach, see Sect. 4.4.

4.1.2 Formulation of the 1D Simplified Model

A simple one-dimensional FSI model is used to describe the flow of an incompressible viscous fluid within an elastic tube. The tube is assumed to possess radial symmetry, and its (circular) cross-sectional size can vary in both time and space, depending on the difference between internal and external pressure. The one-dimensional model is derived from the fully three-dimensional incompressible Navier-Stokes equations, connected with generalized string model with only elastic response function [13, 14, 32, 33]. Since this study focuses on a 2D-1D coupling, it is appropriate to introduce and discuss suitable modifications to ensure consistency with the lower-dimensional formulation. In this context, two possible modifications are considered to better reflect the underlying physics in the 2D case.

This derivation involves making certain simplifying assumptions and integrating the variables over the cross section [13]. The resulting system of coupled evolutionary partial differential equations can be expressed as follows

$$\frac{\partial A}{\partial t} + \frac{\partial Q}{\partial x} = 0\,, \tag{10}$$

$$\frac{\partial Q}{\partial t} + \frac{\partial}{\partial x}\left(\frac{\alpha Q^2}{A}\right) + \frac{A}{\rho}\frac{\partial P}{\partial x} = -K_R \frac{Q}{A}, \tag{11}$$

where A is cross-sectional area, Q is flow rate, P is pressure over cross section, K_R is friction parameter, and α is the so-called momentum flux correction coefficient. In order to close the system, a relation for the average (over the cross section) pressure P in terms of the system variables Q and A has to be provided. Various pressure laws can be found in the literature. In our case, we used the most common relation

$$P = P(A) = p_{ext} + \psi\left(A(x,t), A_0(x), \beta(x)\right), \quad \beta = \frac{\sqrt{\pi} h_0 E}{1-\kappa_E^2}\,, \tag{12}$$

where E is the vessel wall Young modulus, κ_E is the corresponding Poisson ratio, h_0 is the wall thickness, and p_{ext} is the external pressure from surrounding tissues, which has been neglected (set to $p_{ext} = 0$) in this study.

In context of studies of blood flow, particularly in 3D-1D coupled simulations, the most widely adopted form of the function of ψ is

$$\psi_1(A, A_0, \beta) = \beta \frac{\sqrt{A} - \sqrt{A_0}}{A_0}\,, \qquad \text{where} \quad A_0 = \frac{\pi D_0^2}{4}\,. \tag{13}$$

This relationship was derived from the three-dimensional model, including the expression for the elasticity parameter β. However, in our case, for coupling with a 2D flow model, it seems reasonable to consider alternative formulations that could preserve the physical behavior of the 1D model and its dimensional consistency. In this context, it is important to note that in the 2D setting the cross-sectional area A represents the distance between the plates, meaning that $A = D$. Taking into account these constraints, two possible modifications of ψ were proposed for this study:

$$\psi_2(A, A_0, \beta) = \beta\,\frac{A - A_0}{A_0} \quad \text{and} \quad \psi_3(A, A_0, \beta) = \beta\,\frac{A - A_0}{A_0^2} \tag{14}$$

It should be noted that, in general, the physical dimensions (units) of parameter β and cross-sectional size A_0 differ from those used in the original 3D-based relation (13).

It can be shown that after some manipulations the system (10)–(11) can be rewritten in a conservative form as

$$\frac{\partial \boldsymbol{U}}{\partial t} + \frac{\partial \boldsymbol{F}(\boldsymbol{U})}{\partial x} + \boldsymbol{B}(\boldsymbol{U}) = 0\,, \quad x \in (0, L)\,, \quad t > 0\,, \tag{15}$$

which can easily be converted into the quasi-linear (nonconservative) form:

$$\frac{\partial \boldsymbol{U}}{\partial t} + \boldsymbol{H}(\boldsymbol{U})\frac{\partial \boldsymbol{U}}{\partial x} + \boldsymbol{S}(\boldsymbol{U}) = 0 \tag{16}$$

Here, $\boldsymbol{U} = (A, Q)^T$ is the vector of unknowns (conservative variables), $\boldsymbol{F} = (F^A, F^Q)^T$ is the corresponding flux vector, and $(B^A, B^Q)^T$ is a source term:

$$\boldsymbol{F}(\boldsymbol{U}) = \begin{pmatrix} Q \\ \alpha\frac{Q^2}{A} + \int_0^A \frac{A}{\rho}\frac{\partial \psi}{\partial A} dA \end{pmatrix} \tag{17}$$

The Jacobi matrix $\boldsymbol{H}(\boldsymbol{U})$ based on the flux function $\boldsymbol{F}(\boldsymbol{U})$ may be expressed as

$$\boldsymbol{H}(\boldsymbol{U}) = \frac{\partial \boldsymbol{F}}{\partial \boldsymbol{U}} = \begin{pmatrix} 0 & 1 \\ -\alpha\frac{Q^2}{A^2} + \frac{A}{\rho}\frac{\partial \psi}{\partial A} & 2\alpha\frac{Q}{A}. \end{pmatrix} \tag{18}$$

The linearized source term $\boldsymbol{S}(\boldsymbol{U})$ appearing in (16) is given by

$$\boldsymbol{S}(\boldsymbol{U}) = \begin{pmatrix} 0 \\ K_R\frac{Q}{A} + \frac{A}{\rho}\frac{\partial \psi}{\partial A_0}\frac{dA_0}{dx} + \frac{A}{\rho}\frac{\partial \psi}{\partial \beta}\frac{d\beta}{dx} \end{pmatrix}, \tag{19}$$

considering the original source term in the form

$$\boldsymbol{B}(\boldsymbol{U}) = \boldsymbol{S}(\boldsymbol{U}) - \begin{pmatrix} 0 \\ \frac{\partial C_1}{\partial A_0}\frac{dA_0}{dx} + \frac{\partial C_1}{\partial \beta}\frac{d\beta}{dx} \end{pmatrix}, \tag{20}$$

with the coefficient given by

$$C_1(A, A_0, \beta) = \int_0^A c_1^2(A, A_0, \beta) dA\,, \qquad c_1 = \sqrt{\frac{A}{\rho}\frac{\partial \psi}{\partial A}}\,.$$

The eigenvalues λ_i of the matrix $\boldsymbol{H}$ are the solutions of the characteristic equation

$$\det \begin{pmatrix} -\lambda & 1 \\ -\alpha\frac{Q^2}{A^2} + \frac{A}{\rho}\frac{\partial \psi}{\partial A} & 2\alpha\frac{Q}{A} - \lambda \end{pmatrix} = 0\,,$$

leading to polynomial (quadratic) equation

$$\lambda^2 - 2\lambda\alpha\frac{Q}{A} + \alpha\frac{Q^2}{A^2} - \frac{A}{\rho}\frac{\partial\psi}{\partial A} = 0\,, \tag{21}$$

yielding the eigenvalues

$$\lambda_{1,2} = \frac{2\alpha\frac{Q}{A} \pm \sqrt{4\alpha^2\frac{Q^2}{A^2} - 4\alpha\frac{Q^2}{A^2} + 4\frac{A}{\rho}\frac{\partial\psi}{\partial A}}}{2} = \alpha\frac{Q}{A} \pm \sqrt{\frac{A}{\rho}\frac{\partial\psi}{\partial A} + \alpha(\alpha-1)\frac{Q^2}{A^2}}\,. \tag{22}$$

The above-described model is often used in blood flow simulations, under the assumption of unidirectional (pulsatile but nonreversing) flow, where it is common to assume $\alpha = 1$. This choice simplifies the subsequent mathematical analysis and is therefore widely adopted. Also in the presented case of oscillatory/reversing flow, the assumption of $\alpha = 1$ remains reasonable for the Womersley numbers relevant to this study. For a detailed discussion on the influence of the momentum flux correction coefficient α in the 2D-1D coupling, the reader is referred to [15].

By setting $\alpha = 1$, the eigenvalue formula (22) reduces to

$$\lambda_{1,2} = \frac{Q}{A} \pm \sqrt{\frac{A}{\rho}\frac{\partial\psi}{\partial A}}\,, \tag{23}$$

which is also directly visible from the rewritten characteristic equation (21)

$$\underbrace{\lambda^2 - 2\lambda\frac{Q}{A} + \frac{Q^2}{A^2}}_{\left(\lambda-\frac{Q}{A}\right)^2} = \frac{A}{\rho}\frac{\partial\psi}{\partial A} \quad\Longrightarrow\quad \left(\lambda - \frac{Q}{A}\right)^2 = \frac{A}{\rho}\frac{\partial\psi}{\partial A}\,.$$

It is good to note that this relation also implies the requirement of $\frac{\partial\psi}{\partial A} > 0$ for all admissible pressure-area relations.

The eigenvalues have the dimension of velocity (because the ratio Q/A represents the average velocity), and the form of the expression (23) suggests an interpretation of the individual terms as the convective and elastic/wave part of the characteristic velocity. The expressions for the eigenvalues depending on the chosen function ψ are summarized in Table 2. It is clear that while the convective part of the characteristic velocity $\lambda_{conv} = Q/A$ remains the same for all considered forms of ψ (is not explicitly affected by the choice of ψ), the elastic wave part $\lambda_{wave} = \sqrt{\frac{A}{\rho}\frac{\partial\psi}{\partial A}}$ differs according to the specific form of ψ_i.

Although it is formally possible to keep the same numerical value of the elasticity parameter β in all considered forms of ψ_i (despite different dimensions/units of β needed for model consistency), it seems to be wise to readjust this parameter

Table 2 Comparison of $\frac{\partial\psi}{\partial A}$ and corresponding eigenvalues λ_1 and λ_2 for different functions ψ

ψ	$\psi_1 = \beta\frac{\sqrt{A}-\sqrt{A_0}}{A_0}$	$\psi_2 = \beta_2\frac{A-A_0}{A_0}$	$\psi_3 = \beta_3\frac{A-A_0}{A_0^2}$
$\frac{\partial\psi}{\partial A}$	$\frac{\beta}{2A_0\sqrt{A}}$	$\frac{\beta_2}{A_0}$	$\frac{\beta_3}{A_0^2}$
A_0	$\frac{\pi D_0^2}{4}$	D_0	D_0
$\lambda_{1,2}$	$\frac{Q}{A} \pm \sqrt{\frac{\beta}{2\rho A_0}\sqrt{A}}$	$\frac{Q}{A} \pm \sqrt{\frac{\beta_2}{\rho A_0}A}$	$\frac{Q}{A} \pm \sqrt{\frac{\beta_3}{\rho A_0^2}A}$
$\lambda_{1,2}$	$\frac{4}{\pi D}\frac{Q}{D} \pm \sqrt{\frac{\beta D}{\pi^{1/2}\rho D_0^2}}$	$\frac{Q}{D} \pm \sqrt{\frac{\beta}{\pi^{1/2}\rho}\frac{D}{D_0^2}}$	$\frac{Q}{D} \pm \sqrt{\frac{\beta}{\pi^{1/2}\rho}\frac{D}{D_0^2}}$

for each model separately. In order to preserve the physical similarity between the predictions of the 1D model, the elastic wave characteristic velocity λ_{wave} should be kept identical for all forms of ψ_i. This requirement leads to the condition that the terms under the square root in $\lambda_{wave} = \sqrt{\frac{A}{\rho}\frac{\partial\psi}{\partial A}}$ have to be equal (at the reference area A_0), resulting into the compatibility conditions

$$\frac{A}{\rho}\frac{\partial\psi_1}{\partial A} = \frac{A}{\rho}\frac{\partial\psi_2}{\partial A} = \frac{A}{\rho}\frac{\partial\psi_3}{\partial A}. \tag{24}$$

In order to find the corresponding values $\beta_{2,3}$ for $\psi_{2,3}$, we first simplify the needed expression for the function ψ_1

$$\frac{A}{\rho}\frac{\partial\psi_1}{\partial A} = \frac{A}{\rho}\frac{\beta}{2}\frac{1}{A_0\sqrt{A}} = \frac{\beta}{2\rho}\frac{\sqrt{\frac{\pi D^2}{4}}}{\frac{\pi D_0^2}{4}} = \frac{\sqrt{\pi}\beta D}{\pi\rho D_0^2}.$$

Similarly, the expressions emanating from ψ_2 and ψ_3 are rewritten using the 2D channel width (height) D as

$$\frac{A}{\rho}\frac{\partial\psi_2}{\partial A} = \frac{A}{\rho}\frac{\beta_2}{A_0} = \frac{\beta_2}{\rho}\frac{D}{D_0},$$

$$\frac{A}{\rho}\frac{\partial\psi_3}{\partial A} = \frac{A}{\rho}\frac{\beta_3}{A_0^2} = \frac{\beta_3}{\rho}\frac{D}{D_0^2}.$$

Equating now the expressions according to (24) and denoting $\beta = \beta_1$, the parameters β_2 and β_3 (based on the relations summarized in Table 2) can be expressed as a function of the original parameter β derived from a 3D model

$$\beta_3 = \frac{\beta}{\sqrt{\pi}} = \frac{h_0 E}{1-\kappa_E^2}, \quad \beta_2 = \frac{\beta}{\sqrt{\pi}D_0} = \frac{1}{D_0}\frac{h_0 E}{1-\kappa_E^2}. \tag{25}$$

In this sense, the models using ψ_2 and ψ_3 become equivalent, since the resulting formulations differ only by a constant multiplicative factor. The form using ψ_3 (with elasticity parameter β_3) was chosen for the presented numerical simulations.[1]

Taking a closer look to eigenvalues $\lambda_{1,2} = \lambda_{conv} \pm \lambda_{wave}$, considering the numerical values of both parts, it can be seen that both eigenvalues remain real, distinct, having opposite signs. This indicates that the system is strictly hyperbolic and the characteristic analysis ($\lambda_1 > 0$ and $\lambda_2 < 0$) leads to the conclusion that only one boundary condition can be imposed at each end of the 1D tube (channel).

Furthermore, the stability of the explicit numerical discretization scheme is then determined by the Courant-Friedrichs-Lewy (CFL) condition

$$\Delta t \leq C_{max} \frac{\Delta x}{\max|\lambda_{1,2}|}, \tag{26}$$

where the constant C_{max} depends on the chosen discretization and the characteristic velocity scale is given, for the adopted ψ_i, by the corresponding eigenvalues

$$\lambda_{1,2} = \begin{cases} \dfrac{4}{\pi D}\dfrac{Q}{D} \pm \sqrt{\dfrac{\beta}{\pi^{1/2}\rho}\dfrac{D}{D_0^2}} & \text{for} \quad \psi_1\,, \\[2ex] \dfrac{Q}{D} \pm \sqrt{\dfrac{\beta}{\pi^{1/2}\rho}\dfrac{D}{D_0^2}} & \text{for} \quad \psi_2, \psi_3\,. \end{cases}$$

Assuming a typical reference diameter of $D_0 = 0.018\,\text{m}$, it can be observed that the convective part of the eigenvalues for ψ_1 is significantly larger, since the coefficient $\frac{4}{\pi D}$ is greater than the unit coefficient appearing in the expressions for ψ_2 and ψ_3. Although the wave speed contributions are structurally similar in all formulations, the overall magnitude is greater for ψ_1. Consequently, the CFL condition imposes more restrictive time limitations in simulations based on ψ_1, while the use of ψ_2 or ψ_3 allows a longer time step, improving computational efficiency. These relations are used to estimate the time step in the one-dimensional model presented below, although it should be kept in mind that this estimate is based on linear stability analysis, and thus the resulting values may need to be reduced in practical simulations.

These characteristic analysis and observation have an important impact on the computational setup and its implementation in the 2D-1D coupling strategy described in Sect. 4.2.

---

[1] In this form, the physical units of β_3 remain the same as for the original $\beta_1 = \beta$, although the numerical value now differs.

4.1.3 Wave Character of the 1D Model

The governing system (44)–(45) can be rewritten in compact form[2] (considering $\alpha = 1$) as

$$A_t + Q_x = 0\,, \tag{27}$$

$$Q_t + \left(\frac{Q^2}{A}\right)_x + \frac{A}{\rho}P_x = -K_R\frac{Q}{A}\,, \tag{28}$$

with the constitutive pressure-area relation $P(A) = p_{ext} + \psi(A(x,t), A_0, \beta)$. This coupled system of first-order partial differential equations represents a wave equation (single second-order PDE), which can formally be obtained from the above system by taking a derivative with respect to time of the first equation (27) and substituting it into the derivative with respect to space of the second equation (28):

$$A_{tt} + Q_{xt} = 0 \quad \Longrightarrow \qquad Q_{xt} = -A_{tt}\,,$$

$$Q_{tx} + \left(\frac{Q^2}{A}\right)_{xx} + \left(\frac{A}{\rho}P_x\right)_x = -K_R\left(\frac{Q}{A}\right)_x$$

These expressions can be combined into a single, second-order PDE:

$$A_{tt} = \left(\frac{Q^2}{A}\right)_{xx} + \left(\frac{A}{\rho}P_x\right)_x + K_R\left(\frac{Q}{A}\right)_x$$

By expanding the first term on the right-hand side and substituting the pressure-area constitutive relation, the derivative $P_x = \psi_A A_x$ into the second term leads to

$$A_{tt} = \left(\frac{2QQ_xA - Q^2A_x}{A^2}\right)_x + \left(\frac{A}{\rho}\psi_A A_x\right)_x + K_R\left(\frac{Q}{A}\right)_x\,.$$

This equation can be rearranged into the form of a nonlinear wave equation:

$$A_{tt} = \Bigg(\underbrace{\left(\frac{A}{\rho}\psi_A - \frac{Q^2}{A^2}\right)}_{C^2(A,Q)}A_x\Bigg)_x + \underbrace{\left(\left(2Q_x + K_R\right)\frac{Q}{A}\right)_x}_{G(A,Q)}$$

[2] The short notation utilizing subscripts for partial derivatives is adopted for clarity, where $Q_x = \frac{\partial Q}{\partial x}$ is used, for example.

By further expanding the derivatives on the right-hand side, the following equation can be obtained:

$$A_{tt} = \underbrace{\left(\frac{A}{\rho}\psi_A - \frac{Q^2}{A^2}\right)}_{C^2(A,Q)} A_{xx} + \left(\frac{A}{\rho}\psi_A - \frac{Q^2}{A^2}\right)_x A_x +$$
$$+ \underbrace{\left(2Q_x + K_R\right)_x \left(\frac{Q}{A}\right) + \left(2Q_x + K_R\right)\left(\frac{Q}{A}\right)_x}_{G(A,Q)} \tag{29}$$

Apparently, this is a nonlinear second-order PDE, wavelike equation, with wave speed $C(A, Q)$ and forcing term $G(A, Q)$, i.e.,

$$A_{tt} = \left(C^2(A, Q)\, A_x\right)_x + G(A, Q) \tag{30}$$

The forcing term $G(A, Q)$ can be rewritten using the continuity equation (27) by substituting $Q_x = -A_t$, resulting in

$$G(A, Q) = \left(K_R - 2A_t\right)_x \left(\frac{Q}{A}\right) + \left(K_R - 2A_t\right)\left(\frac{Q}{A}\right)_x . \tag{31}$$

It is interesting to note that this wave equation (30) is related to the quasi-linear form (16) of the original first-order system and its characteristic values (23) as

$$C^2(A, Q) = \left(\frac{A}{\rho}\psi_A - \frac{Q^2}{A^2}\right) = \underbrace{\left(\sqrt{\frac{A}{\rho}\psi_A} + \frac{Q}{A}\right)}_{\lambda_1}\underbrace{\left(\sqrt{\frac{A}{\rho}\psi_A} - \frac{Q}{A}\right)}_{-\lambda_2}$$
$$= -\lambda_1\lambda_2 = -\det \boldsymbol{H} .$$

Continuing with this observation and considering the principal invariants (trace and determinant of $\boldsymbol{H}$)

$$\operatorname{tr} \boldsymbol{H} = \lambda_1 + \lambda_2 = 2\frac{Q}{A} \qquad \text{and} \qquad \det \boldsymbol{H} = \lambda_1\lambda_2 = \left(\frac{Q^2}{A^2} - \frac{A}{\rho}\psi_A\right),$$

the wave equation (29) can be written as

$$A_{tt} = -\left(\det \boldsymbol{H}\right)A_{xx} - \left(\det \boldsymbol{H}\right)_x A_x + \left(\operatorname{tr} \boldsymbol{H}\right)Q_{xx} + \left(\operatorname{tr} \boldsymbol{H}\right)_x\left(Q_x + K_R/2\right), \tag{32}$$

where the matrix $\boldsymbol{H}$ of the original system is given by (18) for $\alpha = 1$ as

$$\boldsymbol{H}(A, Q) = \begin{pmatrix} 0 & 1 \\ -\frac{Q^2}{A^2} + \frac{A}{\rho}\psi_A & 2\frac{Q}{A} \end{pmatrix} = \begin{pmatrix} 0 & 1 \\ -\lambda_1\lambda_2 & \lambda_1 + \lambda_2 \end{pmatrix} = \begin{pmatrix} 0 & 1 \\ -\det \boldsymbol{H} & \operatorname{tr} \boldsymbol{H} \end{pmatrix}.$$

Using the continuity equation (27) to replace the spatial derivatives of Q by time derivatives of A (like in (31)), the wave equation (32) can be expressed in terms of the derivatives of A only, as

$$A_{tt} = -\left(\det \boldsymbol{H}\, A_x\right)_x - \left(\operatorname{tr} \boldsymbol{H}\, A_t\right)_x + \left(\operatorname{tr} \boldsymbol{H}\right)_x \left(K_R/2\right). \tag{33}$$

It should be kept in mind that the wave equations (29) and (30) are nonlinear, meaning that the wave speed $C(A, Q)$ and forcing term $G(A, Q)$ depend on the solution $A(x, t)$ and $Q(x, t)$. This translates into the solution-dependent coefficients of the Eq. (33), where the invariants $\operatorname{tr} \boldsymbol{H}(A, Q)$ and $\det \boldsymbol{H}(A, Q)$ include the dependence on the sought $A(x, t)$ and $Q(x, t)$. Interestingly, the wave speed $C(A, Q) = C(\det \boldsymbol{H})$ only depends on the determinant of $\boldsymbol{H}$ and is thus dominated by the elastic wave speed. On the other hand, the forcing source term $G(A, Q) = G(\operatorname{tr} \boldsymbol{H})$ only depends on the mean velocity Q/A and doesn't include any effects of the wall elasticity.

4.1.4 Material Parameters in the 1D Model

The fluid-structure interaction in the 1D FDM and 2D FEM models is determined by the structural characteristics of the material of the elastic wall of the test channel. In the 1D model, the Young modulus E and the Poisson ratio κ_E were combined into the elasticity parameter β in the pressure-area constitutive relation. In the 2D FSI simulations, the linear elasticity model (generalized Hooke's law) is represented by the Piola-Kirchhoff stress tensor defined using the Lamé coefficients (moduli) μ and λ. Here, μ is the shear modulus.[3] The coefficient λ has no particular independent physical interpretation. It can, however, be associated [22] with the bulk (elastic compression/expansion) modulus κ as

$$\kappa = \frac{E}{3(1 - 2\kappa_E)} \qquad \Longrightarrow \qquad \lambda = \kappa - \frac{2}{3}\mu. \tag{34}$$

The parameters E and κ_E used in the presented 1D model are linked to the Lamé parameters μ and λ of the 2D model [22, 34] by

$$E = \frac{\mu(2\mu + 3\lambda)}{\mu + \lambda} \qquad \text{and} \qquad \kappa_E = \frac{\lambda}{2(\mu + \lambda)} \tag{35}$$

[3] Typically denoted G in contrast to Young (tension) modulus E in the engineering literature.

from which the inverse relations can be derived

$$\mu = \frac{E}{2(1+\kappa_E)} \qquad \text{and} \qquad \lambda = \frac{2\kappa_E}{1-2\kappa_E}\mu = \frac{E\,\kappa_E}{(1-2\kappa_E)(1+\kappa_E)}\,. \tag{36}$$

The details on the form of the elastic parameter β can be found in [35] or [36].

4.2 Coupling Strategy for the 2D-1D Model

In order to couple the previously described 2D and 1D FSI models, an explicit coupling approach was chosen and implemented. At first glance, it appears logical to require that the local cross-sectional area A and the flow rate Q at the common interface Γ_3 of the 2D model match the corresponding values of the attached 1D simplified model at each time instant. However, despite its desirability, this alignment cannot be strictly enforced for the coupled model. The primary challenge arises from the mathematical properties of the 1D FSI model, which can be proved to be a hyperbolic system of partial differential equations (PDEs) (see previous Sects. 4.1.2 and 4.1.3). This classification implies, among other issues, that only one boundary condition can be prescribed at each end of the 1D domain. Consequently, it is not possible to prescribe both variables at the common 2D-1D coupling interface. This restriction naturally leads to two fundamental coupling strategies, each allowing only one of the solved quantities (Q or P) to be controlled at the 1D model's interface while the other variable must be set at the opposite uncoupled end of the 1D domain.

In this study, the so-called flow rate continuous coupling is used (see Fig. 5). At each time step, the flow rate Q is first calculated from the 2D model and then consequently used as a boundary condition at the interface for the 1D model. The other way around, in the next time step, the average pressure determined by the 1D model at the coupling interface is imposed as a boundary condition for the 2D model. More information on coupling strategies and their problems can be found in [5, 15, 32, 37].

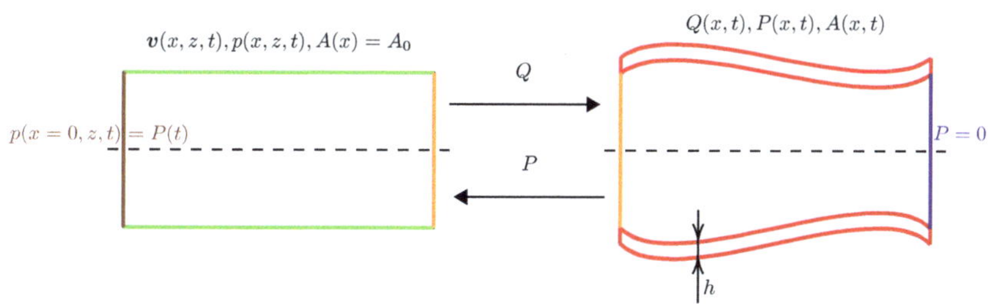

Fig. 5 Flow rate continuous coupling strategy

It is important to note and emphasize that due to the impossibility of prescribing both Q and P simultaneously at the common coupling interface a discontinuity (jump) may occur in one of the quantities. In the chosen approach, only the flow rate Q is passed from the 2D to the 1D model. This ensures that the flow rate will remain continuous across the interface; however, this results in discontinuity of the cross section A at the coupling interface. Therefore, one should be cautious when interpreting the structural (wall) displacements resulting from the 1D model.

Also, pressure can potentially exhibit a minor inconsistency at the common 2D-1D interface due to the explicit time stepping in the coupling procedure, where to obtain new values of flow quantities (at time level $n+1$) in the 2D part of the domain the 1D pressure from the previous time step (time level n) is used at the coupling interface. New pressures are only known at the extreme ends of the coupled 2D-1D domain, that is, at Γ_1 and Γ_6, while the pressure at the common interface Γ_3/Γ_5 needs to be taken from the previous time step. In the present case, this small delay in imposing the pressure drop and possible visual discontinuity of the 2D-1D pressure profile at the common interface is negligible due to the very short time step used in the simulations. This explicit inaccuracy can be either fully removed or at least minimized by sub-iterating the $Q-P$ data transfer between 2D and 1D parts, but such a minor improvement in accuracy comes at nonnegligible additional computational cost. For further details on the coupling approach, see [15, 37].

4.3 Pressure Correction Method

In the presented 2D finite difference simulations, the pressure p and velocity $\boldsymbol{v}$ in the incompressible Navier-Stokes system are linked by means of a pressure correction method. The fundamental concept of the pressure correction method is based on the separation of the calculation of the pressure field from the velocity field. This is achieved by first solving the momentum equation with a known pressure field, such as the pressure obtained in the previous iteration. In the present simulations, the intermediate velocity field $\boldsymbol{v}^*$ is solved from an explicit equation emanating from (8), using the known velocity $\boldsymbol{v}^n$ and pressure p^n from the previous time level n

$$\frac{\boldsymbol{v}^* - \boldsymbol{v}^n}{\Delta t} = -\nabla \cdot (\boldsymbol{v} \otimes \boldsymbol{v})^n - \frac{1}{\rho}\nabla p^n + \nu \Delta \boldsymbol{v}^n \quad \Longrightarrow \quad \boldsymbol{v}^*. \tag{37}$$

This intermediate velocity field $\boldsymbol{v}^*$, however, does not satisfy the continuity equation (the divergence-free constraint (9)). Therefore, it is necessary to update (correct) the newly computed pressure p^{n+1} with respect to p^n and velocity $\boldsymbol{v}^{n+1}$ with respect to $\boldsymbol{v}^*$ in such a way that the velocity field becomes divergence-free again and mass continuity is preserved. This adjustment involves adding the velocity and pressure corrections $\boldsymbol{v}' and p'$ to the intermediate (last known) values

$$\boldsymbol{v}^{n+1} = \boldsymbol{v}^* + \boldsymbol{v}', \qquad p^{n+1} = p^n + p', \tag{38}$$

wherein these final (next time level) values $\boldsymbol{v}^{n+1}$ and p^{n+1} are required to fulfill both the momentum equation

$$\frac{\boldsymbol{v}^{n+1} - \boldsymbol{v}^n}{\Delta t} = -\nabla \cdot (\boldsymbol{v} \otimes \boldsymbol{v})^n - \frac{1}{\rho} \nabla p^{n+1} + \nu \Delta \boldsymbol{v}^n \tag{39}$$

and the divergence-free constraint

$$\nabla \cdot \boldsymbol{v}^{n+1} = 0 \,. \tag{40}$$

The equations for the pressure and velocity corrections are derived by introducing the corrections (38) into Eq. (37)

$$\frac{\boldsymbol{v}^{n+1} - \boldsymbol{v}' - \boldsymbol{v}^n}{\Delta t} = -\nabla \cdot (\boldsymbol{v} \otimes \boldsymbol{v})^n - \frac{1}{\rho} \nabla \left(p^{n+1} - p' \right) + \nu \Delta \boldsymbol{v}^n \tag{41}$$

and subtracting equation (39) from (41). This leads to the relationship between pressure and velocity corrections in the form

$$\boldsymbol{v}' = -\frac{\Delta t}{\rho} \nabla p', \tag{42}$$

from which the velocity correction $\boldsymbol{v}'$ can be calculated, provided that the pressure correction p' is already known. To obtain the equation for the pressure correction, the velocity correction from (38) needs to be substituted into the continuity equation (40), resulting in

$$\nabla \cdot (\boldsymbol{v}^* + \boldsymbol{v}') = 0 \,.$$

Following the substitution from Eq. (42) into (4.3)

$$\nabla \cdot (\boldsymbol{v}^* - \frac{\Delta t}{\rho} \nabla p') = 0 \,,$$

after rearranging, the Poisson equation for pressure correction is obtained in the form

$$\Delta p' = \frac{\rho}{\Delta t} \nabla \cdot \boldsymbol{v}^* \quad \Rightarrow p' \,, \tag{43}$$

where the source term on the right-hand side is proportional to the mass imbalance in the continuity equation considering the intermediate velocity $\boldsymbol{v}^*$.

In summary, by first calculating the p' from (43), introducing it to (42) to obtain $\boldsymbol{v}'$, the intermediate velocity field $\boldsymbol{v}^*$ and the old pressure field p^n can be updated using (38) to obtain the values $\boldsymbol{v}^{n+1}$ and p^{n+1} at the new time level. This procedure was followed in the below presented 2D FDM simulations.

4.4 Discretization of the 2D Governing Equations

The momentum equation (37) for the intermediate velocity $\boldsymbol{v}^* = (v^*, w^*)$ is discretized by a simple central finite difference scheme on an orthogonal equispaced grid with cell sizes Δx and Δz in the $x - z$ coordinate system:

$$\frac{v^*_{i,j} - v^n_{i,j}}{\Delta t} = -\frac{v^n_{i+1,j}v^n_{i+1,j} - v^n_{i-1,j}v^n_{i-1,j}}{2\Delta x} - \frac{v^n_{i,j+1}w^n_{i,j+1} - v^n_{i,j-1}w^n_{i,j-1}}{2\Delta z} -$$
$$-\frac{1}{\rho}\frac{p^n_{i+1,j} - p^n_{i-1,j}}{2\Delta x} + \nu\frac{v^n_{i+1,j} - 2v^n_{i,j} + v^n_{i-1,j}}{\Delta x^2} + \nu\frac{v^n_{i,j+1} - 2v^n_{i,j} + v^n_{i,j-1}}{\Delta z^2}$$

$$\frac{w^*_{i,j} - w^n_{i,j}}{\Delta t} = -\frac{w^n_{i+1,j}v^n_{i+1,j} - w^n_{i-1,j}v^n_{i-1,j}}{2\Delta x} - \frac{w^n_{i,j+1}w^n_{i,j+1} - w^n_{i,j-1}w^n_{i,j-1}}{2\Delta z} -$$
$$-\frac{1}{\rho}\frac{p^n_{i,j+1} - p^n_{i,j-1}}{2\Delta z} + \nu\frac{w^n_{i+1,j} - 2w^n_{i,j} + w^n_{i-1,j}}{\Delta x^2} + \nu\frac{w^n_{i,j+1} - 2w^n_{i,j} + w^n_{i,j-1}}{\Delta z^2}$$

From these discretized equations, the intermediate velocity $\boldsymbol{v}^*_{i,j} = (v^*_{i,j}, w^*_{i,j})$ can be obtained explicitly using just the values at the previous time level n. Once the velocity estimate $\boldsymbol{v}^*_{i,j}$ is known, the pressure correction $p'i, j$ is calculated from the Poisson equation (43), discretized in a similar way as the momentum equations, leading to

$$\frac{p'i+1,j - 2p'i,j + p'i-1,j}{\Delta x^2} + \frac{p'i,j+1 - 2p'i,j + p'i,j-1}{\Delta z^2} =$$
$$= \frac{\rho}{\Delta t}\left(\frac{v^*_{i+1,j} - v^*_{i-1,j}}{2\Delta x} + \frac{w^*_{i,j+1} - w^*_{i,j-1}}{2\Delta z}\right).$$

This system of equations for pressure corrections $p'_{i,j}$ at the internal points of the computational domain is solved iteratively by the Gauss-Seidel method.

The pressure at the time level $(n+1)$ is computed simply by adding the pressure correction, i.e., $p^{n+1}_{i,j} = p^n_{i,j} + p'_{i,j}$. The velocity correction $\boldsymbol{v}'_{i,j} = (v'_{i,j}, w'_{i,j})$ is calculated from the discretized equation (42):

$$v'_{i,j} = \frac{\Delta t}{\rho}\frac{p'_{i+1,j} - p'_{i-1,j}}{2\Delta x}, \qquad w'_{i,j} = \frac{\Delta t}{\rho}\frac{p'_{i,j+1} - p'_{i,j-1}}{2\Delta z}$$

Finally, the new velocity at the time level $(n+1)$ is obtained by applying the correction according to (38), i.e., using $\boldsymbol{v}^{n+1}_{i,j} = \boldsymbol{v}^*_{i,j} + \boldsymbol{v}'i, j$:

$$v^{n+1}_{i,j} = v^*_{i,j} + v'_{i,j}, \qquad w^{n+1}_{i,j} = w^*_{i,j} + w'_{i,j}$$

4.5 Discretization of the 1D Governing Equations

The earlier derived and analyzed system of governing equations for the 1D model, describing the spatiotemporal evolution of the cross-sectional area $A(x, t)$ and flow rate $Q(x, t)$ can be written as

$$\frac{\partial A}{\partial t} + \frac{\partial Q}{\partial x} = 0 \,, \tag{44}$$

$$\frac{\partial Q}{\partial t} + \frac{\partial}{\partial x}\left(\alpha \frac{Q^2}{A}\right) + \frac{A}{\rho}\frac{\partial P}{\partial x} = -K_R \frac{Q}{A} \,, \tag{45}$$

where the adopted model parameters and constitutive laws are given by

$$\begin{aligned} \alpha &= \frac{\int_A \rho u^2 dA}{\rho A \bar{u}^2} \approx 1 \,, \quad K_R = (\zeta + 2) 2\pi \nu \,, \\ P(A) &= P_{\text{ext}} + \beta \, \frac{A - A_0}{A_0^2} \,, \quad P_{\text{ext}} = 0, \quad \beta = \frac{h_0 E}{1 - \kappa_E^2} \,. \end{aligned} \tag{46}$$

Equations (44)–(45) are discretized by a finite difference method, using an explicit central scheme stabilized by a modified Lax-Friedrichs approach. The one-dimensional (in space) computational domain is divided into 100 equal length sub-intervals. For stand-alone 1D model simulations, the time step Δt_1 is chosen such that it satisfies the condition (26), ensuring the numerical stability of the explicit scheme. For the coupled 2D-1D simulations, the time step is chosen such that it satisfies the stability restrictions in both 2D and 1D models.

The standard Lax-Friedrichs approach to discretization of time derivative in (44) is based on the approximation

$$\frac{A_i^{n+1} - A_i^n}{\Delta t_1} = \frac{1}{\Delta t_1}\left(A_i^{n+1} - A_i^n\right) \approx \frac{1}{\Delta t_1}\left(A_i^{n+1} - \frac{1}{2}\left(A_{i-1}^n + A_{i+1}^n\right)\right) ,$$

where the replacement of A_i^n by the spatial average of the neighboring values A_{i-1}^n and A_{i+1}^n, i.e., using $A_i^n \approx \frac{1}{2}(A_{i-1}^n + A_{i+1}^n)$, leads to additional smoothing that stabilizes the solution in the presence of large gradients. In order to avoid over-smoothing of the solution while keeping the needed stabilizing effect of the added numerical diffusion, the convex combination of the original and averaged values is used in the form

$$A_i^n \approx (1 - \xi_A) A_i^n + \xi_A \frac{A_{i-1}^n + A_{i+1}^n}{2} \qquad \text{where} \quad \xi_A \in [0, 1] \,.$$

The adjustable smoothing parameter ξ_A controls the added numerical diffusion and can be fine-tuned depending on the solved case. In the simulations presented later, the value $\xi_A = 0.01$ was chosen and used when needed. Based on this modified Lax-Friedrichs scheme, the cross-sectional area $A(x_i, t_n) \approx A_i^n$ is updated as

$$A_i^{n+1} = (1 - \xi_A)A_i^n + \xi_A \frac{A_{i-1}^n + A_{i+1}^n}{2} - \Delta t_1 \frac{Q_{i+1}^n - Q_{i-1}^n}{2\Delta x_1},$$

with the discretized boundary conditions

$$A_1^{n+1} = 3A_2^{n+1} - 3A_3^{n+1} + A_4^{n+1}, \qquad A_{nx_1+1}^{n+1} = A_0,$$

which represents the inlet boundary condition (at $i = 1$) obtained by quadratic (second-order) extrapolation, while the outlet Dirichlet boundary condition (at $i = nx_1 + 1$) prescribes a fixed-in-time cross-sectional area A_0 corresponding to the considered value of pressure $P = P_{\text{ext}} = 0$. The pressure $P(x_i, t_n) \approx P_i^n$ is computed from the constitutive relation

$$P_i^{n+1} = P_{\text{ext}} + \beta \frac{A_i^{n+1} - A_0}{A_0^2}. \tag{47}$$

The updated pressure P_1^{n+1} at the 1D model inlet is then stored and passed to the two-dimensional simulation as a boundary condition at the interface Γ_3 at the next time step.

For the flow rate $Q(x_i, t_n) \approx Q_i^n$, the discretization is carried out in a way similar to (4.5) to update the values in regular points (for $i = 2, \cdots, nx_1$)

$$\begin{aligned} Q_i^{n+1} &= (1 - \xi_Q)Q_i^n + \xi_Q \frac{Q_{i-1}^n + Q_{i+1}^n}{2} - \\ &- \Delta t_1 \left(\frac{\alpha}{2\Delta x_1} \left(\frac{Q_{i+1}^n Q_{i+1}^n}{A_{i+1}^{n+1}} - \frac{Q_{i-1}^n Q_{i-1}^n}{A_{i-1}^{n+1}} \right) + \frac{A_i^{n+1}}{\rho} \frac{P_{i+1}^{n+1} - P_{i-1}^{n+1}}{2\Delta x_1} + K_R \frac{Q_i^n}{A_i^{n+1}} \right), \end{aligned} \tag{48}$$

where the smoothing coefficient $\xi_Q \in [0, 1]$ can again be fine-tuned for the solved case, independent of ξ_A. The discretized boundary conditions are imposed as

$$Q_1^{n+1} = Q_0, \qquad Q_{nx_1+1}^{n+1} = Q_{nx_1}^{n+1},$$

where at the inlet boundary ($i = 1$) the flow rate is prescribed based on the data Q_0 from the two-dimensional model and at the outlet boundary ($i = nx_1 + 1$) the homogeneous Neumann condition is used, naturally following from the imposed fixed in time cross section $A = A_0$ at the outlet.

4.5.1 Semi-implicit Discretization

The problem being solved becomes stiff as the Young modulus E increases, which together with a higher spatial resolution leads to a gradual loss of robustness, limited stability, and severe restrictions imposed on the admissible time step in the chosen 1D explicit scheme. Therefore, several attempts have been made to improve the robustness and computational efficiency of the adopted numerical scheme.

The system of governing equations (44)–(45) was discretized in space and time using the approach shown in the previous paragraphs. There, a partial improvement was already implemented over the fully explicit discretization. The modification is based on the observation that the system is in fact solved sequentially, when first the $A(x, t_{n+1})$ is found using the known values of $A(x, t_n)$ and $Q(x, t_n)$ from the previous time step and then the $Q(x, t_{n+1})$ is calculated. At this stage, instead of again fully relying on the "old" values $A(x, t_n)$ and $Q(x, t_n)$ from the previous time level, the already known updated values $A(x, t_{n+1})$ and $P(x, t_{n+1})$ are used on the right-hand side of (47) and (48). This simple numerical trick, similar in principle to Gauss-Seidel like modification of a Jacobi solver for linear systems, brings an increase in the 1D model solver robustness. In fact, it results in a partially implicit discretization while solving the system for Q.

In summary, starting from the governing system (44)–(45) rewritten in the form

$$\frac{\partial A}{\partial t} = R^A(A, Q)\,, \tag{49}$$

$$\frac{\partial Q}{\partial t} = R^Q(A, Q)\,, \tag{50}$$

the channel area $A(x, t_{n+1})$ at the new time level is calculated explicitly as

$$\frac{A^{n+1} - A^n}{\Delta t_1} = R^A(A^n, Q^n)\,, \tag{51}$$

and then in the second stage, the new values of $Q(x, t_{n+1})$ are calculated from

$$\frac{Q^{n+1} - Q^n}{\Delta t_1} = R^Q(A^{n+1}, Q^n)\,. \tag{52}$$

Further improvements were built and tested on top of this basic discretization. They resulted in two different versions of a semi-implicit discretization of the governing system.

4.5.2 Fixed-Point Sub-iterations

This approach is a simple extension of the previously described trick with sequential solution of the system for A and Q with partial updates of the unknowns on the right-hand side of the system. The discrete system based on (51)–(52) is solved in a loop, iteratively with index k as

$$\frac{A^{k+1} - A^n}{\Delta t_1} = R^A(A^k, Q^k) \tag{53}$$

$$\frac{Q^{k+1} - Q^n}{\Delta t_1} = R^Q(A^{k+1}, Q^k)\,, \tag{54}$$

starting at $k = 0$ with $A^{k=0} = A^n$ and $Q^{k=0} = Q^n$, iteratively for $k = 0, 1, 2, \ldots$ until $\|A^{k+1} - A^k\| < tol$ and $\|Q^{k+1} - Q^k\| < tol$, respectively, indicating that $A^{n+1} \approx A^{k+1}$ and $Q^{n+1} \approx Q^{k+1}$. This in the limit case leads to implicit, backward-in-time discretization

$$\frac{A^{n+1} - A^n}{\Delta t_1} = R^A(A^{n+1}, Q^{n+1})$$

$$\frac{Q^{n+1} - Q^n}{\Delta t_1} = R^Q(A^{n+1}, Q^{n+1})\,.$$

This iterative procedure results into a more robust numerical model but at the price of an additional computational cost associated with the repeated evaluation of the right-hand side of the discrete equations (53)–(54).

4.5.3 Linearization in Time

One of the drawbacks of all the so far described discretizations of the 1D problem is that in the first stage when calculating the A^{n+1} from (49) or (51), respectively, the right-hand side is treated explicitly. In this case specifically $R^A(A, Q) = -Q_x \approx -Q^n_x$, i.e.,

$$A_t = -Q_x \quad \text{is approximated as} \quad \frac{A_i^{n+1} - A_i^n}{\Delta t_1} = -\frac{Q^n_{i+1} - Q^n_{i-1}}{2\Delta x_1}\,.$$

In order to discretize the right-hand side implicitly, i.e., at the time level $n + 1$, the flow rate gradient Q_x^{n+1} needs to be estimated. This can be done by linearization in time of Q_x based on Taylor expansion at $t = t_n$ as

$$Q_x^{n+1} \approx Q_x^n + \Delta t_1\, Q_{xt}^n\,.$$

The mixed second-order spatiotemporal derivative Q_{xt} can then be obtained by formally taking a derivative with respect to space variable x of (50), leading to

$Q_{xt} = R_x^Q(A, Q)$, which can numerically be approximated as

$$Q_{xt}^n \approx \frac{R_{i+1}^Q(A^n, Q^n) - R_{i-1}^Q(A^n, Q^n)}{2\Delta x_1} . \tag{55}$$

Using this expansion, the explicit update of area A^{n+1} from (51) can be changed to

$$\frac{A^{n+1} - A^n}{\Delta t_1} = R^A(A^n, Q^{n+1}) ,$$

which in the discrete version gives

$$\frac{A_i^{n+1} - A_i^n}{\Delta t_1} = -\frac{Q_{i+1}^n - Q_{i-1}^n}{2\Delta x_1} - \Delta t_1 \frac{R_{i+1}^Q(A^n, Q^n) - R_{i-1}^Q(A^n, Q^n)}{2\Delta x_1} . \tag{56}$$

This implicit-like treatment of the first part (the cross section A) in the governing system leads to much more robust numerical algorithm, with only minimal additional computational cost.

It is good to note that the added implicit correction term on the right-hand side of (56) can easily (cheaply) be approximately evaluated from the remembered right-hand side of (52) from the previous time step, by only calculating its gradient using (55).

5 Numerical Results

The overview of the selected numerical results presented in this section is split into two parts: First, both the FEM and FDM codes are validated using the benchmark based on the Womersley analytical solution for a periodically oscillating flow in an infinitely long straight pipe (between two parallel plates in 2D). This allows us to adjust the numerical methods and establish the spatiotemporal resolving properties of both numerical solvers. The second part of the presented results focuses on a mutual comparison of the full 2D-2D FSI problem (solved by FEM) with the simplified coupled 2D-1D problem (solved by FDM).

5.1 Comparison of Flow Field with Womersley Solution

Despite the fact that this paper primarily focuses on FSI problems, the validation of computational codes is performed first on a simple unsteady fluid flow problem, where the analytical solution for a periodically oscillating flow is known.

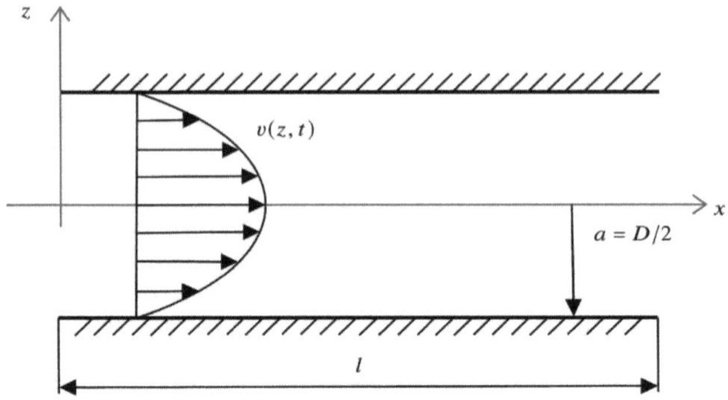

Fig. 6 Schematic sketch of the flow between two parallel infinite flat plates

For this purpose, the well-known Womersley solution is used for the viscous fluid flow between two parallel flat plates separated by a distance $D = 2a$.

The coordinate system axes z and x are defined as perpendicular and parallel to these plates, as illustrated in Fig. 6. The flow is driven by a time-periodic pressure gradient (or pressure drop), written here in a complex form

$$\frac{\partial p}{\partial x} = \frac{p_2 - p_1}{l} = -A_\omega \exp(i\omega t),$$

where A_ω represents the amplitude of pressure gradient oscillations, $t \in I = [0, T]$ denotes the time (in seconds s), and p_1 and p_2 are the pressures at the (positions 1 and 2) inlet and outlet, respectively, separated by a distance l. The variable ω represents the angular frequency of oscillations (in $radians/s$), defined as

$$\omega = 2\pi f = \frac{\nu Wo^2}{a^2},$$

where f is the frequency in $cycles/s$, ν the kinematic viscosity of the fluid, and Wo is the (nondimensional) Womersley number.

Analytical solutions to the incompressible Navier-Stokes equations for the velocity profile, volumetric flow rate, and velocity gradient have been derived, for example, in [38–40]. According to these references, the real part of the solution for velocity v, $v(z,t)$: $\Omega \times I \to \mathbb{R}$ is given by

$$\begin{aligned} v(z,t) =& \frac{A_\omega}{\omega\rho\gamma}\Big([\sinh\Phi_1(z)\sin\Phi_2(z) + \sinh\Phi_2(z)\sin\Phi_1(z)]\cos(\omega t) + \\ &+ \big[\gamma - \cosh\Phi_1(z)\cos\Phi_2(z) - \cosh\Phi_2(z)\cos\Phi_1(z)\big]\sin(\omega t)\Big), \end{aligned} \tag{57}$$

where $\Phi_1(z)$ and $\Phi_2(z)$ are defined as

$$\Phi_1(z) = \frac{Wo}{\sqrt{2}}\left(1+\frac{z}{a}\right), \quad \Phi_2(z) = \frac{Wo}{\sqrt{2}}\left(1-\frac{z}{a}\right),$$

$$\gamma = \cosh\left(\sqrt{2}Wo\right) + \cos\left(\sqrt{2}Wo\right).$$

The time-periodic volumetric flow rate can be obtained (by integrating velocity) as

$$Q(t) = \frac{2A_\omega a^3}{\mu}\left(\frac{(\Theta_1-\Theta_2)}{\sqrt{2}Wo^3}\cos(\omega t) - \frac{(\Theta_1-\Theta_2)+\sqrt{2}Wo}{\sqrt{2}Wo^3}\sin(\omega t)\right), \tag{58}$$

where Θ_1 and Θ_2 are defined as

$$\Theta_1 = \frac{\sinh\left(\sqrt{2}Wo\right)}{\gamma}, \qquad \Theta_2 = \frac{\sin\left(\sqrt{2}Wo\right)}{\gamma}.$$

Furthermore, the velocity gradient at the wall (wall-normal derivative) is given by

$$\begin{aligned}\frac{dv(a,t)}{dz} =& \frac{-A_\omega\, Wo}{\sqrt{2}\,\omega\,\rho a\,\gamma}\Big\{\left[\sinh(\sqrt{2}Wo)+\sin(\sqrt{2}Wo)\right]\cos(\omega t)+ \\ &+\left[\sinh(\sqrt{2}Wo)-\sin(\sqrt{2}Wo)\right]\sin(\omega t)\Big\} = -\frac{dv(-a,t)}{dz}.\end{aligned} \tag{59}$$

5.1.1 FDM

To systematically evaluate the accuracy of the finite difference method (FDM) used in the 2D flow model, a series of simulations was performed with varying spatial resolutions, defined by the number of grid cells in the longitudinal (axial) and transversal (radial) directions, defined by nx and nz, respectively. Equivalently, the resolution can be characterized by varying the grid sizes Δx and Δz. The case setup remained otherwise unchanged for all these simulations, ensuring that any differences in the resulting errors could solely be attributed to varying spatial resolutions.

The velocity error in each simulation was evaluated at each time step using an integral norm of the difference between the numerical (discrete) solution v_d and the analytical (exact) solution v_e, calculated over the inlet cross section (channel height):

$$err(v) = \left(\int_{-a}^{a}\left(v_d(z)-v_e(z)\right)^2 dz\right)^{1/2} \approx \sqrt{\sum_{j=1}^{nz}\left(v_d(z_j)-v_e(z_j)\right)^2\Delta z} \tag{60}$$

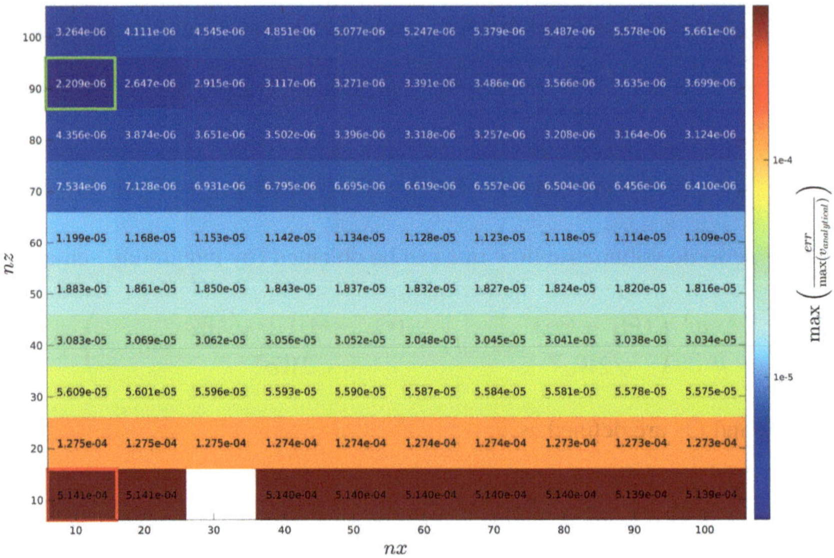

Fig. 7 Heatmap of the maximum relative norm of the error of velocity profile at the inlet, normalized by the amplitude of oscillating velocity over the last period of oscillation

Figure 7 presents a heatmap of the maximum relative norm of the error of the velocity profile evaluated at the inlet boundary Γ_1, normalized by the amplitude of the analytical velocity, calculated during the final period. The figure highlights the impact of spatial resolution on the solution accuracy. The minimum and maximum errors are marked with green and red frames for clarity. The results indicate that refining the grid in the longitudinal x-direction (by increasing nx) has only a minor impact on the velocity profile, which is evident from the nearly uniform error levels across increasing nx for fixed nz. In contrast, increasing the resolution in the transversal z-direction (by increasing nz) significantly reduces the error, reflecting the importance of accurately capturing the non-monotone structure of the velocity profile.

The resulting error varies not only as a function of the grid spacing but also as a function of time as the solution (numerical and analytical) evolves. This time variation of the error for the presented FDM simulations is shown in Fig. 9. Similarly, the flow rate error can be evaluated at each time instant taking a difference of the integral of the discrete velocity profile and the corresponding analytical exact value obtained from (58). The velocity gradient (wall normal derivative) of the velocity of the discrete solution can also be compared with the theoretical value given by the formula (59). The flow rate and velocity gradient error vs. time plots for the presented simulations are shown in Figs. 10 and 8, respectively.

Figure 8 shows the normalized velocity gradient over time for various numerical grids. The results indicate that the overall accuracy is strongly dependent on spatial discretization, with coarser grids leading to pronounced deviations from

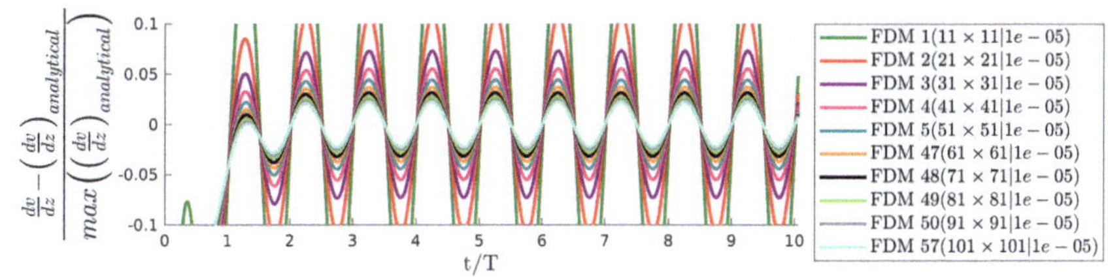

Fig. 8 Relative error of the velocity gradient, normalized by the amplitude of velocity gradient

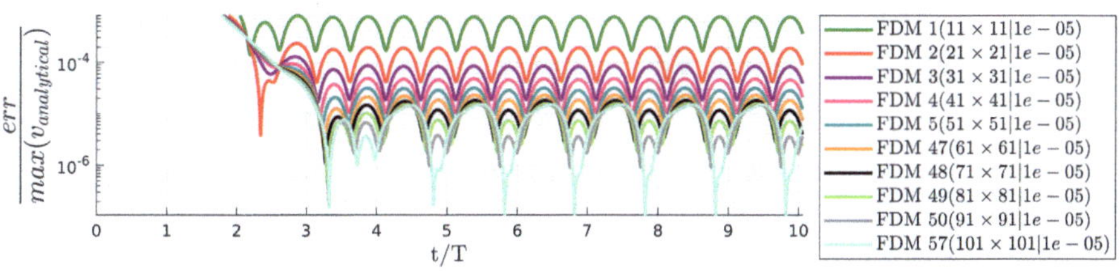

Fig. 9 Relative norm of the error of velocity profile at the inlet, normalized by the amplitude v_{max} of the oscillating velocity

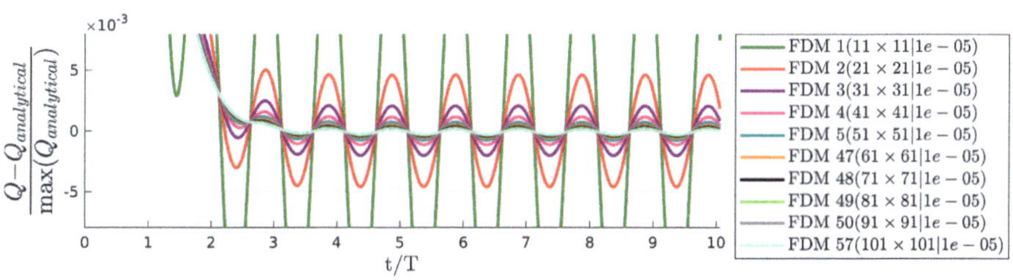

Fig. 10 Relative error of the volume flow rate of the fluid, normalized by the amplitude Q_{max} of the oscillating volume flow rate

the analytical solution. Although refining the grid in the x-direction has a minor effect on the solution, a finer resolution in the z-direction significantly influences (improves) the accuracy.

The assessment of the resulting norm of the error of velocity profile (Fig. 9) and volumetric flow rate (Fig. 10) is not as straightforward as that of the velocity gradient. Here, it can be observed that a finer spatial resolution does not necessarily lead to better (more accurate) results. This phenomenon arises because all simulations in this section were performed with the same time step Δt. Thus, when increasing the spatial resolution while keeping a fixed time step, numerical errors may become more pronounced. As a result, refining only the (spatial) grid does not guarantee an improvement in the accuracy of velocity profiles and volumetric flow rates. These results, however, indicate where a timestep adjustment may be required and whether the associated improvement in accuracy justifies the additional computational cost.

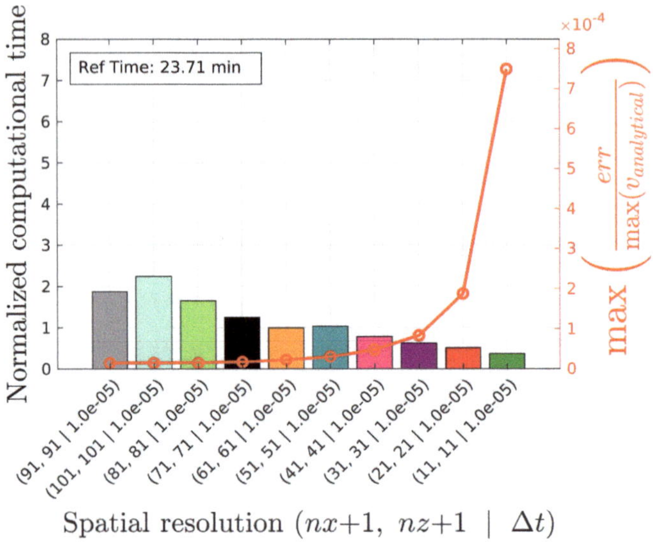

Fig. 11 Computational time and solution error chart for different spatial resolutions

It is worth noting that the velocity error shown in Fig. 9 for the FDM method is in most cases (grid resolutions) much smaller than the same error shown in Fig. 12 for the FEM method. The results of the FDM method even on very coarse grids easily match the accuracy of the FEM method on much finer grids. This facilitates the choice of the computational grid for the 2D-1D coupled FDM simulations.

To achieve a balance between accuracy and computational efficiency (cost) in the later performed coupled 2D-1D simulations, the configuration with $nx = 50 \times 2$, $nz = 50$ was selected for further use. The accuracy tests presented above demonstrate that the velocity gradient exhibits a significant improvement with a finer resolution in the z direction. The chosen resolution provides sufficient accuracy in this regard while avoiding excessive computational cost. The same holds for velocity profiles and volumetric flow rates, where further refinement of the grid does not necessarily return proportionate improvements.

To assess the relative computational cost associated with different spatial resolutions, the provided bar chart (Fig. 11) presents the normalized computational time for various grid configurations. The reference computational time, indicated in the legend, corresponds to $nx = 60 \times 2$, $nz = 60$, and all other times are scaled accordingly.

5.1.2 FEM

The numerical simulations for the coupled 2D-2D problem are performed using the finite element library `deal.II` [41]. The implementation leverages the hp framework provided by `deal.II` to extend the pressure finite element, which is defined only in the fluid domain, into the structure domain using a finite element with no degrees of freedom. For an additional validation, we refer to [28] and [29] where the method and solver were validated on well-known stationary and nonstationary benchmarks from [42]. The code for the numerical simulations is based on [43].

Similar to the validation of the finite difference method, we validate the finite element method computations. Table 3 presents the mesh and timestep sizes used in the simulations, including the number of cells in each direction. For example, a mesh with $nx = 72$ and $nz = 12$ will result in $(72 \cdot 2 + 1) \cdot 13 = 1885$ degrees of freedom for the pressure and $(72 \cdot 2 \cdot 2 + 1)(12 \cdot 2 + 1) = 14450$ degrees of freedom for both velocity components together.
For the validation of the FEM simulations, we use the same error norms as in the FDM validation. The simulations were performed for several spatial resolutions and three different time steps to assess both spatial and temporal accuracy.

Figure 12 shows the norm of the relative error of the velocity profile at the inlet normalized by the amplitude v_{max} of the oscillating velocity. The velocity error is mainly affected by the time step Δt, being lowest for the shortest time step.

Figure 13 illustrates the relative error of the velocity gradient, clearly demonstrating convergence under mesh refinement. As expected, the three most accurate cases correspond to the finest mesh (with smallest Δz, i.e., highest nz). In contrast to the accuracy of the whole velocity profile, the timestep size Δt appears to be sufficiently small, as it has only a small visible impact on the error of the velocity gradient at the wall.

Figure 14 presents the relative error in the volume flow rate, where both the timestep size and the mesh size visibly influence the results. The three largest errors correspond to the largest timestep sizes. As Δt decreases, the error is reduced, with the smallest Δt yielding the three most accurate results.

Figure 15 shows an overview of the relative computational cost of the individual FEM simulations performed. As expected, the cost scales proportionally to the number of elements in the computational grid. Apparently, the solution error is only marginally affected by spatial resolution. It seems that the quadratic finite elements approximate very well the sought solution (most of the time close to a parabolic velocity profile) from very low spatial resolution already. However, the error in the FEM simulations is much more affected by the time step Δt. The error scales down based on the time step, with the corresponding error being virtually the same for all simulations performed at a certain time step, independent of the spatial resolution of the grid. This is also confirmed by the time plots of the errors in Figs. 12 and 14, where the graphs corresponding to the same time step almost perfectly coincide, independent of the spatial resolution.

Our implementation supports multiple time integration schemes, including backward Euler, Crank-Nicolson, and shifted Crank-Nicolson, with the latter allowing for a tunable degree of implicitness. Rather than being a limitation, the observations discussed below highlight a fundamental aspect of implicit time integration. Specifically, while the implicit discretization used in the FEM method effectively

Table 3 Spatial grid and timestep resolutions used for the FEM simulations

Case	1	2	3	4	5	6	7	8	9	10	11	12
v dofs	14,450	14450	14,450	28,322	28,322	28,322	56,066	56,066	56,066	28,850	28,850	28,850
p dofs	1885	1885	1885	3625	3625	3625	7105	7105	7105	3757	3757	3757
nx	72×2	72×2	72×2	72×2	72×2	72×2	72×2	72×2	72×2	144×2	144×2	144×2
nz	12	12	12	24	24	24	48	48	48	12	12	12
Δt	0.005	0.0025	0.01	0.005	0.0025	0.01	0.005	0.0025	0.01	0.005	0.0025	0.01

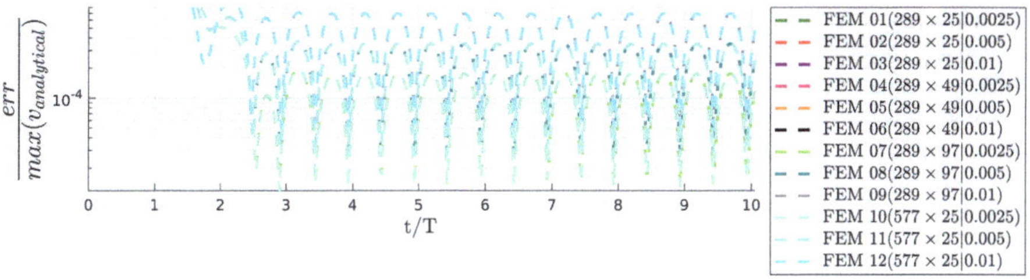

Fig. 12 Relative norm of the error of velocity profile at the inlet, normalized by the amplitude v_{max} of the oscillating velocity

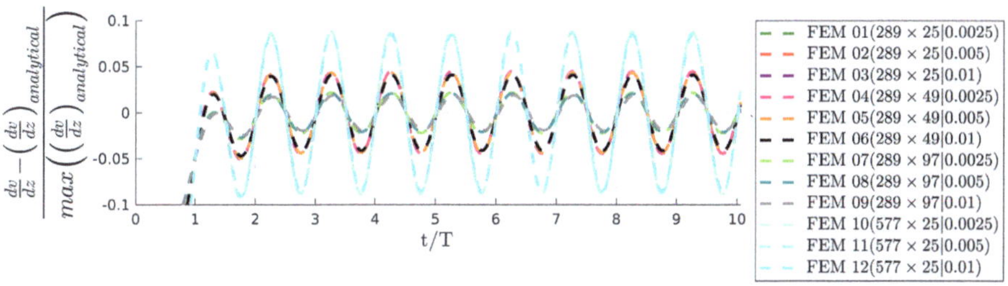

Fig. 13 Relative error of the velocity gradient, normalized by the amplitude of velocity gradient

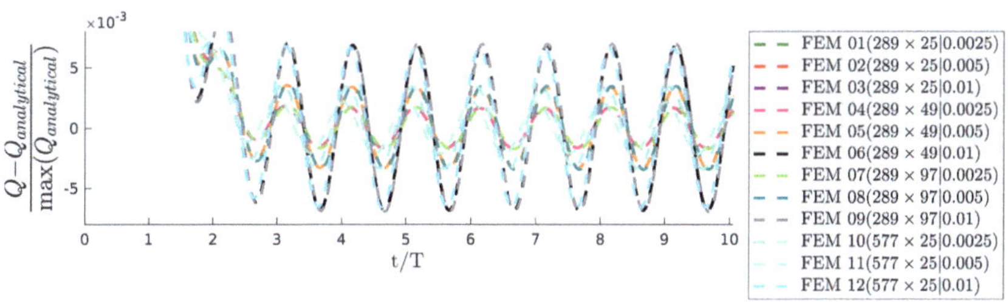

Fig. 14 Relative error of the volume flow rate of the fluid, normalized by the amplitude Q_{max} of the oscillating volume flow rate

removes stability-related constraints on the time step, it still influences the temporal accuracy of the solution. In the unsteady test case considered here, the primary source of numerical error is a temporal phase shift between computed and analytical solutions, rather than insufficient spatial resolution. In this context, it is worth noting that the FDM simulations summarized in Figs. 7, 8, 9, and 10 were performed with the time step $\Delta t = 10^{-5}s$, which is two to three orders of magnitude smaller than the corresponding time steps used for the FEM simulations, ranging from $\Delta t = 2.5 \cdot 10^{-3} - 1.0 \cdot 10^{-2}s$. This highlights the practical advantage of implicit time integration in FEM: significantly larger time steps can be employed without compromising stability. However, for problems where high temporal accuracy is required, as in this unsteady case, both methods may ultimately be constrained by similar accuracy demands, potentially requiring comparable timestep sizes.

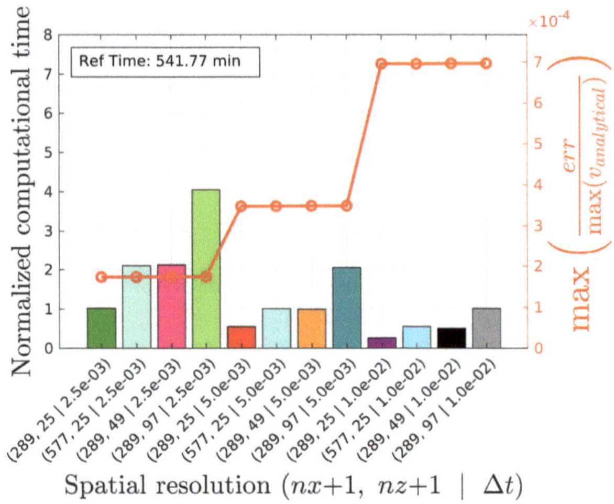

Fig. 15 Computational time and solution error chart for different spatial resolutions and time steps

A more detailed direct comparison of the computation time and efficiency of the two approaches (FDM vs. FEM) is not offered, as they were implemented in completely different ways and solved on different hardware. The FDM solver was implemented in MATLAB and solved on a regular PC, while the FEM solver, which is based on the `deal.II` library, is suitable for massively parallel computations on high-performance supercomputers, which were not used here. The results shown were also calculated on a regular PC.

5.2 *Results for the Coupled FSI Problems*

The numerical tests of the coupled FSI problem focus on the computational verification of the coupling strategy of the 2D-1D simplified problem. From the above-presented discussion of the problem, it follows that the elastic response of the 1D model depends mainly on the form of the pressure-area relation $P(A)$ and on the geometric and material parameters involved. The elasticity coefficient β emanates from the original formulation (12)

$$\beta = \frac{\sqrt{\pi} h_0 E}{1 - \kappa_E^2},$$

where E is the wall Young modulus, κ_E is the corresponding Poisson ratio, and h_0 is the wall thickness. From this formula, it is possible to estimate the behavior of the elastic deformation. For the pressure-area law in the form using $\psi_3(A)$,

$$P(x,t) = p_{ext} + \psi_3\left(A(x,t)\right) = p_{ext} + \beta_3 \frac{A(x,t) - A_0}{A_0^2}, \tag{61}$$

which after rearranging the terms leads to

$$A(x,t) - A_0 = \frac{A_0^2}{\beta_3}\Big(P(x,t) - p_{ext}\Big) = \frac{A_0^2}{h_0}\frac{1-\kappa_E^2}{E}\Big(P(x,t) - p_{ext}\Big), \quad (62)$$

showing how the difference of the local pressure $P(x,t)$ from the external one translates into the change of the local area $A(x,t)$ with respect to the equilibrium state A_0. Considering a fixed reference geometry with a given ratio A_0^2/h_0, the mechanical response (deformation) of the 1D model will mainly depend on the value of the Young modulus E. This *inverse proportionality of the deformation on the Young modulus* suggests that with higher E the channel walls are more stiff and the deformation (subject to the given pressure difference) will be smaller than in the case of lower values of E.

Based on the eigenvalue (and timestep) analysis, it is possible to expect that also the mathematical model will exhibit a "stiff" behavior, as the elastic wave speed

$$\lambda_{wave} = \sqrt{\frac{A}{\rho}\frac{\partial\psi}{\partial A}} = \sqrt{\frac{A}{\rho}\frac{\beta_3}{A_0^2}} \approx \sqrt{\frac{h_0}{\rho A_0}\frac{E}{1-\kappa_E^2}} \propto \sqrt{E}$$

scales proportionally to $\sqrt{E}$ and thus the time step for numerical simulations scales according to $\Delta t \propto \frac{\Delta x}{\sqrt{E}}$.

The physiological uncertainty and the large variability of the values of the Young modulus E in the human respiratory system led us to primarily focus the presented simulations on the effects of the material parameter E and the accuracy and stability limitations originating from the choice of the discretization parameters Δx and Δt.

5.2.1 Relation Between the FEM and FDM Structure Models

It has been shown how the 1D FSI model used in the FDM simulations can be rewritten in the form of a nonlinear wave equation and what the wave speeds and forcing terms are in that case. A similar trick can be used to analyze the structural model in the FEM simulations.

Starting from the coupled mathematical model used for the fully 2D FEM simulations, i.e., from the system (3) equipped with the constitutive relations for the stress in the fluid (4) and structure (5), respectively. The structural force balance part of the system (3) can be written as[4]

$$\rho_s\,\partial_t \boldsymbol{v}_s = \mathrm{div}\Pi\,,$$
$$\boldsymbol{v}_s = \partial_t \boldsymbol{u}_s$$

[4] The hats denoting the reference configuration were dropped for simplicity.

simply postulating that the force due to inertia (proportional to structural acceleration and density) is balanced by the acting forces represented by the divergence of the structural stress tensor Π. The structural velocity $\boldsymbol{v}_s$ is simply defined as a time derivative of the corresponding displacement $\boldsymbol{u}_s$. The structural stress is related to the displacement gradient by the Piola-Kirchhof linear elasticity model by

$$\Pi := 2\mu\,\boldsymbol{\varepsilon} + \lambda\,\mathrm{tr}(\boldsymbol{\varepsilon})\mathbb{I}$$

with $\boldsymbol{\varepsilon} = \frac{1}{2}\big(\nabla \boldsymbol{u}_s + \nabla \boldsymbol{u}_s^T\big)$ being the symmetric part of the deformation (displacement) gradient tensor. The coupling and interaction with the fluid flow are ensured by the boundary conditions on the fluid-structure interface, requesting, among others, the continuity of the normal stresses across the common interface.

By combining all these above relations, a single equation for the spatiotemporal evolution of the structural displacement $\boldsymbol{u}_s$ can be obtained:

$$\rho_s\, \partial_{tt} \boldsymbol{u}_s = \mathrm{div}\Big(\mu(\nabla \boldsymbol{u}_s + \nabla \boldsymbol{u}_s^T) + \lambda\,(\nabla \boldsymbol{u}_s) : \mathbb{I}\Big) \tag{63}$$

This equation relates the acceleration of the structure to its deformation. As noted earlier, the Lamé moduli μ and λ, respectively, can be expressed in terms of the Young modulus E and Poisson ratio κ_E using (36),

$$\mu = \frac{E}{2(1+\kappa_E)} \qquad \text{and} \qquad \lambda = \frac{2\kappa_E}{1-2\kappa_E}\,\mu = \frac{E\,\kappa_E}{(1-2\kappa_E)(1+\kappa_E)}\,,$$

leading after substitution into (63) and division by the structural density ρ_s to the equation for the structural displacement

$$\partial_{tt} \boldsymbol{u}_s = \frac{E}{2\rho_s(1+\kappa_E)}\,\mathrm{div}\left(\big(\nabla \boldsymbol{u}_s + \nabla \boldsymbol{u}_s^T\big) + \frac{2\kappa_E}{1-2\kappa_E}\,(\nabla \boldsymbol{u}_s) : \mathbb{I}\right). \tag{64}$$

The left-hand side of (64) contains the second temporal derivative of the displacement $\boldsymbol{u}_s$, i.e., the structural acceleration, while on the right-hand side appear the second spatial derivatives of the same displacement $\boldsymbol{u}_s$. More details on such second-order evolution equation for the structural displacement can be found in [21], for example. It is a kind of wave equation, with the wave velocity $C_{2D}(E, \kappa_E, \rho_s)$

$$C_{2D}^2(E, \kappa_E, \rho_s) = \frac{E}{2\rho_s(1+\kappa_E)} = \frac{\mu}{\rho_s}\,,$$

due to structural elasticity. In fact, the quotient on the right-hand side is just the shear modulus μ divided by the structural density, indicating that the elastic response is primarily due to bending (change of curvature) of the structure, rather than due to tension considered in the 1D structural model (coming from a generalized string or independent ring models) characterized by the Young (tension) modulus E.

The structural displacement equation (64) can be directly compared to its one-dimensional counterpart derived in Sect. 4.1.3 for the 1D simplified FDM model, i.e., to Eq. (30)

$$A_{tt} = \left(C^2(A, Q)\, A_x\right)_x + G(A, Q) = \left(\left(\frac{A}{\rho}\psi_A - \frac{Q^2}{A^2}\right) A_x\right)_x + G(A, Q)\,,$$

where the wave velocity can be expressed including the pressure-area relation ψ_3 as

$$C^2(A, Q) = \left(\frac{A}{\rho}\psi_A - \frac{Q^2}{A^2}\right) = \left(\frac{A}{\rho}\frac{E}{1-\kappa_E^2}\frac{h_0}{A_0^2} - \frac{Q^2}{A^2}\right).$$

Considering now that $A \doteq A_0$ and keeping in mind that the average velocity Q/A is small, temporarily even vanishing during the respiratory cycle, an estimate of the elastic wave speed for the 1D model can be given in the form

$$C_{1D}^2(E, \kappa_E, \rho) \doteq \frac{1}{\rho}\frac{E}{1-\kappa_E^2}\frac{h_0}{A_0} = \frac{E}{2\rho(1+\kappa_E)}\frac{2\,h_0}{A_0(1-\kappa_E)}\,.$$

Interestingly, the density ρ appearing in the 1D model is the density of the fluid and not structure, like in the 2D model; thus, $\rho = \rho_f$, which leads to

$$C_{1D}^2(E, \kappa_E, \rho) = \frac{E}{2\rho(1+\kappa_E)}\frac{2\,h_0}{A_0(1-\kappa_E)} = \underbrace{\frac{E}{2\rho_s(1+\kappa_E)}}_{C_{2D}^2}\frac{2\,h_0}{A_0(1-\kappa_E)}\frac{\rho_s}{\rho_f}\,.$$

For the incompressible structure case considered here, the Poisson ratio should be[5] $\kappa_E \approx 1/2$. The geometry of the channel is determined by the wall thickness $h_0 = 0.0014$ m and channel width (cross section) $A_0 = 0.018$ m, which means that the relation between the corresponding elastic wave velocities in 2D and 1D models is approximately

$$C_{1D}^2 = \underbrace{\frac{2\,h_0}{A_0(1-\kappa_E)}}_{\approx 0.2}\underbrace{\frac{\rho_s}{\rho_f}}_{\approx 10^3} C_{2D}^2 \approx 200\, C_{2D}^2\,, \tag{65}$$

considering the density of fluid (air) $\rho_f = 1.024\,\mathrm{kg\cdot m^{-3}}$ and structure $1000\,\mathrm{kg\,m^{-3}}$.

[5] The 2D structural model is singular for $\kappa_E = 1/2$, therefore the value $\kappa_E = 0.48$ was chosen for most simulations.

Mainly because of the high structure-fluid density ratio the 1D model will behave like having the elastic Young modulus at least two orders of magnitude higher. Therefore, it is to be expected that the displacements predicted by the 1D elastic model will be about two orders of magnitude smaller than those resulting from the 2D model for the same Young modulus E. This rough estimate, however, can be altered by the second part of the divergence term in (64), taking into account the compression/expansion effects of the structure.

However, for the intended use, of the 1D model, as a boundary condition replacement for multidimensional models, important will be the pressure feedback (pushback) from the 1D model. In such a case, the pressure-area relation comes to play again, converting the structural displacement back to pressure force. In this way, the final effect (expressed in terms of pressure) of the different structural displacements can be reduced.

5.2.2 FDM

In the 2D-1D coupled system used in the FDM simulations, two substantially different mathematical models are connected. One of the key differences is the role that pressure plays in each of these models. In the 2D model based on the system of incompressible Navier-Stokes equations, the pressure field $p(x, z, t)$ only occurs in the gradient term ∇p, and, therefore, the whole field can be shifted arbitrarily (by an additive constant field), without any influence on the resulting velocity field. The role of the pressure p in the 2D model is to enforce the divergence-free constraint, which is realized by means of the pressure-correction method used for the pressure-velocity coupling. In the simplified 1D model, however, the pressure $P(x, t)$ is determined from an artificial state-like equation relating the local pressure $P(x)$ to local cross section $A(x)$. The role of the pressure P in the 1D model is to act as a force returning the elastic walls of the channel to the static (no flow) equilibrium position, represented by the cross section A_0. Therefore, this pressure $P(A)$ cannot be shifted arbitrarily. This double role of pressure is essential in the adopted computational setup.

In the fully 2D FEM FSI setup, the flow is driven by a prescribed pressure drop (average pressure gradient) between the inlet boundary Γ_1 and the outlet Γ_6. By imposing such pressure difference, the flow develops, based on the applied external (pressure) force. This setup was used and validated in the simulations of the time-periodic oscillating flow between two parallel plates (2D planar channel) shown and discussed previously.

For the simplified coupled 2D-1D FDM simulations, a very similar pressure difference-driven setup was chosen and tested. Also in this case, the predefined time-periodic pressure difference between inlet and outlet boundaries Γ_1 and Γ_6 was imposed. The pressure at the common 2D-1D interface Γ_3/Γ_5 is determined via coupling conditions as described in Sect. 4.2. In this way, the 1D model co-determines the pressure drop for the 2D rigid part of the model and can thus affect the resulting flow rate and velocity profiles. The pressure conditions for this 2D-1D configuration can be summarized as

$$
\begin{aligned}
p\big|_{\Gamma_1}(t) &= l\ A_\omega \cos(\omega t)\ , \\
p\big|_{\Gamma_3}(t) &= P\big|_{\Gamma_5}(t)\ , \\
P\big|_{\Gamma_6}(t) &= 0 = p_{ext}\ .
\end{aligned}
\tag{66}
$$

In these pressure-driven simulations, the required time-dependent flow rate can be achieved, without knowing (needing to know) a priori the form of the inlet velocity profile.

5.2.3 FDM Simulation Results

The results shown in this section are aimed at demonstrating the effects of the attached 1D model on the flow in the 2D portion of the model. It is a prototype of the intended use of the simplified 1D model as a replacement of the often-used elementary boundary conditions. It should allow to incorporate the feedback effects of a larger network of elastic tubes, typically found in respiratory problem simulations, without excessive additional computational cost.

The selected test case is pressure (pressure drop)-driven; therefore, the pressure evolution along the channel is the key factor determining the flow. In the case of a rigid channel, where the flow is fully developed, the pressure is a linear function of the axial coordinate x at each time instant. For the semielastic model represented here by the coupled 2D-1D problem, the pressure exhibits some deviation from this simple linear form, due to change of the cross section along the channel (in the elastic part). This behavior is shown in Fig. 16, presenting the pressure deviation from the linear profile of the rigid channel (provided by the analytical solution).

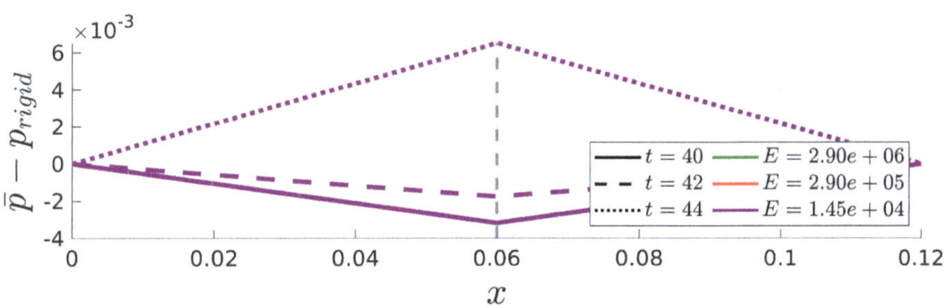

Fig. 16 Difference between the average pressure $\bar{p}$ in semielastic 2D-1D model and the pressure p_{rigid} in a 2D fully rigid model as a function of axial position x for various times t and Young moduli E. The vertical gray line marks the interface between the rigid and elastic parts of the channel

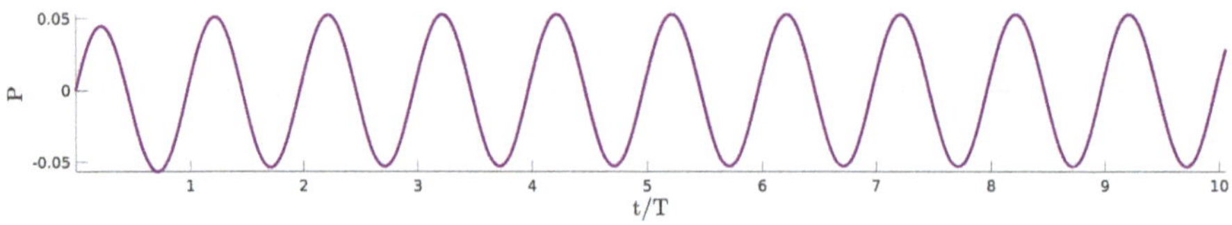

Fig. 17 Mean pressure vs. time, at the coupling interface for different values of Young modulus: $E = 2.9 \cdot 10^6$ (green), $E = 2.9 \cdot 10^5$ (red), and $E = 1.45 \cdot 10^4$ (purple)

Due to quite rigid channel walls, in the 1D elastic part, and their minimal deformation, the pressure in both parts of the model remains (almost) linear at all times, although the slopes of the plots can differ in the rigid and elastic part. In all cases, the pressure is approximately matched (equal) at the coupling interface at all times, as expected from the chosen explicit flow rate continuous coupling strategy described in Sect. 4.2. The pressure (deviation) differs in time as expected, reaching a zero value at the far left and right boundaries Γ_1 and Γ_6, where the pressure is a priori known and prescribed. On the coupling interface Γ_3/Γ_3, however, the deviation is maximal. Interestingly, while the variation in time is apparent, changes in the Young modulus E of the elastic walls do not cause visible differences in the axial pressure profiles. This unexpected behavior is documented and commented on later.

In contrast to the fully rigid case, the pressure difference between the two ends (inlet/outlet boundaries) of the 2D rigid portion of the channel is now partially determined by the feedback from the 1D elastic model. While the pressure at the left boundary Γ_1 is prescribed solely as a function of time, the pressure on the coupling interface Γ_3 is now governed by the attached 1D model; see Fig. 17.

In the case of the fully rigid channel, the pressure at the coupling interface, located at the distance $l/2$ from both ends, should be exactly equal to the average of the pressures at the ends of the channel. This means that in the present case when $p_{ext} = 0$ was set at the far right end Γ_6, the pressure at the interface Γ_3 should be at all times equal to $\frac{1}{2}p(t)\big|_{\Gamma_1}$. However, the actual evolution of the pressure at this interface in the semielastic test case is affected by the coupled 1D model, leading to the altered pressure history shown in Fig. 17.

There is no visible difference in the pressure history for various Young moduli E, and all the corresponding graphs coincide.

The resulting pressure drop over the rigid 2D part of the channel (between Γ_1 and Γ_3) is not fully controlled by the prescribed boundary condition but is partially affected by the pressure in the 1D elastic part of the domain. By this mechanism, the pressure drop along the 2D part of the domain is not exactly equal to the one for the fully rigid case (for which analytical solution is known), and the resulting velocity profile at the inlet can differ visibly. This is shown in Fig. 18, where the differences of velocity profiles at the boundary Γ_1 are drawn at several time instants. Also here, the velocity profiles plotted for various Young moduli E coincide, showing

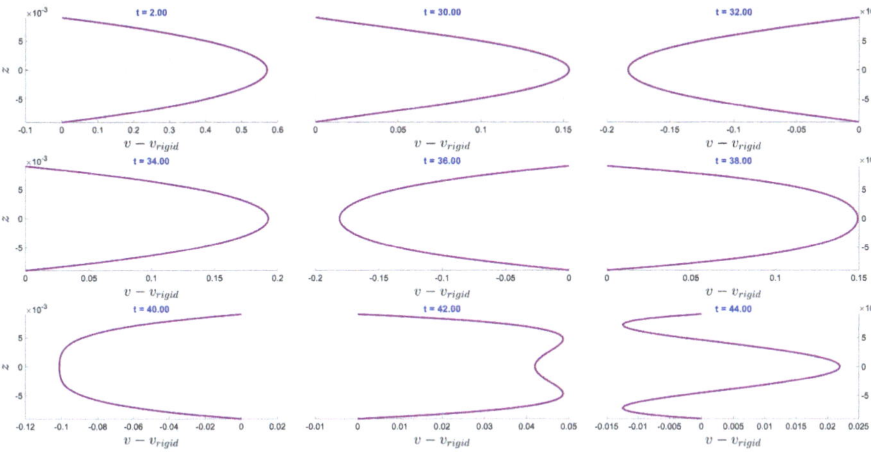

Fig. 18 Difference between the velocity profiles predicted by the semielastic 2D-1D model (FDM solution) and the fully rigid 2D model (analytical solution) at the inlet Γ_1, shown at different times t. Colored lines correspond to different values of Young modulus: $E = 2.9 \cdot 10^6$ (green), $E = 2.9 \cdot 10^5$ (red), and $E = 1.45 \cdot 10^4$ (purple)

no apparent differences. This aligns with the findings regarding the axial pressure profiles in Fig. 16.

Due to the pressure deviations at the coupling interface, the resulting pressure drop between the ends Γ_1 and Γ_3 is altered, leading to changes in the resulting flow rate measured at the inlet cross section Γ_1 as shown in Fig. 19 allowing to compare the flow rate history in the fully rigid channel case with those of the semielastic 2D-1D coupled model.

Once again, the different Young modulus E of the channel does not change the pressure and thus the flow rate feedback of the 1D model. However, as it can be seen from Fig. 20, the maximal wall displacement is scaled linearly with E as expected, based on the relation (62).

In summary, the attached 1D model of the elastic channel provides a small but nonnegligible feedback effect that alters the flow in the 2D part of the model. In contrary to our initial expectations, the effect turned out to be virtually independent of the elasticity (Young) modulus E, despite the model predicting very different wall displacements. This behavior, at least for the adopted pressure-driven setup, can probably be attributed to the pressure-displacement algebraic relation (62). The imposed (given) pressure oscillations lead to different wall displacements (for different moduli E), but those displacements still end up causing the same pressure pushback to the fluid flow, due to the use of the same, just inverse linear (solution-independent) displacement-pressure law (61).

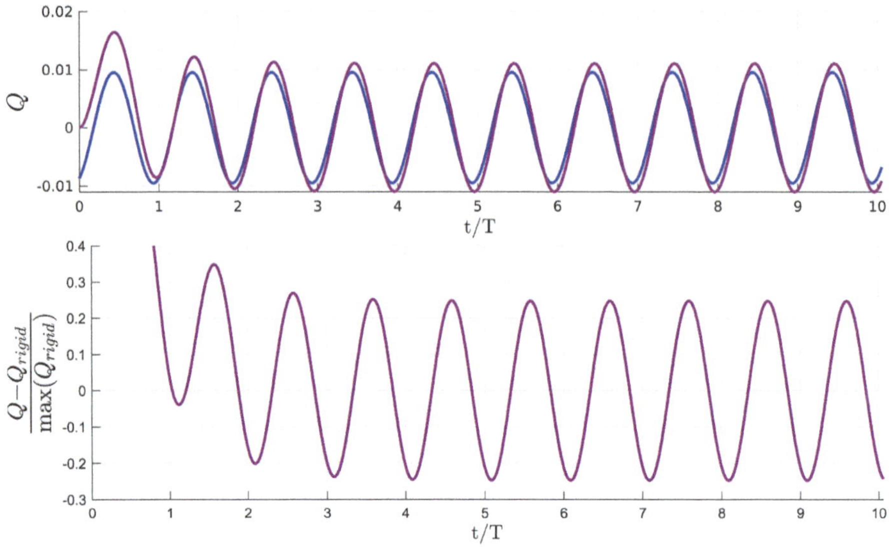

Fig. 19 Flow rate history at the inlet for the fully rigid case blue and for the semielastic cases with different values of Young modulus: $E = 2.9 \cdot 10^6$ (green), $E = 2.9 \cdot 10^5$ (red), and $E = 1.45 \cdot 10^4$ (purple)

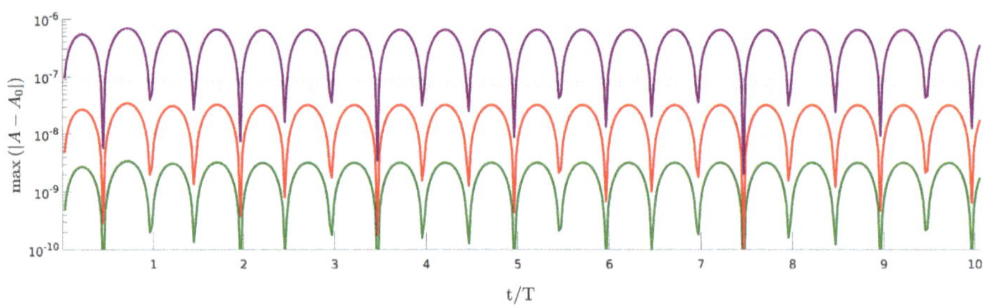

Fig. 20 Maximum wall displacement vs. time history for different values of Young modulus: $E = 2.9 \cdot 10^6$ (green), $E = 2.9 \cdot 10^5$ (red), and $E = 1.45 \cdot 10^4$ (purple)

5.2.4 FEM

The presented finite element approach to the fluid-structure interaction modeling and simulation represents in this work the standard reference solution with which the simplified 2D-1D coupled model is compared. The results of the FEM simulations presented here are chosen to demonstrate the type of outputs one can obtain for different cases and parameter settings.

5.2.5 FEM Simulation Results

From the 2D finite element ALE formulation, the resulting fields of flow quantities (velocity and pressure) can be presented together with the corresponding artificial displacements for both the structure and the fluid regions. An example of such fields for one of the simulations is shown in Figs. 21 and 22 corresponding to two different time instants in the same simulation. At the time $t = 42.2\,\text{s}$ (shown in Fig. 21),

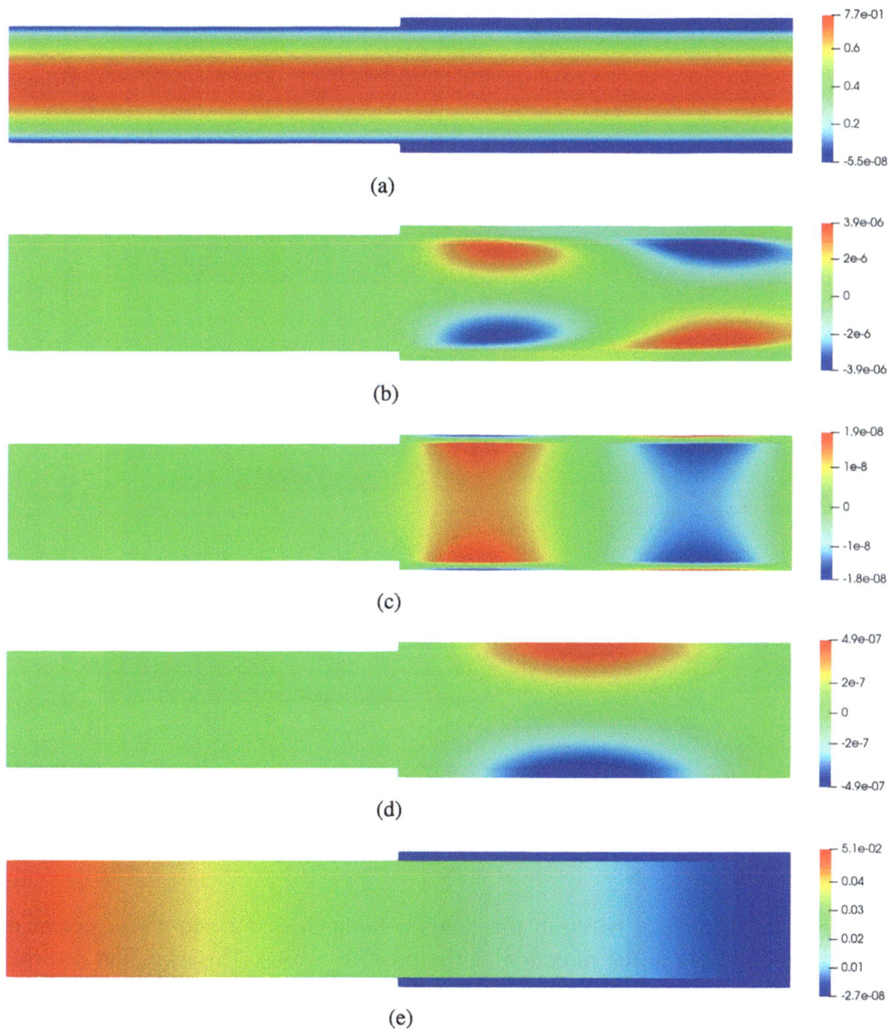

Fig. 21 FEM solution fields at time $t = 42.2\,\text{s}$. (**a**) Horizontal (x component) velocity. (**b**) Vertical (z component) velocity. (**c**) Horizontal (x component) displacement. (**d**) Vertical (z component) displacement. (**e**) Pressure

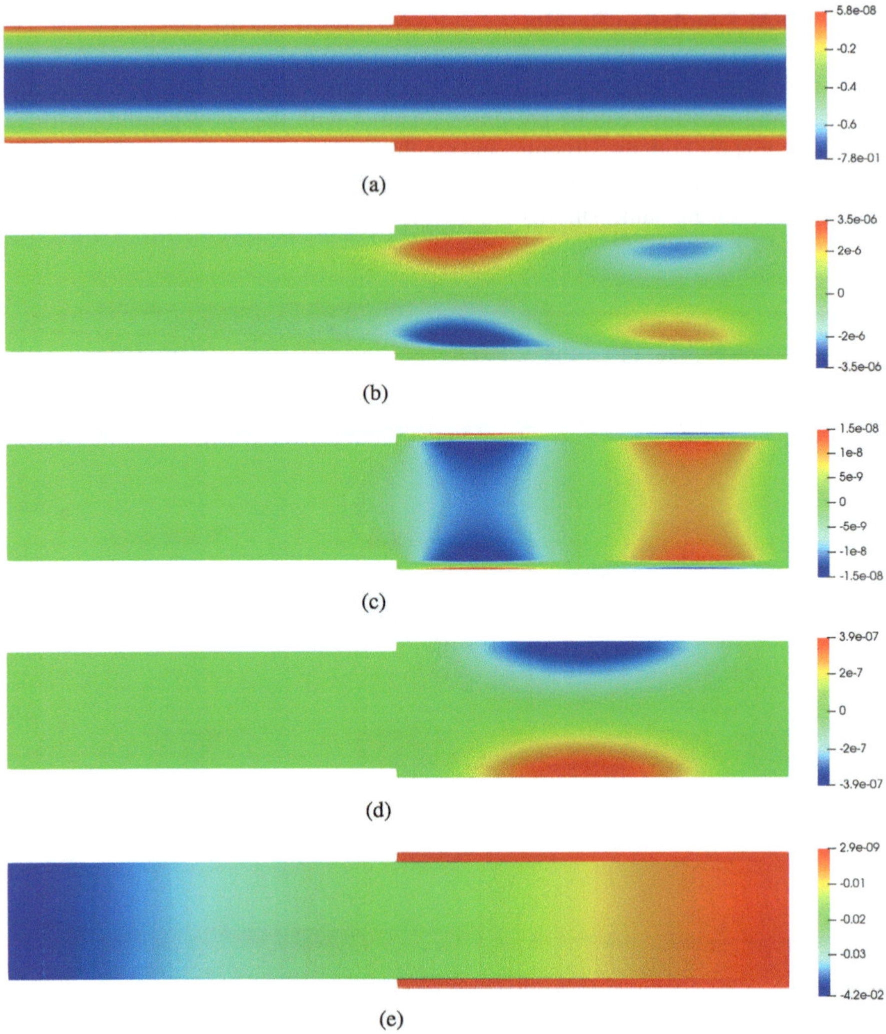

Fig. 22 FEM solution fields at time $t = 44.5$ s. (**a**) Horizontal (x component) velocity. (**b**) Vertical (z component) velocity. (**c**) Horizontal (x component) displacement. (**d**) Vertical (z component) displacement. (**e**) Pressure

the pressure difference between the domain boundaries Γ_6 and Γ_1 is negative, generating the flow in the direction of the positive x axis. This situation reverses at the later time $t = 44.5$ s (shown in Fig. 22), where both the pressure difference and the resulting velocity change their sign.

In both cases, the results show the expected behavior, with axial flow being dominant, when the transversal (radial) velocity component is about five orders of magnitude smaller. The vertical (radial) displacement reaches its maximum approximately at the midpoint of the elastic channel segment, a direct consequence of the zero-displacement boundary conditions applied at both ends. Since the ALE (Arbitrary Lagrangian-Eulerian) approach is used, where the solution is computed on a fixed reference domain rather than on a deforming, time-dependent domain, some clarification regarding the interpretation of the results is necessary. The governing equations for both the fluid and the structure are numerically transformed and solved on a static reference configuration. This approach facilitates the handling of problems involving moving boundaries and large deformations by decoupling the computational mesh motion from material motion. As a result, the deformations depicted in Figs. 21d and 22d are presented within the reference geometry, illustrating the artificial displacement field computed during the simulation. These artificial displacements represent the mesh deformation necessary to accommodate the physical movement of the material within the fixed computational domain. However, in this specific application, the magnitude of these displacements is minimal, rendering the distinction between the reference and current configurations negligible. In Fig. 3, the current (deformed) domain has deliberately been scaled by a specific factor to enhance visibility. This enlargement facilitates a clearer depiction of the deformation patterns resulting from the Arbitrary Lagrangian-Eulerian (ALE) method. Conversely, Figs. 21c and 22c reveal the artificial longitudinal displacement, manifesting as stretching and compression of the channel wall. These visualizations underscore the mesh adjustments implemented to accurately capture the dynamic behavior of the system. Furthermore, within the ALE framework, the computed velocity field extends across both structural and fluid domains. In the structural domain, this velocity field corresponds to the physical movement of the material. In the fluid domain, the velocity field imparts an additional component, representing the grid movement. This coupling ensures that the fluid flow accounts for the moving boundaries and mesh deformations, providing a consistent and accurate representation of the fluid-structure interaction.

Under reverse-flow conditions (Fig. 22), the field patterns and magnitudes remain essentially unchanged, differing only in the sign of both velocity and displacement. In Fig. 23, the current (deformed) domain is depicted at time instants $t = 42.2\,\mathrm{s}, 43.6\,\mathrm{s}$, and $t = 45\,\mathrm{s}$. The visualizations represent the absolute magnitude of the deformation field. Given the minimal deformations inherent in this application, the deformation values have been magnified by a factor of 4000 to enhance visibility and facilitate a clearer understanding of the deformation patterns. The simulations used a Poisson ratio of $\kappa_E = 0.49$ and Lamè coefficients $\mu = 973154$ and $\lambda = 47684564$ on a mesh of 4608 elements with 78829 degrees of freedom.

Further insights into the results of the 2D FEM simulations can be gained by comparing the corresponding graphs with those generated by the 2D-1D FDM simulations. Among these outputs, the pressure along the channel centerline (shown in Fig. 24) is particularly important, as the pressure-feedback coupling is intended to be the primary output of the simplified 1D model.

The altered pressure field due to the elasticity of the part of the channel will affect the corresponding pressure drop for the rigid part and thus the flow field generated in it. This is visible in the resulting velocity profiles at the inlet that differ from

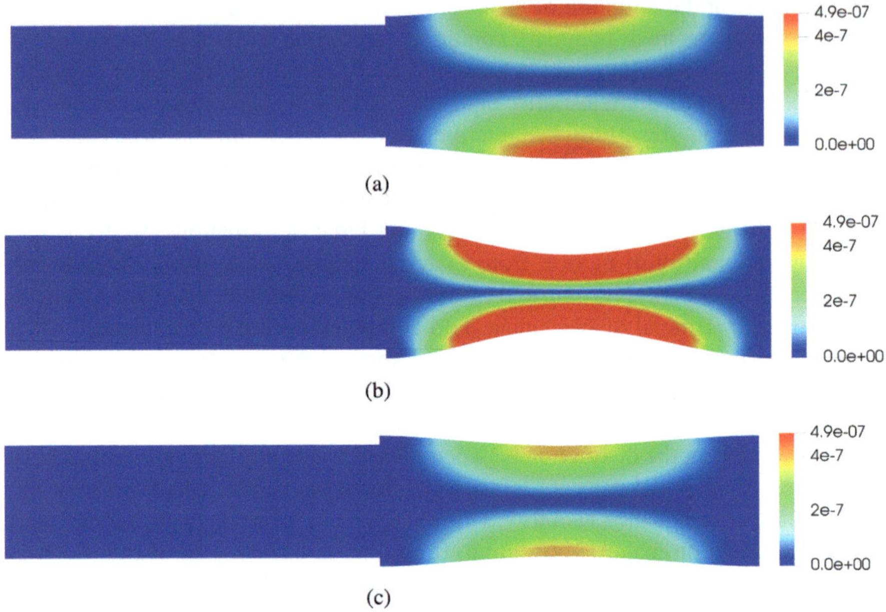

Fig. 23 Visualization of the current (deformed) domain with the absolute magnitude of the deformation field magnified by a factor of 4000 at 3 different time points. (**a**) Absolute magnitude of the deformation field at time $t = 42.2$ s. (**b**) Absolute magnitude of the deformation field at time $t = 43.6$ s. (**c**) Absolute magnitude of the deformation field at time $t = 44.5$ s

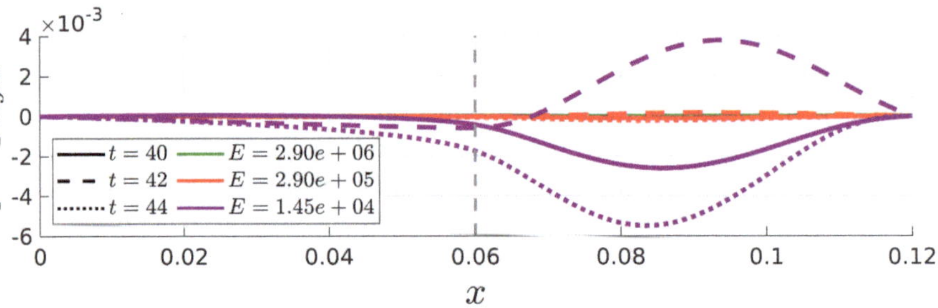

Fig. 24 Difference between the mean pressure $\bar{p}$ and the pressure analytical solution for rigid channel p_{rigid} as a function of position x for various times t and Young moduli E. The vertical gray line marks the interface between solid and elastic parts of the tube

those corresponding to the fully rigid case, represented by the Womersley analytical solution. The corresponding differences of velocity profiles at the inlet boundary Γ_1 are shown in Fig. 25.

The evolution of velocity profiles over time illustrates how the wall elasticity influences deviations from the analytical solution during fluid-structure interaction. This behavior is specific for the pressure-driven setup chosen for this series of

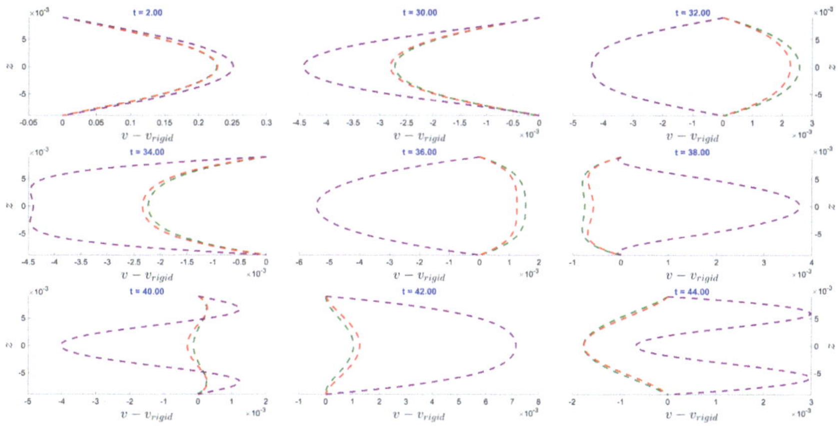

Fig. 25 Difference between numerical and analytical velocity profiles at the inlet along the tube axis z, shown at different times t. Colored lines correspond to different values of Young modulus: $E = 2.9 \cdot 10^6$ (green), $E = 2.9 \cdot 10^5$ (red), and $E = 1.45 \cdot 10^4$ (purple)

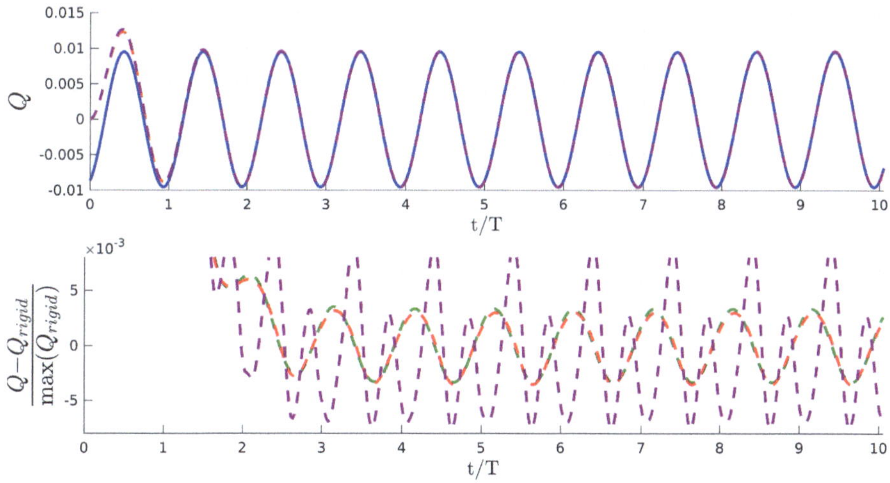

Fig. 26 Flow rate history for different values of Young modulus: $E = 2.9 \cdot 10^6$ (green), $E = 2.9 \cdot 10^5$ (red), and $E = 1.45 \cdot 10^4$ (purple)

simulations, where the flow rate is not known a priori and is obtained as a part of the solution.

Due to velocity profile deviations caused by the channel elasticity, the achieved flow rates also differ, for the given (prescribed) pressure difference, compared to the fully rigid channel case. This effect is apparent from the time plots of the resulting flow rates at the channel inlet shown in Fig. 26.

Except for the initial transition period, the resulting flow rate closely follows that of the fully rigid case. When the differences are plotted separately as in the bottom

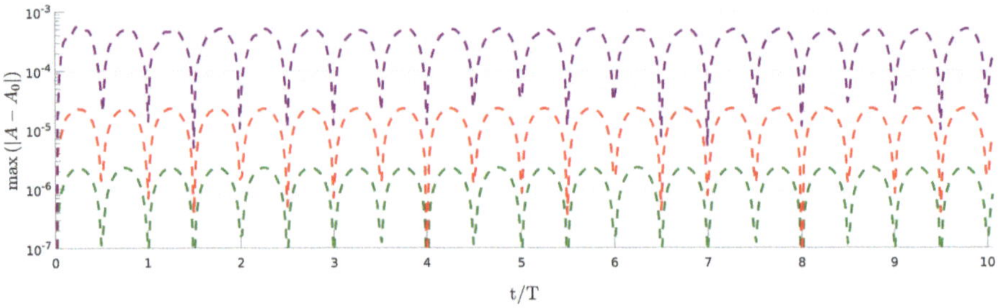

Fig. 27 Maximal wall displacement for different values of Young modulus: $E = 2.9 \cdot 10^6$ (green), $E = 2.9 \cdot 10^5$ (red), and $E = 1.45 \cdot 10^4$ (purple)

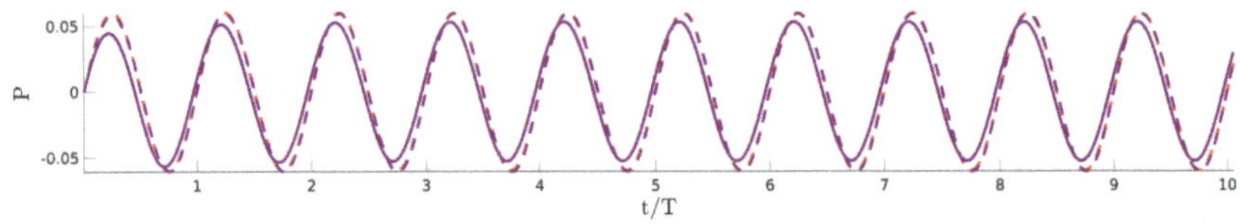

Fig. 28 Comparison of the mean pressure at the coupling interface Γ_3 obtained by FDM (solid lines) and FEM (dashed lines) for different Young moduli: $E = 2.9 \cdot 10^6$ (green), $E = 2.9 \cdot 10^5$ (red), and $E = 1.45 \cdot 10^4$ (purple)

panel of Fig. 26, it becomes apparent that the magnitude of the deviations changes periodically in time, being mildly dependent on the Young modulus E of the channel wall. But also, in this case, the reduction of the Young modulus by a factor of ten has almost no effect, while reducing the modulus further by a factor of 200 finally brings some more pronounced solution differences.

The maximum wall displacement in the FEM simulations scales approximately linearly with the Young modulus E, as documented in Fig. 27 showing the time history of the displacement magnitude for the three values considered of E. It is interesting to note that for the higher Young moduli the wall displacements differ substantially, but the pressure, flow rate, and velocity profiles are almost identical. A similar trend was also observed in the FEM simulations where, however, the magnitudes of the deviations were different.

5.2.6 FEM vs. FDM Result Comparison

The individual results shown and discussed separately for the FDM and FEM models in previous sections are now visualized together to point out and discuss their mutual differences.

The key parameter providing the feedback and mutual coupling of the rigid and elastic part of the channel in both 2D-2D FEM and 2D-1D FDM simulations is the pressure at the common interface Γ_3. The time history of this interfacial pressure is shown in Fig. 28.

In the fully rigid channel, this (mean) pressure should follow exactly a sinusoidal curve with amplitude $A_\omega \cdot l/2 = 1 \cdot 0.06 = 0.06\,\mathrm{Pa}$, because the interface is located at position $x = l/2 = 0.06\,\mathrm{m}$ from the far right boundary Γ_6, where the pressure is set to $p_{ext} = 0$. The presented comparison shows that the amplitude of the pressure predicted by the FDM is smaller, while the FEM model remains very close to the fully rigid reference case. The position of the local maxima/minima of the pressure history reveals a visible phase delay of the FDM solution.

As already noted earlier, the effect of the variation of the Young modulus E does not translates into a visible difference in pressure feedback, despite the wall displacements being different by an order of magnitude.

Based on the above-presented difference between FDM and FEM solutions in the pressure history at the coupling interface, the expected variation of the resulting flow rate is observed in Fig. 29.

The flow rate is generated by the difference in pressure between the prescribed value at the boundary Γ_1 and the pressure obtained at the coupling interface Γ_3. The lower pressure predicted by the FDM model leads to higher pressure drop and thus to higher flow rate. A similar mechanism applies for the FEM model, where the resulting flow rate becomes smaller. Also, in this case, the variation of the Young modulus of the elastic wall has virtually no effect on the predicted flow rates in both models.

The comparison of maximal wall displacement history for both the FDM and the FEM is shown in Fig. 30.

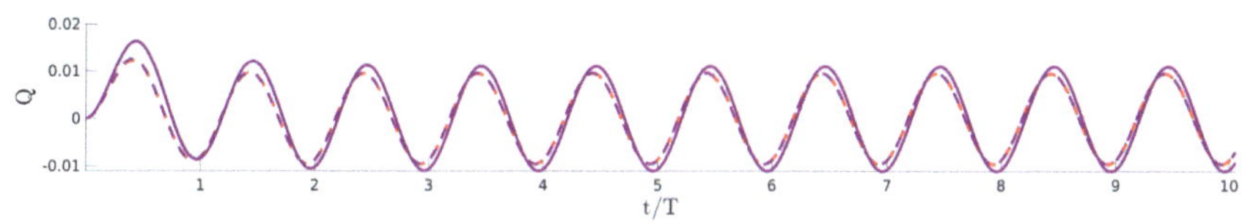

Fig. 29 Comparison of the flow rate at the inlet Γ_1 obtained by FDM (solid lines) and FEM (dashed lines) for different Young moduli: $E = 2.9 \cdot 10^6$ (green), $E = 2.9 \cdot 10^5$ (red), and $E = 1.45 \cdot 10^4$ (purple)

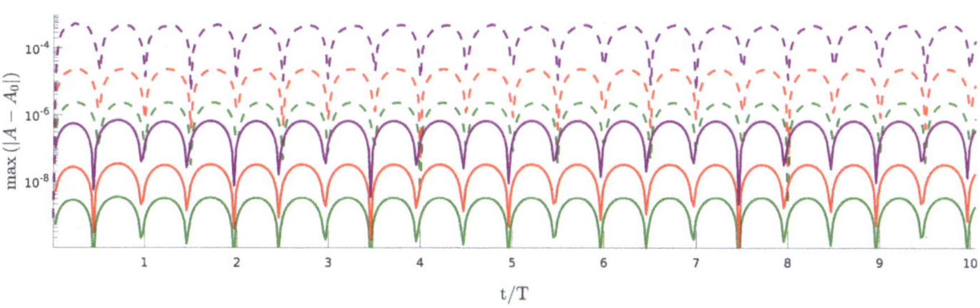

Fig. 30 Comparison of the maximal displacement obtained by FDM (solid lines) and FEM (dashed lines) for different Young moduli: $E = 2.9 \cdot 10^6$ (green), $E = 2.9 \cdot 10^5$ (red), and $E = 1.45 \cdot 10^4$ (purple)

It confirms what was expected, based on the analysis detailed in Sect. 5.2.1, that the displacements predicted by the FDM are at least two orders of magnitude smaller than those found in the results of FEM simulations. Due to the limitations of the 1D model setup, regarding the displacement boundary conditions, the displacement profiles also differ substantially in shape.

6 Conclusion

The finite difference and finite element codes were tested and validated for the case of a periodically oscillating flow, close to the one found in the respiratory flows. In the case of the rigid channel, both codes have proved to be sufficiently accurate and efficient in properly resolving the important flow features represented by the analytical solution. The accuracy, measured as an error with respect to the exact analytical solution, was found to depend more on the time step rather than being affected by an insufficient spatial resolution. This effect was more pronounced with the FEM simulations, for which the typically used time step was larger than in the explicit FDM simulations (where a much lower time step must be used due to stability restrictions).

The semielastic channel case was chosen to assess the effects of elasticity of walls on the flow and pressure conditions in the whole system. Under the chosen simulation parameters, closely resembling those of the human respiratory system, the effects of elasticity were rather subtle. The displacements predicted by both models were close to negligible, but they have still visibly (measurably) affected the flow. In this specific pressure-driven setup, the flow rate was not prescribed but was obtained as a part of the solution. The very small amplitude of the imposed pressure boundary condition (small imposed pressure drop) generates a physiologically reasonable flow velocity at which one should not expect large deformations of the trachea. From this point of view, the presented results met our expectations.

The channel wall displacement shape predicted by the 2D FEM model apparently differs from the one predicted by the 1D simplified model. It has to because both models have (must have) different boundary conditions for the elastic structure. In the 2D FEM model, the channel cross section is fixed (prescribed) at both sides of the elastic part. Therefore, the magnitude of the wall displacement is maximal at some position between the two fixed ends. The displacement shape reminds one of a string with fixed ends. However, in the simplified 1D FDM model, only the far exit cross-sectional area is fixed, while at the (coupling) 2D-1D interface the area has to be determined from the 1D model, subject to the prescribed flow rate. This leads to displacements that are maximal at the coupling interface, decaying to zero at the fixed far end of the channel. The wall behaves like a moving (almost rigid and flat) plate that rotates/oscillates around the fixed end point.

In the pressure-driven flow, where the pressure oscillations at one end of the semielastic channel generate the reciprocating flow, the effects of wall elasticity are very limited. Although wall displacements do depend on the elasticity (Young) modulus of the wall, the pushback effect through the force exerted on the fluid remains indifferent for a wide range of Young moduli. This was to some extent observed in both the 2D-2D FEM and the 2D-1D FDM simulations. This behavior can most likely be attributed to the pressure-driven setup used in all simulations, where the pressure disturbances in the whole channel are governed and limited by the prescribed time-dependent pressure boundary condition. In this study, we examine the impact of varying the wall's elasticity modulus on the fluid-structure interaction (FSI) dynamics. Our simulations reveal that, within a certain range of Young's modulus values, the force and stress balance at the fluid-structure interface remains effectively preserved. Consequently, the mutual interaction patterns between the fluid and the structure exhibit minimal variation across different simulations, regardless of the specific Young's modulus employed. The elasticity effect should rather be seen in the difference in wall displacements that leads to an increase (or decrease) of the channel volume, allowing the fluid to temporarily accumulate. This may lead to a phase shift and a hysteresis effect in the time history of velocity and flow rate.

This assessment of the effects of elasticity has only a limited validity for the chosen model setup and explored range of parameters. For simulations of larger part of the respiratory system, including the whole tracheobronchial tree and going even deeper into the lungs, the structural properties of the considered channels change dramatically, exhibiting lower Young modulus and thinner walls. Under such conditions, the deformation of the walls can significantly alter flow patterns, leading to variations in pressure-feedback mechanisms. An open question is the use of the 2D-1D coupled model in a case when the flow will be generated by a prescribed time-periodic wall displacement (typical situation for respiratory flows). In such setup, the displacement will generate a pressure disturbance leading to a force driving the flow. Thus, the flow rate is expected to depend much more on the Young modulus than in the present prescribed pressure-driven scenario.

Further analysis and modifications of the 1D simplified model will be needed, together with additional simulations, in order to fully explore and improve its potential applicability in realistic simulations of respiratory problems.

Acknowledgments The financial support for the present work was provided by the *Czech Technical University, Faculty of Mechanical Engineering*, under the *Grant No. SGS25/123/OHK2/3T/12*, and partly by the Czech Academy of Sciences under the *Praemium Academiae of Š. Nečasová* and RVO:67985840. From Kassel's side, this work was embedded in the project *BiTWerk* funded by the *University of Kassel* and in the *DFG Research Training Group 2749 Multiscale Clocks.*

Competing Interests The authors have no conflicts of interest to declare that are relevant to the content of this chapter.

References

1. Lancmanová, A., Bodnár, T.: Numerical simulations of human respiratory flows - A review. Discover Applied Sciences **7**(242) (2025). https://doi.org/10.1007/s42452-025-06617-x
2. Lancmanová, A., Bodnár, T.: Effects of higher order discretization on the predictions of steady turbulent airflow in human respiratory system. Journal of Computational And Applied Mathematics **466**, 116593 (2025). https://doi.org/10.1016/j.cam.2025.116593
3. Robertson, A., Sequeira, A., Kameneva, M.: Hemorheology. In: Galdi et al. [4], chap. III., pp. 63–120
4. Galdi, G., Rannacher, R., Robertson, A., Turek, S. (eds.): Hemodynamical Flows - Modeling, Analysis and Simulation, *Oberwolfach Seminars*, vol. 37. Birkäuser (2008)
5. Lancmanová, A., Bodnár, T., Sequeira, A.: On the development of a numerical model for the simulation of air flow in the human airways. In: Topical Problems of Fluid Mechanics 2023, pp. 112–118. Institute of Thermomechanics CAS, Prague (2023). https://doi.org/10.14311/TPFM.2023.016
6. Lancmanová, A., Bodnár, T.: Numerical study of the steady airflow in the human respiratory system during inhaling and exhaling. In: Topical Problems of Fluid Mechanics 2024, pp. 124–131. Institute of Thermomechanics CAS, Prague (2024). https://doi.org/10.14311/TPFM.2024.017
7. Weibel, E.R.: Morphometry of the Human Lung. Springer-Verlag, Berlin (1963)
8. Mammana, M., Bonis, A., Verzeletti, V., Dell'Amore, A., Rea, F.: Tracheal tissue engineering: Principles and state of the art. Bioengineering **11**(2), 198 (2024)
9. Furlow, P., Mathisen, D.: Surgical anatomy of the trachea. Annals of Cardiothoracic Surgery **7**(2), 255–260 (2018)
10. Minnich, D., Mathisen, D.: Anatomy of the trachea, carina, and bronchi. Thoracic Surgery Clinics **17**(4), 571–585 (2007)
11. Peiró, J., Veneziani, A.: Reduced models of the cardiovascular system, pp. 347–394. Springer Milan, Milano (2009)
12. Formaggia, L., Quarteroni, A., Veneziani, A.: Multiscale models of the vascular system, pp. 395–446. Springer Milan, Milano (2009)
13. Formaggia, L., Lamponi, D., Quarteroni, A.: One-dimensional models for blood flow in arteries. Journal of Engineering Mathematics **47**, 251–276 (2003)
14. Formaggia, L., Quateroni, A., Veneziani, A.: Cardiovascular Mathematics Modeling and simulation of the circulatory system. Springer-Verlag Italia, Milano (2009)
15. Lancmanová, A., Bodnár, T., Friedmann, E.: Influence of momentum flux correction coefficient in 2D-1D coupling. In: A. Sequeira, A. Silvestre, S. Valtchev, J. Janela (eds.) Numerical Mathematics and Advanced Applications ENUMATH 2023, Volume 2, Lecture Notes in Computational Science and Engineering, pp. 104–114. Springer Cham (2025). https://doi.org/10.1007/978-3-031-86169-7_11
16. Donea, J., Giuliani, S., Halleux, J.P.: An arbitrary Lagrangian-Eulerian finite element method for transient dynamic fluid-structure interactions. Computer Methods in Applied Mechanics and Engineering **33**(1–3), 689–723 (1982)
17. Nobile, F., Vergara, C.: Partitioned algorithms for fluid-structure interaction problems in haemodynamics. Milan Journal of Mathematics **80**(2), 443–467 (2012)
18. Dunne, T.: An Eulerian approach to fluid–structure interaction and goal-oriented mesh adaptation. International Journal for Numerical Methods in Fluids **51**(9–10), 1017–1039 (2006)
19. Formaggia, L., Quarteroni, A., Veneziani, A.: Cardiovascular Mathematics: Modeling and simulation of the circulatory system, vol. 1. Springer-Verlag Milan (2009)
20. Richter, T., Wick, T.: Finite elements for fluid–structure interaction in ALE and fully Eulerian coordinates. Computer Methods in Applied Mechanics and Engineering **199**(41–44), 2633–2642 (2010)
21. Gürarslan, G.: Numerical modelling of linear and nonlinear diffusion equations by compact finite difference method. Applied Mathematics and Computation **216**, 2472–2478 (2010)

22. Irgens, F.: Continuum Mechanics. Springer (2008)
23. Stein, K., Tezduyar, T., Benney, R.: Mesh moving techniques for fluid-structure interactions with large displacements. Journal of Applied Mechanics **70**(1), 58–63 (2003)
24. Wick, T.: Fluid-structure interactions using different mesh motion techniques. Computers & Structures **89**(13–14), 1456–1467 (2011)
25. Dunne, T.: Adaptive finite element approximation of fluid-structure interaction based on Eulerian and arbitrary Lagrangian-Eulerian variational formulations. Ph.D. thesis, University of Heidelberg (2007)
26. Wick, T.: Adaptive finite element simulation of fluid-structure interaction with application to heart-valve dynamics. Ph.D. thesis, University of Heidelberg (2011)
27. Richter, T.: Fluid-structure Interactions: Models, Analysis and Finite Elements, vol. 118. Lecture Notes in Computational Science and Engineering. Springer (2017)
28. Drobny, A., Friedmann, E.: Numerical simulation of viscoelastic fluid–structure interaction benchmarks and their application to the human eye. SN Applied Sciences **4**(11), 299 (2022)
29. Drobny, A.: Mathematical modelling and adaptive finite element simulation of viscoelastic fluid-structure interaction systems and chemical processes with applications to ophthalmology. Ph.D. thesis, Kassel, Universität Kassel, Fachbereich Mathematik und Naturwissenschaften, Institut für Mathematik (2023)
30. Heywood, J.G., Rannacher, R.: Finite-element approximation of the nonstationary Navier–Stokes problem. Part IV: Error analysis for second-order time discretization. SIAM Journal on Numerical Analysis **27**(2), 353–384 (1990)
31. Amestoy, P.R., Duff, I.S., L'Excellent, J.Y.: Multifrontal parallel distributed symmetric and unsymmetric solvers. Computer Methods in Applied Mechanics and Engineering **184**(2–4), 501–520 (2000)
32. Quarteroni, A., Veneziani, A., Vergara, C.: Geometric multiscale modeling of the cardiovascular system, between theory and practice. Computer Methods in Applied Mechanics and Engineering. **302** (2016)
33. Sherwin, S., Franke, V., Peiro, J., Parker, K.H.: One-dimensional modelling of a vascular network in space-time. Journal of Engineering Mathematics. **47**, 217–250 (2003)
34. Gurtin, M., Fried, E., Anand, L.: The mechanics and thermodynamics of continua. Cambridge University Press (2010)
35. Formaggia, L., Veneziani, A.: Reduced and multiscale models for the human cvs (2003). Technical Report, PoliMI, Milan
36. Quarteroni, A., Tuver, M., Veneziani, A.: Computational vascular fluid dynamics: problems, models and methods. Computing and Visualisation in Science **2**, 163–197 (2000). https://doi.org/10.1007/s007910050039
37. Lancmanová, A., Bodnár, T., Sequeira, A.: On the 2D-1D coupling in numerical simulations of simplified air flow model in human airways. In: 10th edition of the International Conference on Computational Methods for Coupled Problems in Science and Engineering (COUPLED PROBLEMS 2023). CIMNE (2023). https://doi.org/10.23967/c.coupled.2023.022
38. Kurzweg, U.: Enhanced heat conduction in oscillating viscous flows within parallel-plate channels. Journal of Fluid Mechanics **156**, 291–300 (1985)
39. Landau, L.D., Lifshitz, E.M.: Fluid Mechanics (Volume 6 of A Course of Theoretical Physics). Pergamon Press (1959)
40. Loudon, C., Tordesillas, A.: The use of the dimensionless Womersley number to characterize the unsteady nature of internal flow. Journal of Theoretical Biology **191**, 63–78 (1998)
41. Arndt, D., Bangerth, W., Davydov, D., Heister, T., Heltai, L., Kronbichler, M., Maier, M., Pelteret, J.P., Turcksin, B., Wells, D.: The deal.II finite element library: Design, features, and insights. Computers & Mathematics with Applications **81**, 407–422 (2021). https://doi.org/10.1016/j.camwa.2020.02.022. URL https://arxiv.org/abs/1910.13247

42. Turek, S., Hron, J.: Proposal for numerical benchmarking of fluid-structure interaction between an elastic object and laminar incompressible flow. In: H.-J. Bungartz and M. Schäfer (Eds.). Fluid-Structure Interaction: Modelling, Simulation, Optimisation, pp. 371–385. Springer-Verlag Berlin Heidelberg (2006)
43. Wick, T.: Solving monolithic fluid-structure interaction problems in arbitrary Lagrangian Eulerian coordinates with the deal.ii library. Archive of Numerical Software (ANS), Vol. 1, pp. 1–19 (2013)

Non-Newtonian Compressible Fluids with Stochastic Right-Hand Side

Pavel Ludvík and Václav Mácha

Abstract Fluids with shear rate-dependent viscosity form a special class of non-Newtonian fluids that play a crucial role in continuum dynamics. We consider a compressible barotropic flow of such fluids, governed by a generalized Navier-Stokes system. Due to the nonlinearity in the elliptic term, obtaining the existence of weak solutions is challenging. To address this, we introduce the concept of a measure-valued solution, whose existence is established in a very general setting using an abstract theorem on the existence of a Young measure. We also discuss the proof of this key result.

1 Introduction

The compressible non-Newtonian fluids are fluids whose viscosity is not constant, and it may vary with the gradient of velocity and/or pressure. This allows us to capture certain physical phenomena, for example, die swell, delayed die swell, shear thinning/thickening, and so on. The mathematical description of such fluids was deduced by Málek and Rajagopal [1].

The existence of a weak solution to such a system without white noise is quite delicate as there is a lack of compactness of the velocity. The first papers in this direction are due to Mamontov [2, 3] who considered an exponential growth of the viscosity together with an isothermal pressure ($p(\varrho) = a\varrho$ for some $a > 0$). The existence of a weak solution was further treated by Feireisl, Liao, and Málek [4] who worked with a special form of the stress tensor and further by Zhikhov and Pastukhova [5] for quite general setting—unfortunately, it turned out to be wrong—we refer to the introduction of [4].

P. Ludvík
Faculty of Science, Palacký University Olomouc, Olomouc, Czech Republic
e-mail: pavel.ludvik@upol.cz

V. Mácha (✉)
Institute of Mathematics of the Czech Academy of Sciences, Praha, Czech Republic
e-mail: macha@math.cas.cz

T. Bodnár et al. (eds.), *Stochastics in Fluids*, Lecture Notes in Mathematical Fluid Mechanics, https://doi.org/10.1007/978-3-032-02299-8_3

To overcome this existing issue, the notion of *dissipative weak solution* was introduced. Its existence was proven by Abbattiello, Feireisl, and Novotný [6] for isentropic pressure and by Basarić [7] for an isothermal pressure.

There are several reasons for working with the stochastic force term in such system—it can contain some numerical as well as physical uncertainties. This is the main aim of this article—we introduce a notion of *measure-valued solution* for the nonlinear compressible Navier-Stokes equation with the stochastic force term, and we provide a proof of its existence.

Concerning the notion of the measure-valued solution for the stochastic equations, we refer to the article by Hofmanová, Koley, and Sarkar [8] where this notion was given for the compressible Euler equations. Further, the stochastic version of the dissipative weak solution (martingale dissipative solution) was introduced by Hofmanová, Zhu, and Zhu in [9] who worked with the incompressible Euler system.

This paper is devoted to a nonlinear system of the form

$$\begin{aligned} \mathrm{d}(\varrho \mathbf{u}) + \operatorname{div}(\varrho \mathbf{u} \otimes \mathbf{u})\mathrm{d}t + \operatorname{div} \mathbb{S} \mathrm{d}t - \nabla p(\varrho)\mathrm{d}t &= G(\varrho, \varrho \mathbf{u})\mathrm{d}W \\ \mathrm{d}\varrho + \operatorname{div}(\mathbf{u})\mathrm{d}t &= 0 \end{aligned} \tag{1}$$

considered on $(0, T) \times \Omega$ where $T > 0$ and $\Omega \subset \mathbb{R}^3$ is a bounded Lipschitz domain. The unknowns are the velocity $\mathbf{u} : (0, T) \times \Omega \to \mathbb{R}^3$ and the density $\varrho : (0, T) \times \Omega \to [0, \infty)$. Furthermore, we assume that $\mathbb{S} \in \mathbb{R}^{3\times 3}_{sym}$ satisfies

$$\mathbb{S} : D\mathbf{u} = F^*(\mathbb{S}) + F(D\mathbf{u}) \tag{2}$$

for a suitable convex nonnegative potential $F : \mathbb{R}^{3\times 3}_{sym} \to \mathbb{R}$—here, F^* denotes a convex conjugate of F, and $D\mathbf{u}$ denotes the symmetric part of the gradient of $\mathbf{u}$, i.e., $2D\mathbf{u} = \left(\nabla \mathbf{u} + (\nabla \mathbf{u})^T\right)$. We just recall that, assuming F is a proper convex lower-semicontinuous function, (2) is equivalent to

$$\mathbb{S} \in \partial F(D\mathbf{u})$$

or

$$D\mathbf{u} \in \partial F^*(\mathbb{S})$$

where ∂F denotes the subdifferential of F. Taking

$$F(D\mathbf{u}) = \frac{\mu}{2}|D\mathbf{u}|^2 + \frac{\lambda}{2}|\operatorname{div} \mathbf{u}|^2,$$

one obtains a system describing the Newtonian fluid. Such a system has been extensively studied—the existence of a weak solution to a deterministic equation is provided by nowadays standard Lions-Feireisl theory (see [10] and [11]). Besides that, also the stochastic case has been examined—we refer to book by Breit, Feireisl, and Hofmanová [12] and to a pioneering paper by Breit and Hofmanová [13]. There are many more results dealing with the analysis of the Newtonian fluid, and it is impossible to present an exhaustive list; thus, we just presented some of the most significant representatives.

We would also like to point out that the above implicit relation between $\mathbb{S}$ and $D\mathbf{u}$ contains a lot of models, for example, the power law fluids, i.e., fluids whose stress tensor satisfies, for example,

$$\mathbb{S}(D\mathbf{u}) = \left(1 + |D\mathbf{u}|^2\right)^{(p-2)/2} D\mathbf{u}$$

which is attained by the choice

$$F(D\mathbf{u}) = \frac{1}{p}\left(1 + |D\mathbf{u}|^2\right)^{p/2}.$$

The system (1) is complemented with the homogenous Dirichlet boundary condition

$$\mathbf{u}\restriction_{\partial\Omega} = 0, \;\; \Omega \text{ is of class } C^{2+\nu}, \; \nu > 0 \tag{3}$$

which captures the case when the fluid sticks to the boundary of Ω.

Let us specify the assumptions on data. We start with the definition of the stochastic forcing term. Let $(O, \mathfrak{F}, (\mathfrak{F}_t)_{t\geq 0}, \mathbb{P})$ be a stochastic basis with a complete right-continuous filtration and W be a cylindrical $(\mathfrak{F}_t)$-Wiener process in a separable Hilbert space $\mathfrak{U}$ formally given by

$$W = \sum_{k=1}^{\infty} e_k W_k(t),$$

where $\{W_k\}_{k=1}^{\infty}$ is a sequence of mutually independent real-valued Brownian motions relative to $(\mathfrak{F}_t)_{t\geq 0}$ and $\{e_k\}_{k=1}^{\infty}$ is an orthonormal basis of $\mathfrak{U}$.

Next, we assume that

$$G(\varrho, \mathbf{u})\,\mathrm{d}W = \sum_{k=1}^{\infty} \mathbf{G}_k(\varrho, \mathbf{u})\,\mathrm{d}W_k, \quad \mathbf{G}_k(\varrho, \mathbf{u}) = \varrho\mathbf{F}_k(\varrho, \mathbf{u}) \tag{4}$$

where

$$\mathbf{F}_k : \Omega \times [0, \infty) \times \mathbb{R}^3 \to \mathbb{R}^3, \mathbf{F}_k \in C^1(\Omega \times [0, \infty) \times \mathbb{R}^3)$$

satisfies

$$\|\mathbf{F}_k\|_{L^\infty(\Omega\times[0,\infty)\times\mathbb{R}^3)} + \|\nabla_{\mathbf{u}}\mathbf{F}_k\|_{L^\infty(\Omega\times[0,\infty)\times\mathbb{R}^3)} \leq f_k, \quad \sum_{k=1}^{\infty} f_k^2 < \infty,$$

and there is some positive constant α such that $\mathbf{F}_k(\varrho, \mathbf{u})$ is zero whenever $\varrho < \alpha$, $\varrho > \frac{1}{\alpha}$, or $|\mathbf{u}| > \frac{1}{\alpha}$—this last assumption is here just for simplicity as there is a known method allowing to remove it (we refer to [12]). We define $\mathbb{F} = \{\mathbf{F}_k\}_{k=1}^{\infty}$.

The initial condition is $\varrho(0, \cdot) = \varrho_0(\cdot)$ and $\varrho(0, \cdot)\mathbf{u}(0.\cdot) = (\varrho\mathbf{u})_0(\cdot)$, and the initial data are supposed to satisfy

$$\int_\Omega P(\varrho_0)\,\mathrm{d}x < \infty, \quad \int_\Omega \frac{1}{2}\frac{(\varrho\mathbf{u})_0^2}{\varrho_0}\,\mathrm{d}x < \infty, \text{ and } \int_\Omega (\varrho\mathbf{u})_0\,\mathrm{d}x < \infty$$

$\mathbb{P}$-almost surely. Here, P stands for the pressure potential introduced below (8). However, this initial datum is given only by its law—we refer to (23) and (24) for the precise statement of the initial condition.

Next, the potential F is supposed to be of class C^1, to satisfy $F(0) = 0$, and we assume the existence of an N-function $g : [0, \infty) \to [0, \infty)$ (see [14, Chapter 8]) such that

$$F(D\mathbf{u}) \geq g(|D\mathbf{u}|). \tag{5}$$

Up to our knowledge, the generalized compressible Navier-Stokes equation with a stochastic right-hand side as presented above has not been studied yet.

The main aim of this paper is to provide a notion of a solution which is on one hand sufficiently general so one can still deduce its existence and, on the other hand, there is a chance that it still possesses the weak-strong uniqueness property—this is going to be the content of the forthcoming paper.

To achieve this, we use a theorem on the existence of Young measures for a sequence of function-valued random variables, as introduced in [12]. We revisit this theorem under relaxed assumptions and provide a detailed proof in Sect. 2.

Section 3 focuses on the analysis of the Galerkin approximation, which serves as the starting point for the subsequent limit process, leading to the existence of a properly defined measure-valued solution to (1). The existence of solution to this approximation is established using the methods presented in [12].

The limit in the Galerkin approximation is presented in Sect. 4, which also provides the proper definition of the measure-valued solution. The existence of such a solution is summarized in this section as well, stated in a theorem that we consider the main result of this paper.

2 Remarks on Young Measures

In our analysis below, we make use of two standard theorems that appear in the literature in various forms. We state the specific variants needed for our purposes here.

Theorem 2.1 (de la Vallée Poussin Criterion, see [15, Theorem 4.5.9.]) *Let $(X, \mathcal{A}, \mu)$ be a measure space with a finite measure. A family $\mathcal{F}$ of μ-integrable functions is equi-integrable if and only if there exists a nondecreasing function $g : [0, +\infty) \to [0, +\infty)$ such that*

$$\lim_{t \to +\infty} \frac{g(t)}{t} = \infty \text{ and } \sup_{f \in \mathcal{F}} \int_X g(|f(x)|)\, \mathrm{d}\mu(x) \le \infty.$$

In such a case, one can choose a convex nondecreasing function g.

Theorem 2.2 (Dunford-Pettis Theorem, see [15, Theorem 4.7.18.]) *Let $(X, \mathcal{A}, \mu)$ be a measure space with a finite measure, and let $\mathcal{F}$ be a family of μ-measurable functions. The set $\mathcal{F}$ is equi-integrable precisely when it is weakly relatively compact in $L^1(X)$.*

The main tool yielding the measure-valued solution is the concept of the Young measures (see e.g. [16, Theorem 6.2]) adapted to the stochastic setting. The theorem following can be found as Theorem 2.8.1 in [11] (see also [8, Lemma 2.2]):

Lemma 2.1 *Let $N, M \in \mathbb{N}$, $\Omega \subset \mathbb{R}^N$, let $(O, \mathfrak{F}, \mathbb{P})$ be a complete probability space, and let $\{\mathbf{U}_n\}_{n=1}^{\infty}$, $\mathbf{U}_n : \Omega \times O \to \mathbb{R}^M$, be a sequence of random variables such that*[1]

$$\mathbb{E}\left[\|\mathbf{U}_n\|^p_{L^p(\Omega)}\right] \le C, \text{ for a certain } p \in (1, \infty).$$

Then, there exists a new subsequence $\{\tilde{\mathbf{U}}_n\}_{n=1}^{\infty}$ (not relabeled), defined on the standard probability space $\left([0,1], \overline{\mathcal{B}[0,1]}, \mathcal{L}\right)$ ($\mathcal{L}$ stands for the Lebesgue measure), and a parametrized family $\left\{\tilde{\nu}_{\omega,x}\right\}_{(\omega,x)\in O\times\Omega}$ of random probability measures on $\mathbb{R}^M$, regarded as a random variable taking values in $(L^\infty_{\omega^*}(\Omega; \mathcal{P}(\mathbb{R}^M)), w^*)$, such that $\mathbf{U}_n$ has the same law as $\tilde{\mathbf{U}}_n$, i.e., $\mathbf{U}_n \sim_d \tilde{\mathbf{U}}_n$, and the following property holds: for any Carathéodory function $G = G(x, Z)$, $x \in \Omega$, $Z \in \mathbb{R}^M$, such that

$$|G(x, Z)| \le C(1 + |Z|^q),\ 1 \le q < p, \text{ uniformly in } x,$$

implies $\mathcal{L}$-a.s.

$$G(\cdot, \tilde{\mathbf{U}}_n) \rightharpoonup \overline{G} \text{ in } L^{p/q}(\Omega)$$

[1] Hereinafter, C is reserved for a constant which may vary from line to line but remains independent of solutions. Any specific dependencies will be explicitly highlighted when necessary.

where

$$\overline{G}(x) = \langle \tilde{\nu}_{\omega,x}; G(x,\cdot)\rangle := \int_{\mathbb{R}^M} G(x,z)\,\mathrm{d}\tilde{\nu}_{\omega,x}(z)$$

for almost all $x \in \Omega$.

Here, we present a stronger theorem together with its proof, namely the following:

Theorem 2.3 *Let* $N, M \in \mathbb{N}$, $\Omega \subset \mathbb{R}^N$ *of finite Lebesgue measure* $\mathcal{L}$, *let* $(O, \mathfrak{F}, \mathbb{P})$ *be a complete probability space, and let* $\{\mathbf{U}_n\}_{n=1}^{\infty}$, $\mathbf{U}_n : \Omega \times O \to \mathbb{R}^M$, *be a sequence of random variables such that*

$$\sup_{n\in\mathbb{N}} \int_O \int_\Omega g(|\mathbf{U}_n|)\mathrm{d}\mathcal{L}(x)\mathrm{d}\mathbb{P}(\omega) \le C,$$

for some $g : [0,\infty) \to [0,\infty)$ *which is a nondecreasing function such that* $\lim_{t\to\infty} \frac{g(t)}{t} = \infty$.

Then, there exists a subsequence $\{\tilde{\mathbf{U}}_n\}_{n=1}^{\infty}$ *defined on the standard probability space* $\left([0,1], \overline{\mathcal{B}[0,1]}, \mathcal{L}\right)$ *and a parametrized family* $\left\{\tilde{\nu}_{\omega,x}\right\}_{(\omega,x)\in O\times\Omega}$ *as in the previous lemma such that for every Carathéodory function* $G = G(x, Z)$ *such*

$$G(x, \tilde{\mathbf{U}}_n) \text{ is weakly convergent in } L^1(\Omega)\ \mathcal{L}\text{-almost surely}$$

it holds that

$$G(x, \tilde{\mathbf{U}}_n) \rightharpoonup \langle \tilde{\nu}_{\omega,x}; G(x,\cdot)\rangle$$

in $L^1(\Omega)$ $\mathcal{L}$*-almost surely.*

Proof Let $(\nu_n)_{\omega,x} := \delta_{\mathbf{U}(\omega,x)}$ and let $\mathcal{L}[\nu_n]$ be its law. Next, let

$$B_{R,1} = \left\{\nu \in L^\infty(\Omega; \mathcal{P}(\mathbb{R}^M), w^*),\ \int_\Omega \int_{\mathbb{R}^M} |\xi|\,\mathrm{d}\nu_x(\xi)\mathrm{d}\mathcal{L}(x) \le R\right\}.$$

This set is relatively compact in $L^\infty(\Omega; \mathcal{P}(\mathbb{R}^M), w^*)$. This can be justified by the observation that in [17, Theorem 5], without changing the proof, we can replace $\mathbb{R}$ by $\mathbb{R}^M$ and the assumption that (X, λ) is a finite measure space with the assumptions that (X, λ) is a σ-finite space and the corresponding $L^1(X)$ is separable. Consequently, by the Chebyshev inequality,

$$\mathcal{L}[\nu_n](B_{R,1}^c) = \mathbb{P}\left(\nu_n \notin B_{R,1}\right) = \mathbb{P}\left(\int_\Omega |\mathbf{U}_n|\,\mathrm{d}x > R\right) \le \frac{1}{R}\mathbb{E}\left[\|\mathbf{U}_n\|_{L^1(\Omega)}\right] \le \frac{C}{R}.$$

The family of laws $\{\mathcal{L}[\nu_n]\}_{n=1}^{\infty}$ is therefore tight in $L^\infty(\Omega; \mathcal{P}(\mathbb{R}^M), w^*)$.

Next, we show that also the family of laws $\{\mathcal{L}[\mathbf{U}_n]\}_{n=1}^{\infty}$ is tight on $L^1(\Omega, w)$. For the fixed $R > 0$, we consider the set

$$\left\{\mathbf{V} \in L^1(\Omega),\ \|g(\mathbf{V})\|_{L^1(\Omega)} \leq R\right\},$$

which is equi-integrable in $L^1(\Omega, w)$ by Theorem 2.1. Thus, it is also relatively weakly compact by Theorem 2.2.

Therefore,

$$\mathcal{L}[\mathbf{U}_n]\left(\left\{\mathbf{V} \in L^1(\Omega),\ \|g(\mathbf{V})\|_{L^1(\Omega)} > R\right\}\right) \leq \frac{1}{R}\mathbb{E}\left[\|g(\mathbf{U}_n)\|_{L^1(\Omega)}\right] \leq \frac{C}{R},$$

by Chebyshev inequality, and we thus obtain the desired tightness.

Thus, the family of joint laws $\{\mathcal{L}[\mathbf{U}_n, \nu_n],\ n \in \mathbb{N}\}$ is tight in $L^1(\Omega, w) \times L^{\infty}(\Omega; \mathcal{P}(\mathbb{R}^M), w^*)$, and we may apply Jakubowski-Skorokhod's theorem (see [18, Theorem 2] or [12, Theorem 2.7.1]) to obtain a complete probability space $\left([0,1], \overline{\mathcal{B}[0,1]}, \mathcal{L}\right)$ and Borel random variables $[\tilde{\mathbf{U}}_n, \tilde{\nu}_n]$ with the law $\mathcal{L}[\mathbf{U}_n, \nu_n]$, for $n \in \mathbb{N}$, and converging $\tilde{\mathbb{P}}$-a.s. to a Borel random variable $[\tilde{\mathbf{U}}, \tilde{\nu}]$ possessing values in $L^1(\Omega, w) \times L^{\infty}(\Omega; \mathcal{P}(\mathbb{R}^M), w^*)$.

Now, let us define

$$\begin{aligned} Q := \Bigg\{(\mathbf{V}, \nu) \in L^1(\Omega) \times L^{\infty}(\Omega; \mathcal{P}(\mathbb{R}^M), w^*),\ &\int_{\Omega} \psi(x) \int_{\mathbb{R}^M} \phi(\xi) \mathrm{d}\nu_x(\xi) \mathrm{d}x \\ &= \int_{\Omega} \psi(x)\phi(\mathbf{V}) \mathrm{d}x \text{ for all } \psi \in L^1(\Omega), \phi \in C_b(\mathbb{R}^M)\Bigg\}. \end{aligned}$$

Then, from the definition of ν_n, $n \in \mathbb{N}$, and the equality of laws of $[\mathbf{U}_n, \nu_n]$ and $[\tilde{\mathbf{U}}_n, \tilde{\nu}_n]$ follows

$$\begin{aligned} 1 = \mathbb{P}\Bigg(&\int_{\Omega} \psi(x) \int_{\mathbb{R}} \phi(\xi) \mathrm{d}\nu_{n,x}(\xi) \mathrm{d}x = \\ &\int_{\Omega} \psi(x)\phi(\mathbf{U}_n) \mathrm{d}x \text{ for all } \psi \in L^1(\Omega), \phi \in C_b(\mathbb{R}^M)\Bigg) \\ &= \mathcal{L}[\mathbf{U}_n, \nu_n](Q) = \mathcal{L}[\tilde{\mathbf{U}}_n, \tilde{\nu}_n](Q) \\ &= \mathcal{L}\Bigg(\int_{\Omega} \psi(x) \int_{\mathbb{R}^M} \phi(\xi) \mathrm{d}\tilde{\nu}_{n,x}(\xi) \mathrm{d}x \\ &= \int_{\Omega} \psi(x)\phi(\tilde{\mathbf{U}}_n) \mathrm{d}x \text{ for all } \psi \in L^1(\Omega), \phi \in C_b(\mathbb{R}^M)\Bigg). \end{aligned}$$

We can conclude that $\tilde{\nu}_{n,\omega,x}(\cdot) = \delta_{\hat{\mathbf{U}}_n(\omega,x)}(\cdot)$ for all ω from a set of full probability measure $\mathcal{L}$.

Finally, let us take ω from a subset of O of full measure such that $\tilde{\nu}_{n,\omega,x}(\cdot) = \delta_{\hat{\mathbf{U}}_n(\omega,x)}(\cdot)$ together with the weak convergence of $[\tilde{\mathbf{U}}_n, \tilde{\nu}_n]$ to $[\tilde{\mathbf{U}}, \tilde{\nu}]$. If we denote

$$\overline{\phi(\tilde{\mathbf{U}})} := \langle \tilde{\nu}_{\omega,x}; \phi(\cdot)\rangle = \int_{\mathbb{R}^M} \phi(\xi)\mathrm{d}\tilde{\nu}_{\omega,x}(\xi)\mathrm{d}x,$$

we have, for all $\psi \in L^1(\Omega)$, $\phi \in C_b(\mathbb{R}^M)$,

$$\begin{aligned}\lim_{n\to\infty}\int_\Omega \psi(x)\phi(\tilde{\mathbf{U}}_n)\mathrm{d}x &= \lim_{n\to\infty}\int_\Omega \psi(x)\int_{\mathbb{R}^M}\phi(\xi)\mathrm{d}\tilde{\nu}_{n,x}(\xi)\mathrm{d}x \\ &= \int_\Omega \psi(x)\int_{\mathbb{R}^M}\phi(\xi)\mathrm{d}\tilde{\nu}_x(\xi)\mathrm{d}x = \int_\Omega \psi(x)\overline{\phi(\tilde{\mathbf{U}})}\mathrm{d}x.\end{aligned}$$

We can conclude that $\phi(\tilde{\mathbf{U}}_n) \rightharpoonup \overline{\phi(\tilde{\mathbf{U}})}$ in $L^1(\Omega, w)$ for every $\phi \in C_b(\mathbb{R}^M)$.

On the other hand, we can apply the theorem stating the existence of a Young measure (see [16, Theorem 6.2] or [12, Theorem 2.8.2.]). Since the probability measures are determined by their action on continuous bounded functions, the obtained Young measure must coincide with $\tilde{\nu}_{\omega,x}$ for every $\omega \in \Omega$. Knowing this, we can extend our previous conclusion from $\phi \in C_b(\mathbb{R}^M)$ to any Carathéodory function $G = G(x, Z)$ such that $G(x, \tilde{\mathbf{U}}_n)$ is weakly convergent in $L^1(\Omega)$ almost surely. This gives the demanded claim. □

As stated in [8, Lemma 2.3], by relaxing our requirements, we can still obtain a reasonable outcome in terms of the concentration defect measures. We present the lemma along with its proof (which is omitted in the cited reference), which is a stochastic modification of the proof of [19, Lemma 2.1].

Let us recall that a sequence of functions generating a Young measure (i.e., satisfying the conclusion of Theorem 2.3) is weakly convergent in L^1 (see, for example, [20, Remark 3.131]).

Lemma 2.2 *Let $N, M \in \mathbb{N}$, $\Omega \subset \mathbb{R}^N$ of finite Lebesgue measure, let $(O, \mathfrak{F}, \mathbb{P})$ be a complete probability space, and let $\{\mathbf{U}_n\}_{n=1}^\infty$, $\mathbf{U}_n : \Omega \times O \to \mathbb{R}^M$, be a sequence of random variables generating a Young measure $\{\nu_{\omega,x}\}_{x\in\Omega}$. Let $G : \mathbb{R}^M \to [0,\infty)$ be a continuous function such that*

$$\sup_{n\in\mathbb{N}} \mathbb{E}\left[\|G(\mathbf{U}_n)\|^p_{L^1(\Omega)}\right] < \infty, \textit{ for a certain } p \in (1,\infty),$$

and let F be continuous function satisfying

$$F : \mathbb{R}^M \to \mathbb{R}, \quad |F(z)| \le G(z), \textit{ for all } z \in \mathbb{R}^M.$$

Let us denote $\mathbb{P}$-a.s.

$$F_\infty := \tilde{F} - \langle \nu_{\omega,x}; F(\cdot)\rangle\,\mathrm{d}x, \quad G_\infty := \tilde{G} - \langle \nu_{\omega,x}; G(\cdot)\rangle\,\mathrm{d}x$$

where $\tilde{F}, \tilde{G} \in \mathcal{M}(\Omega)$ (the space of bounded Radon measures on Ω) are w^*-limits of $\{F(\mathbf{U}_n)\}_{n=1}^{\infty}$, $\{G(\mathbf{U}_n)\}_{n=1}^{\infty}$, respectively, in $\mathcal{M}(\Omega)$. Then $|F_\infty| \leq G_\infty$ $\mathbb{P}$-a.s.

Proof First, we express $\tilde{F}$ and $\tilde{G}$ as the w^*-limits of the sequences $\{F(\mathbf{U}_n)\}_{n=1}^{\infty}$ and $\{G(\mathbf{U}_n)\}_{n=1}^{\infty}$ in $\mathcal{M}(\Omega)$, respectively, while also incorporating the truncation parameter $T > 0$. Hence,

$$\langle \tilde{F}; \phi \rangle = \lim_{n\to\infty} \int_{\Omega\cap\{|\mathbf{U}_n| \leq T\}} F(\mathbf{U}_n)\,\phi\,\mathrm{d}x + \lim_{n\to\infty} \int_{\Omega\cap\{|\mathbf{U}_n| > T\}} F(\mathbf{U}_n)\,\phi\,\mathrm{d}x,$$

$$\langle \tilde{G}; \phi \rangle = \lim_{n\to\infty} \int_{\Omega\cap\{|\mathbf{U}_n| \leq T\}} G(\mathbf{U}_n)\,\phi\,\mathrm{d}x + \lim_{n\to\infty} \int_{\Omega\cap\{|\mathbf{U}_n| > T\}} G(\mathbf{U}_n)\,\phi\,\mathrm{d}x,$$

for every $\phi \in C_b(\Omega)$ $\mathbb{P}$-a.s.

Because F *and* G are bounded continuous on the set $\{x \in \Omega,\ |x| \leq T\}$, we can employ a Young measure $\{\nu_{\omega,x}\}_{x\in\Omega}$ to obtain

$$\lim_{n\to\infty} \int_{\Omega\cap\{|\mathbf{U}_n| \leq T\}} F(\mathbf{U}_n)\,\phi\,\mathrm{d}x = \int_{\Omega\cap\{|\mathbf{U}_n| \leq T\}} \langle \nu_{\omega,x}; F(\cdot) \rangle\,\phi\,\mathrm{d}x,$$

$$\lim_{n\to\infty} \int_{\Omega\cap\{|\mathbf{U}_n| \leq T\}} G(\mathbf{U}_n)\,\phi\,\mathrm{d}x = \int_{\Omega\cap\{|\mathbf{U}_n| \leq T\}} \langle \nu_{\omega,x}; G(\cdot) \rangle\,\phi\,\mathrm{d}x,$$

for every $\phi \in C_b(\Omega)$ $\mathbb{P}$-a.s.

Then, we get, for almost all $\omega \in O$ and every $\phi \in C_b(\Omega)$,

$$\lim_{T\to\infty} \left(\lim_{n\to\infty} \int_{\Omega\cap\{|\mathbf{U}_n| \leq T\}} F(\mathbf{U}_n)\,\phi\,\mathrm{d}x \right) = \int_{\Omega} \langle \nu_{\omega,x}; F(\cdot) \rangle\,\phi\,\mathrm{d}x,$$

$$\lim_{T\to\infty} \left(\lim_{n\to\infty} \int_{\Omega\cap\{|\mathbf{U}_n| \leq M\}} G(\mathbf{U}_n)\,\phi\,\mathrm{d}x \right) = \int_{\Omega} \langle \nu_{\omega,x}; G(\cdot) \rangle\,\phi\,\mathrm{d}x.$$

By combining the above considerations, we arrive at the conclusion that

$$\langle F_\infty; \phi \rangle = \lim_{T\to\infty} \left(\lim_{n\to\infty} \int_{\Omega\cap\{|\mathbf{U}_n| > T\}} F(\mathbf{U}_n)\,\phi\,\mathrm{d}x \right),$$

$$\langle G_\infty; \phi \rangle = \lim_{T\to\infty} \left(\lim_{n\to\infty} \int_{\Omega\cap\{|\mathbf{U}_n| > T\}} G(\mathbf{U}_n)\,\phi\,\mathrm{d}x \right),$$

for every $\phi \in C_b(\Omega)$ $\mathbb{P}$-a.s.

Since $|F| \leq G$, it follows that

$$|\langle F_\infty; \phi \rangle| \leq \langle G_\infty; \phi \rangle,$$

for every $\phi \in C_b(\Omega)$ $\mathbb{P}$-a.s. which completes the proof of $|F_\infty| \leq G_\infty$ $\mathbb{P}$-a.s. □

In what follows, we give the proof of the existence of a measure-valued solution to (1) endowed with the Dirichlet boundary condition (3). We do not expect any complication when treating other types of well-known boundary conditions like the full-slip boundary condition or the periodic boundary condition.

3 Galerkin Approximation

Since $L^2(\Omega)$ has a countable Schauder basis consisting of smooth functions with zero on the boundary, we consider n-dimensional spaces $X_n \subset C_0^\infty(\Omega)$ with the property

$$L^2(\Omega) = \overline{\bigcup_n X_n}^{\|\cdot\|_{L^2}}. \tag{6}$$

Let Π_n be the orthonormal projection $\Pi_n : L^2(\Omega) \to X_n$. In this section, we tackle the problem of the existence of a pair $(\varrho_n, \mathbf{u}_n) \in C([0, T], L^1) \times C([0, T], X_n)$ satisfying the system

$$\begin{aligned} \mathrm{d}\varrho_n + \operatorname{div}(\varrho_n \mathbf{u}_n)\,\mathrm{d}t &= 0 \\ \mathrm{d}\Pi_n(\varrho_n \mathbf{u}_n) + \Pi_n\big((\operatorname{div} \varrho_n \mathbf{u}_n \otimes \mathbf{u}_n) + \nabla p(\varrho_n)\big)\,\mathrm{d}t &= \Pi_n\big(\operatorname{div} \mathbb{S}_n\big)\,\mathrm{d}t \\ &\quad + \Pi_n(\varrho \Pi_n(\mathbb{F}(\varrho, \mathbf{u})))\,\mathrm{d}W. \end{aligned} \tag{7}$$

It is worthwhile to mention that X_n is of finite dimension, and thus all norms are equivalent there. As a matter of fact, we use $\|\cdot\|_{X_n}$ for any functional which acts as a norm on X_n.

Definition 3.1 Let Λ be a Borel probability measure on $C^{2+\nu}(\Omega) \times X^n$. Then, $((O, \mathfrak{F}, (\mathfrak{F}_t)_{t\geq 0}, \mathbb{P}), \varrho_n, \mathbf{u}_n, W)$ is called a *martingale solution* to (7) with initial datum Λ if the following is satisfied:

1. $(O, \mathfrak{F}, (\mathfrak{F}_t)_{t\geq 0}, \mathbb{P})$ is a stochastic basis with a complete right-continuous filtration.
2. W is a cylindrical Wiener process.
3. The density ϱ_n is $(\mathfrak{F}_t)$ progressively measurable and satisfies

$$\varrho_n \in L^p((0, T) \times \Omega) \text{ for any } p \in (1, \infty),\ \varrho_n \geq \underline{\varrho} > 0 \quad \mathbb{P}\text{-a.s.}$$

4. The velocity $\mathbf{u}_n$ is $(\mathfrak{F}_t)$ progressively measurable and satisfies

$$\mathbf{u}_n \in C([0,T],\ X_n) \quad \mathbb{P}\text{-a.s.}$$

5. There exists an $\mathfrak{F}_0$ measurable random variable $(\varrho_0, \mathbf{u}_0)$ such that

$$\mathcal{L}(\varrho_n(0), \mathbf{u}_n(0)) = \Lambda.$$

6. The approximate continuity equation

$$\partial_t \varrho_n + \operatorname{div}(\varrho_n \mathbf{u}_n) = 0$$

holds in $(0,T) \times \Omega$ $\mathbb{P}$-a.s. and $\varrho(0) = \varrho_0$ $\mathbb{P}$-a.s.

7. The approximate momentum equation

$$\begin{aligned}
-\int_0^T \partial_t \Phi \int_\Omega \varrho_n \mathbf{u}_n \varphi \,\mathrm{d}x\mathrm{d}t - \Phi(0) \int_\Omega \varrho_0 \mathbf{u}_0 \cdot \varphi \,\mathrm{d}x \\
= \int_0^T \Phi \int_\Omega (\varrho_n \mathbf{u}_n \otimes \mathbf{u}_n : \nabla\varphi + p(\varrho_n) \operatorname{div}\varphi) \,\mathrm{d}x\mathrm{d}t \\
- \int_0^T \Phi \int_\Omega \mathbb{S}(D\mathbf{u}_n) : \nabla\varphi \,\mathrm{d}x\mathrm{d}t \\
+ \int_0^T \Phi \int_\Omega \varrho_n \Pi_n[\mathbb{F}(\varrho_n, \mathbf{u}_n)] \cdot \varphi \,\mathrm{d}x\mathrm{d}W
\end{aligned}$$

holds for all $\Phi \in C_c^\infty([0,T))$ and all $\varphi \in X_n$ $\mathbb{P}$-a.s.

8. The energy (in)equality

$$\begin{aligned}
-\int_0^T \partial_t \Phi \int_\Omega \left(\frac{1}{2}\varrho|\mathbf{u}|^2 + P(\varrho)\right) \mathrm{d}x\mathrm{d}t - \Phi(0) \int_\Omega \left(\frac{1}{2}\varrho|\mathbf{u}|^2 + P(\varrho)\right) \mathrm{d}x \\
+ \int_0^T \Phi \int_\Omega \mathbb{S}(D\mathbf{u}) : \nabla\mathbf{u} \,\mathrm{d}x\mathrm{d}t = \frac{1}{2}\sum_{k=1}^\infty \int_0^T \Phi \int_\Omega \varrho\, |\Pi_n(\mathbf{F}_k(\varrho, \mathbf{u}))|^2 \,\mathrm{d}x\mathrm{d}t \\
\int_0^T \Phi \int_\Omega \varrho \Pi_n\left[\mathbb{F}(\varrho, \mathbf{u})\right] \cdot \mathbf{u} \,\mathrm{d}x\mathrm{d}W \qquad (8)
\end{aligned}$$

holds for all $\Phi \in C_c^\infty([0,T))$ $\mathbb{P}$-a.s. Here, P stands for the usual pressure potential, i.e.,

$$P(\varrho) = \varrho \int_1^\varrho \frac{p(z)}{z^2} \,\mathrm{d}z.$$

Let us mention that the energy inequality and the Burkholder-Davis-Gundy inequality yield

$$\mathbb{E}\left[\left|\sup_{\tau\in[0,T]}\int_\Omega \frac{1}{2}\varrho_n|\mathbf{u}_n|^2 + P(\varrho_n)\,\mathrm{d}x\right|^r\right] + \mathbb{E}\left[\left|\int_0^T\int_\Omega \mathbb{S}:\nabla\mathbf{u}_n\,\mathrm{d}x\mathrm{d}t\right|^r\right]$$
$$\leq C\mathbb{E}\left[\left|\int_\Omega \frac{1}{2}\varrho_0|\mathbf{u}_0|^2 + P(\varrho_0)\,\mathrm{d}x\right|^r + 1\right] \tag{9}$$

for any $r > 1$.

The starting point of the existence of the measure-valued solution is the following theorem. Its proof is recalled for reader's convenience in the next section.

Theorem 3.1 *Let Λ be a Borel probability measure on $C^{2+\nu}(\Omega)\times X^n$ such that*

$$\Lambda[0<\underline{\varrho}\leq\varrho,\ \|\varrho\|_{C^{2+\nu}(\Omega)}\leq\overline{\varrho}] = 1, \qquad \int \|\mathbf{v}\|_{X^n}^r\,\mathrm{d}\Lambda[\varrho,\mathbf{v}]\leq\overline{u}$$

for some deterministic constants $\underline{\varrho}$, $\overline{\varrho}$, and$\overline{u}$ and some $r > 2$. Then there exists the martingale solution specified in Definition 3.1.

3.1 Proof of Theorem 3.1

Let $\{\mathbf{w}_k\}_{k=1}^\infty$ be the orthogonal basis of $W_0^{3,2}(\Omega)$ formed by the smooth eigenvectors of the problem

$$-\Delta^3\mathbf{w} = \lambda\mathbf{w},$$
$$\mathbf{w}\restriction_{\partial\Omega} = 0.$$

(see [21, Theorem 4.11 and Remark 4.14 in the Appendix] for the existence of such basis). We define $X_n = \mathrm{span}\{\mathbf{w}_k, k = 1,\ldots,n\}$ as the sequence of spaces satisfying (6).

Fix $n\in\mathbb{N}$. In the first step, we show the solvability of the system consisting of the following:

- The continuity equation

$$\partial_t\varrho + \mathrm{div}(\varrho[\mathbf{u}]_R) = \varepsilon\Delta\varrho \tag{10}$$

  endowed with the boundary condition

$$\frac{\partial\varrho}{\partial\mathbf{n}} = 0 \text{ on } \partial\Omega$$

- The momentum equation

$$
\begin{aligned}
\Pi_n[\mathrm{d}(\varrho\mathbf{u})] + \Pi_n[\operatorname{div}(\varrho[\mathbf{u}]_R \otimes \mathbf{u})]\,\mathrm{d}t - \mu\Pi_n[\Delta^3\mathbf{u}]\,\mathrm{d}t \\
- \Pi_n[\operatorname{div}\mathbb{S}(D\mathbf{u})]\,\mathrm{d}t + \Pi_n[\chi(\|\mathbf{u}_n - R\|_{X_n})\nabla p(\varrho)]\,\mathrm{d}t \\
- \varepsilon\Pi_n[\Delta(\varrho\mathbf{u})]\,\mathrm{d}t = \Pi_n[\varrho\Pi_n[\mathbb{F}(\varrho, \varrho\mathbf{u})]]\,\mathrm{d}W,
\end{aligned}
\tag{11}
$$

where $\mu > 0$, $\varepsilon > 0$, and $R > 0$ are given parameters. Here, $\chi(z)$ is a smooth nonnegative decreasing function which equals 1 for $z < 0$ and is equal to 0 for $z > 1$ and $[\mathbf{u}]_R = \chi_R(\|\mathbf{u}\|_{X_n} - R)\mathbf{u}$—for the precise meaning, we refer also to [12, Chapter 4]. We look for the solution in the form

$$
\mathbf{u} = \sum_{i=1}^{N} c_i \boldsymbol{\omega}_i,
$$

where $c_i \in \mathbb{R}$ are unknowns assuming the stochastic basis $(O, \mathfrak{F}, (\mathfrak{F}_t)_{t\geq 0}, \mathbb{P})$ and the cylindrical process W are given. Systems (10) and (11) are uniquely solvable for initial conditions satisfying

$$
\begin{aligned}
\varrho_0 \in C^{2+\nu}(\Omega), \quad \mathbf{u}_0 \in X_n, \quad \mathbb{P}\{0 < \underline{\varrho} \le \varrho_0,\ \|\varrho_0\|_{C^{2+\nu}} \le \overline{\varrho}\} = 1, \\
\mathbb{E}[\|\mathbf{u}_0\|^r_{X_n}] \le \overline{u},
\end{aligned}
$$

where $\underline{\varrho}, \overline{\varrho}, \overline{u}$, and $r > 2$. Moreover, we choose ϱ_0 and $\mathbf{u}_0$ such that $\mathcal{L}(\varrho_0, \mathbf{u}_0) = \Lambda$ and ϱ_0 and $\mathbf{u}_0$ are $\mathfrak{F}_0$ measurable. The abovementioned existence of the solution is shown (with slight modification) in [12, Theorem 4.1.12]. In the rest of the proof, we pass with parameters μ and ε to zero and with R to infinity.

First, we tend with μ to zero. The energy balance

$$
\begin{aligned}
-\int_0^T \partial_t\Phi \int_\Omega \frac{1}{2}\varrho|\mathbf{u}|^2 + P(\varrho)\,\mathrm{d}x\mathrm{d}t - \Phi(0)\int_\Omega \frac{1}{2}\varrho|\mathbf{u}|^2 + P(\varrho)\,\mathrm{d}x \\
+ \int_0^T \Phi \int_\Omega \mathbb{S}(D\mathbf{u}) : \nabla\mathbf{u} + \mu|\nabla^3\mathbf{u}|^2 + \varepsilon\varrho|\nabla\mathbf{u}|^2 + \varepsilon P''(\varrho)|\nabla\varrho|^2\,\mathrm{d}x\mathrm{d}t \\
= \frac{1}{2}\sum_{k=1}^{\infty}\int_0^T \Phi \int_\Omega \varrho|\Pi_n[F_k(\varrho, \mathbf{u})]|^2\,\mathrm{d}x\mathrm{d}t \\
+ \int_0^T \Phi \int_\Omega \varrho\Pi_n[\mathbb{F}(\varrho, \mathbf{u})]\cdot\mathbf{u}\,\mathrm{d}t\mathrm{d}W
\end{aligned}
$$

holds for all $\Phi \in C_c^\infty([0,T))$ $\mathbb{P}$-almost surely—see [12, Proposition 4.1.14]. The Burkholder-Davis-Gundy inequality and the Gronwall lemma yield

$$\begin{aligned}
&\mathbb{E}\left[\left|\sup_{\tau\in[0,T]}\int_\Omega \frac{1}{2}\varrho|\mathbf{u}|^2 + P(\varrho)\,\mathrm{d}x\right|^r\right] \\
&\quad + \mathbb{E}\left[\left|\int_0^T\int_\Omega \mathbb{S}:\nabla\mathbf{u} + \varepsilon\varrho|\nabla\mathbf{u}|^2 + \mu|\nabla^3\mathbf{u}|^2 + \varepsilon P''(\varrho)|\nabla\varrho|^2\,\mathrm{d}x\mathrm{d}t\right|^r\right] \\
&\qquad \le C\,\mathbb{E}\left[\left|\int_\Omega \frac{1}{2}\varrho_0|\mathbf{u}_0|^2 + P(\varrho_0)\,\mathrm{d}x\right|^r + 1\right], \qquad (12)
\end{aligned}$$

whenever $r \ge 2$. This (together with (2)) enables us to deduce the following estimates:

$$\begin{aligned}
\mathbb{E}\left[\left|\int_0^T\int_\Omega F(D\mathbf{u})\,\mathrm{d}x\mathrm{d}t\right|^r\right] &\le C, \\
\mathbb{E}\left[\left\|\varrho|\mathbf{u}|^2\right\|^r_{L^\infty((0,T);L^1(\Omega))}\right] &\le C, \qquad (13)\\
\mathbb{E}\left[\|P(\varrho)\|^r_{L^\infty((0,T);L^1(\Omega))}\right] &\le C
\end{aligned}$$

Since $[\mathbf{u}]_R$ is bounded, (10) yields the existence of $C > 0$ such that

$$\begin{aligned}
&\operatorname{esssup}_{t\in[0,T]}\left(\|\varrho(t)\|_{C^{2+\nu}(\Omega)} + \|\partial_t\varrho\|_{C^\nu(\Omega)}\right) \le C, \\
&\frac{1}{C} \le \varrho(x) \le C, \quad \text{for all } (t,x)\in[0,T)\times\Omega. \qquad (14)
\end{aligned}$$

This can be found, for example, at [22] as Proposition 7.39 or at [12] as Theorem A.2.5.

Consequently, $(13)_2$ yields

$$\mathbb{E}\left[\|\mathbf{u}\|^r_{L^\infty((0,T);X_n)}\right] \le C.$$

We deduce similarly to [12, Chapter 4.1.2.9] that

$$\mathbb{E}\left[\|\mathbf{u}\|^r_{C^\beta([0,T);X_n)}\right] \le C$$

for some $\beta > 0$.

Everything is prepared for the use of the Jakubowski-Skorokhod theorem (see [12, Theorem 2.7.1]). We define

$$\begin{aligned} \mathcal{X} &= \mathcal{X}_{\varrho_0} \times \mathcal{X}_{\mathbf{u}_0} \times \mathcal{X}_{\varrho} \times \mathcal{X}_{\mathbf{u}} \times \mathcal{X}_W \\ &= C^{2+\nu}(\Omega) \times X_n \times C^{\iota}([0,T], C^{2+\iota}(\Omega)) \times C^{\kappa}([0,T]; X_n) \times C([0,T]; \mathfrak{U}_0) \end{aligned}$$

for some $\iota, \kappa > 0$. Let ϱ_μ and $\mathbf{u}_\mu$ be a solution to (10) and (11) for some $\mu > 0$. The laws of $\{(\varrho_0, \mathbf{u}_0, \varrho_\mu, \mathbf{u}_\mu, W)\}_{\mu>0}$ are tight in $\mathcal{X}$.

Indeed, the ball $B_L \subset C^{\nu}([0,T]; C^{2+\nu}(\Omega))$ is compact in $\mathcal{X}_\varrho$ for every $0 < \iota < \nu$ and thus

$$\mathcal{L}_{\varrho_\mu}(B_L) = \mathbb{P}(\varrho_\mu \in B_L) = 1$$

assuming L is sufficiently large due to (14) (note the constants there are not random variables). Consequently, the laws of $\{\varrho_\mu\}_{\mu>0}$ are tight in $\mathcal{X}_\varrho$.

Similarly, the ball $B_M \subset C^{\beta}([0,T]; X_n)$ centered at zero with radius $M > 0$ is compact in the space $C^{\kappa}([0,T]; X_n)$ for all $0 < \kappa < \beta$. With the help of the Chebyshev inequality, we get

$$\begin{aligned} \mathcal{L}_{\mathbf{u}_\mu}(B_M) &= \mathbb{P}(\mathbf{u}_\mu \in B_M) = 1 - \mathbb{P}(\|\mathbf{u}_\mu\|_{C^\beta([0,T];X^n)} > M) \\ &= 1 - \frac{1}{M^r}\mathbb{E}\left(\|\mathbf{u}_\mu\|^r_{C^\beta([0,T];X_n)}\right) \geq 1 - \frac{C}{M^r}, \end{aligned}$$

and, consequently, the laws of $\{\mathbf{u}_\mu\}_{\mu>0}$ are tight in $\mathcal{X}_{\mathbf{u}}$.

Finally, the law of $\{W\}$ is tight in $\mathcal{X}_W$ since it is a singleton. The same reasoning gives that the law of $\{\varrho_0\}$ is tight in $C^{2+\nu}(\Omega)$ and the law of $\{\mathbf{u}_0\}$ is tight in X_n.

According to the Jakubowski-Skorokhod theorem, we get the existence of a stochastic basis $(\tilde{O}, \tilde{\mathfrak{F}}, (\tilde{\mathfrak{F}}_t)_{t\geq 0}, \tilde{\mathbb{P}})$ and random variables $\tilde{\varrho}_0$, $\tilde{\mathbf{u}}_0$, $\tilde{\varrho}_\mu$, $\tilde{\mathbf{u}}_\mu$, and $\tilde{W}_\mu$ which coincide in law to (respectively) ϱ_0, $\mathbf{u}_0$, ϱ_μ, $\mathbf{u}_\mu$, and W. Moreover, there exists a pentad $(\tilde{\varrho}_0, \tilde{\mathbf{u}}_0, \tilde{\varrho}, \tilde{\mathbf{u}}, \tilde{W})$ such that

$$(\tilde{\varrho}_0, \tilde{\mathbf{u}}_0, \tilde{\varrho}_\mu, \tilde{\mathbf{u}}_\mu, \tilde{W}_\mu) \to (\tilde{\varrho}_0, \tilde{\mathbf{u}}_0, \tilde{\varrho}, \tilde{\mathbf{u}}, \tilde{W})$$

$\mathbb{P}$-almost surely in the topology of $\mathcal{X}$. In particular, $\tilde{\varrho}$ and $\tilde{\mathbf{u}}$ satisfy

$$\partial_t \tilde{\varrho} + \operatorname{div}(\tilde{\varrho}[\tilde{\mathbf{u}}]_R) = \varepsilon \Delta \tilde{\mathbf{u}} \tag{15}$$

$\mathbb{P}$-almost surely with the initial condition $\tilde{\varrho}(0) = \tilde{\varrho}$. Next, it satisfies

$$
\begin{aligned}
&- \int_0^T \partial_t \Phi \int_\Omega \tilde{\varrho}\tilde{\mathbf{u}}\, \boldsymbol{\varphi}\, \mathrm{d}x\mathrm{d}t - \Phi(0) \int_\Omega \tilde{\varrho}_0 \tilde{\mathbf{u}}_0\, \boldsymbol{\varphi}\, \mathrm{d}x \\
&\qquad = \int_0^T \Phi \int_\Omega \big(\tilde{\varrho}[\tilde{\mathbf{u}}]_R \otimes \tilde{\mathbf{u}} + \chi(\|\tilde{\mathbf{u}} - R\|_{X_n}) p(\varrho)\mathbb{I}\big) : \nabla \boldsymbol{\varphi}\, \mathrm{d}x\mathrm{d}t \\
&\qquad\qquad - \int_0^T \Phi \int_\Omega (\mathbb{S}(D\tilde{\mathbf{u}}) : \nabla\boldsymbol{\varphi} - \varepsilon \tilde{\varrho}\tilde{\mathbf{u}}\, \Delta\boldsymbol{\varphi})\, \mathrm{d}x\mathrm{d}t \\
&\qquad\qquad\qquad + \int_0^T \Phi \int_\Omega \tilde{\varrho}\Pi_n[\mathbb{F}(\tilde{\varrho}, \tilde{\mathbf{u}})] \cdot \boldsymbol{\varphi}\, \mathrm{d}x\mathrm{d}W
\end{aligned}
\tag{16}
$$

for each $\Phi \in C_c([0, T))$ and $\boldsymbol{\varphi} \in X_n$ $\mathbb{P}$-almost surely—the strong convergence of $\tilde{\varrho}_\mu$ and $\tilde{\mathbf{u}}_\mu$ is sufficient to obtain the convergence of the stochastic process—we refer to [12, Lemma 2.6.6 and proof of Proposition 4.3.15].

Next, we pass with $\varepsilon \to 0$ similarly as above. In particular, let ϱ_ε and $\mathbf{u}_\varepsilon$ be the solutions to (15) and (16) constructed above and emanating from the initial conditions ϱ_0 and $\mathbf{u}_0$. The energy inequality obtained similarly to (12) has the form

$$
\begin{aligned}
&\mathbb{E}\left[\left| \sup_{\tau\in[0,T]} \int_\Omega \frac{1}{2}\varrho_\varepsilon |\mathbf{u}_\varepsilon|^2 + P(\varrho_\varepsilon)\, \mathrm{d}x \right|^r\right] \\
&\qquad + \mathbb{E}\left[\left| \int_0^T \int_\Omega \mathbb{S} : \nabla\mathbf{u}_\varepsilon + \varepsilon\varrho_\varepsilon |\nabla\mathbf{u}_\varepsilon|^2 + \varepsilon P''(\varrho_\varepsilon)|\nabla\varrho_\varepsilon|^2\, \mathrm{d}x\mathrm{d}t \right|^r\right] \\
&\qquad\qquad \le C\, \mathbb{E}\left[\left| \int_\Omega \frac{1}{2}\varrho_0 |\mathbf{u}_0|^2 + P(\varrho_0)\, \mathrm{d}x \right|^r + 1\right]
\end{aligned}
\tag{17}
$$

whenever $r \ge 2$ and one gets also the existence of C such that

$$
\frac{1}{C} \le \varrho(t, x) \le C, \quad \text{for all } (t, x) \in [0, T) \times \Omega. \tag{18}
$$

Consequently, (17) and (18) yield

$$
\begin{aligned}
\mathbb{E}\left[\left| \int_0^T \int_\Omega F(D\mathbf{u}_\varepsilon)\, \mathrm{d}x\mathrm{d}t \right|^r\right] &\le C, \\
\mathbb{E}\left[\|\mathbf{u}_\varepsilon\|^r_{L^\infty((0,T);X_n)} \right] &\le C
\end{aligned}
$$

and also

$$
\mathbb{E}\left[\|\mathbf{u}\|^r_{C^\beta([0,T];X_n)} \right] \le c \tag{19}
$$

for some $\beta > 0$. We consider the sequence (labeled by $\varepsilon > 0$)

$$(\varrho_0, \mathbf{u}_0, \varrho_\varepsilon, \mathbf{u}_\varepsilon, W)$$

in the space

$$\begin{aligned}\mathcal{X} = \mathcal{X}_{\varrho_0} \times \mathcal{X}_{\mathbf{u}_0} \times \mathcal{X}_{\varrho} \times \mathcal{X}_{\mathbf{u}} \times \mathcal{X}_W \\ = C^{2+\nu}(\Omega) \times X_n \times L^q(((0,T)\times\Omega), w) \\ \times C^\kappa([0,T]; X_n) \times C([0,T]; \mathfrak{U}_0)\end{aligned}$$

for an arbitrary $q \in (1, \infty)$. Again, this sequence is tight:

We recall that $B_L = \{\mathbf{u} \in L^q((0,T)\times\Omega),\ \|\mathbf{u}\|_{L^q((0,T)\times\Omega)} \le L\}$ is compact in the weak topology for every $M > 0$ and thus the estimate (18) yields

$$\mathcal{L}_{\varrho_\varepsilon} = \mathbb{P}(\varrho_\varepsilon \in B_L) = 1$$

once L is sufficiently large.

Similarly as before, the ball $B_M \subset C^\beta([0,T]; X_n)$ is compact in $C^\kappa([0,T]; X_n)$ whenever $\kappa < \beta$ and the Chebyshev inequality with (19) give

$$\begin{aligned}\mathcal{L}_{\mathbf{u}_\varepsilon}(B_M) = \mathbb{P}(\mu_\varepsilon \in B_M) = 1 - \mathbb{P}(\|\mu_\varepsilon\|_{C^\beta([0,T];X_n)} > M) \\ \ge 1 - \frac{1}{M^r}\mathbb{E}\left(\|\mu_\varepsilon\|^r_{C^\beta([0,T];X_n)}\right) \ge 1 - \frac{C}{M^r}\end{aligned}$$

which implies the desired tightness.

The set $\{\varrho_0, \mathbf{u}_0, W\}$ is a singleton in $\mathcal{X}_{\varrho_0} \times \mathcal{X}_{\mathbf{u}_0} \times \mathcal{X}_W$ and therefore it is tight.

As above, the Jakubowski-Skorokhod theorem gives the existence of a stochastic basis $(\tilde{O}, \tilde{\mathfrak{F}}, (\tilde{\mathfrak{F}}_t)_{t\ge0}, \tilde{\mathbb{P}})$ and random variables $\tilde{\varrho}_0, \tilde{\mathbf{u}}_0, \tilde{\varrho}_\varepsilon, \tilde{\mathbf{u}}_\varepsilon$, and $\tilde{W}$ which coincide in law (respectively) to $\varrho_0, \mathbf{u}_0, \varrho_\varepsilon, \mathbf{u}_\varepsilon$, and W such that

$$(\tilde{\varrho}_0, \tilde{\mathbf{u}}_0, \tilde{\varrho}_\varepsilon, \tilde{\mathbf{u}}_\varepsilon, \tilde{W}) \to (\tilde{\varrho}_0, \tilde{\mathbf{u}}_0, \tilde{\varrho}, \tilde{\mathbf{u}}, \tilde{W}) \text{ in } \mathcal{X}$$

for some pentad $(\tilde{\varrho}_0, \tilde{\mathbf{u}}_0, \tilde{\varrho}, \tilde{\mathbf{u}}, \tilde{W})$. The pentad satisfies the following:

- The continuity equation

$$\partial_t\tilde{\varrho} + \operatorname{div}(\tilde{\varrho}[\tilde{\mathbf{u}}]_R) = 0 \tag{20}$$

in a weak sense $\mathbb{P}$-almost surely with initial condition $\tilde{\varrho}(0) = \tilde{\varrho}_0$ $\mathbb{P}$-almost surely. The zero on the right-hand side follows from (17) which yields

$$\mathbb{E}\left[\left|\int_0^T\int_\Omega \varepsilon|\nabla\tilde{\varrho}_\varepsilon|^2\,\mathrm{d}x\mathrm{d}t\right|^r\right] \le C,$$

since $P''(\tilde{\varrho}_\varepsilon) \geq C$ as long as $\tilde{\varrho}_\varepsilon$ and $\tilde{\varrho}_\varepsilon^{-1}$ are bounded. Therefore,

$$\mathbb{E}\left[\left|\int_0^T \int_\Omega \varepsilon \nabla \tilde{\varrho}_\varepsilon \nabla \varphi \,\mathrm{d}x\mathrm{d}t\right|\right] \leq \sqrt{\varepsilon}\,\mathbb{E}\left[\int_0^T \int_\Omega \varepsilon |\nabla \tilde{\varrho}_\varepsilon|^2 \,\mathrm{d}x\mathrm{d}t\right] \to 0,$$

whenever $\varphi \in C^\infty([0,T]\times\Omega)$ is such that $\|\nabla\varphi\|_{L^2((0,T)\times\Omega)} = 1$. Consequently, the right-hand side of (15) tends to zero almost surely.

It is also worthwhile to mention that we can multiply (15) by $(1+\log\tilde{\varrho}_\varepsilon)$ and (20) by $(1+\log\tilde{\varrho})$ to learn

$$\begin{aligned}&\partial_t\left(\tilde{\varrho}_\varepsilon \log\tilde{\varrho}_\varepsilon\right) + \operatorname{div}(\tilde{\varrho}_\varepsilon \log\tilde{\varrho}_\varepsilon [\tilde{\mathbf{u}}_\varepsilon]_R)\\ &\qquad\qquad + \tilde{\varrho}_\varepsilon \operatorname{div}\tilde{\mathbf{u}}_\varepsilon = \varepsilon\left(\Delta(\tilde{\varrho}_\varepsilon\log\tilde{\varrho}_\varepsilon) - \frac{|\nabla\tilde{\varrho}_\varepsilon|^2}{\tilde{\varrho}_\varepsilon}\right)\\ &\partial_t\left(\tilde{\varrho}\log\tilde{\varrho}\right) + \operatorname{div}\left(\tilde{\varrho}\log\tilde{\varrho}\tilde{\mathbf{u}}\right) + \tilde{\varrho}\operatorname{div}\tilde{\mathbf{u}} = 0.\end{aligned}$$

Integrating the above equations over Ω and subtracting them, we deduce

$$\int_\Omega \overline{\tilde{\varrho}\log\tilde{\varrho}}(t,\cdot)\,\mathrm{d}x \leq \int_\Omega \tilde{\varrho}\log\tilde{\varrho}(t,\cdot)\,\mathrm{d}x$$

for almost all $t \in (0,T)$ where $\overline{f(\tilde{\varrho})}$ denotes the weak limit of $f(\tilde{\varrho}_\varepsilon)$. This is sufficient to claim that $\tilde{\varrho}_\varepsilon \to \tilde{\varrho}$ almost everywhere and, consequently, $\tilde{\varrho}_\varepsilon \to \tilde{\varrho}$ strongly in $L^p((0,T)\times\Omega)$ for every $p \in [1,\infty)$ almost surely.

- The momentum equation

$$\begin{aligned}&-\int_0^T \partial_t\Phi \int_\Omega \tilde{\varrho}\tilde{\mathbf{u}}\,\boldsymbol{\varphi}\,\mathrm{d}x\mathrm{d}t - \Phi(0)\int_\Omega \varrho_0\mathbf{u}_0\,\boldsymbol{\varphi}\,\mathrm{d}x\\ &\quad = \int_0^T \Phi\int_\Omega \left(\tilde{\varrho}[\tilde{\mathbf{u}}]_R\otimes\tilde{\mathbf{u}} + \chi(\|\tilde{\mathbf{u}} - R\|_{X_n})p(\varrho)\mathbb{I}\right) : \nabla\boldsymbol{\varphi}\,\mathrm{d}x\mathrm{d}t\\ &\quad - \int_0^T \Phi\int_\Omega \mathbb{S}(D\tilde{\mathbf{u}}) : \nabla\boldsymbol{\varphi}\,\mathrm{d}x\mathrm{d}t + \int_0^T \Phi\int_\Omega \tilde{\varrho}\Pi_n[\mathbb{F}(\tilde{\varrho},\tilde{\mathbf{u}})]\cdot\boldsymbol{\varphi}\,\mathrm{d}x\mathrm{d}W\end{aligned}$$

holds for each $\Phi \in C_c([0,T))$ and $\boldsymbol{\varphi} \in X_n$ $\mathbb{P}$-almost surely.

- We can neglect certain positive terms on the left-hand side of (17) (recall that convex continuous function is weakly sequentially lower semicontinuous) to deduce

$$\begin{aligned}&\mathbb{E}\left[\left|\sup_{\tau\in[0,T]}\int_\Omega \frac{1}{2}\varrho|\mathbf{u}|^2 + P(\varrho)\,\mathrm{d}x\right|^r\right] + \mathbb{E}\left[\left|\int_0^T\int_\Omega \mathbb{S} : \nabla\mathbf{u}\,\mathrm{d}x\mathrm{d}t\right|^r\right]\\ &\qquad\qquad \leq C\mathbb{E}\left[\left|\int_\Omega \frac{1}{2}\varrho_0|\mathbf{u}_0|^2 + P(\varrho_0)\,\mathrm{d}x\right|^r + 1\right] \qquad (21)\end{aligned}$$

with R-independent constant C on the right-hand side. It is worthwhile to mention that the above solution is unique. The stochastic terms in the above system are treated as above—i.e., the strong convergence of both $\tilde{\varrho}_\varepsilon$ and $\tilde{\mathbf{u}}_\varepsilon$ is sufficient.

It remains to tend with R to infinity. This is done by a suitable use of the stopping time argument (see [12, Chapter 4.2]). In particular, let ϱ_R and $\mathbf{u}_R$ be the unique solution obtained in the previous paragraph, and let d_t denote the material derivative with the velocity field $[\mathbf{u}]_R$. Then, (20) yields

$$\mathrm{d}_t \varrho_R = -\varrho_R \operatorname{div}[\mathbf{u}_R]_R,$$

and we deduce

$$\underline{\varrho} e^{-\int_0^t \|\operatorname{div}[\mathbf{u}_R]_R\|_{L^\infty(\Omega)}\,\mathrm{d}t} \leq \varrho(t,x) \leq \overline{\varrho} e^{\int_0^t \|\operatorname{div}[\mathbf{u}_R]_R\|_{L^\infty(\Omega)}\,\mathrm{d}t},$$

where $\underline{\varrho} \leq \varrho_0(x) \leq \overline{\varrho}$ for every $x \in \Omega$. Since all norms on X_n are equivalent, the Young inequality and (5) yield

$$\|\operatorname{div}[\mathbf{u}_R]_R\|_{L^\infty(\Omega)} \leq \|D\mathbf{u}_R\|_{L^\infty(\Omega)} \sim \|D\mathbf{u}_R\|_{L^1(\Omega)} \leq Cg^*(1) + \int_\Omega F(D\mathbf{u}_R)\,\mathrm{d}x.$$

We deduce from (21) that

$$\mathbb{E}\left[e^{-\int_0^T \int_\Omega F(D\mathbf{u}_R)\,\mathrm{d}x\mathrm{d}t} \sup_{t\in(0,T)} \|\mathbf{u}_R\|^2_{L^2(\Omega)}\right] \leq C.$$

We define three events,

$$A = \left[e^{-\int_0^T \int_\Omega F(D\mathbf{u}_R)\,\mathrm{d}x\mathrm{d}t} \sup_{t\in(0,T)} \|\mathbf{u}_R\|^2_{L^2(\Omega)} \leq a_R\right],$$

$$B = \left[\int_0^T \int_\Omega F(D\mathbf{u}_R)\,\mathrm{d}x\mathrm{d}t \leq b_R\right],$$

$$S = \left[\sup_{t\in(0,T)} \|\mathbf{u}_R\|^2_{L^2(\Omega)} \leq a_R e^{b_R}\right],$$

where a_R and b_R are sequences satisfying $\lim a_R = \lim b_R = \infty$ and $a_R e^{b_R} = R$. Next, because $S \supset A \cap B$, the Chebyshev inequality together with (21) gives

$$\mathbb{P}(S) \geq \mathbb{P}(A) + \mathbb{P}(B) - 1 \geq 1 - \frac{C}{a_R} - \frac{C}{b_R} \to 1. \tag{22}$$

We define the stopping time

$$\tau_R = \inf\{t \in [0, T],\ \|\mathbf{u}_R\|_{X_n} > R\}.$$

Recall that $[\mathbf{u}_R(t,\cdot)]_R = \mathbf{u}_R(t,\cdot)$ whenever $t < \tau_R$. We have just deduced that $\mathbb{P}(\sup_{R\in\mathbb{N}} \tau_R = T) = 1$ due to (22).

4 Measure-Valued Martingale Solution

First, we construct the laws of initial conditions Λ_n in such a way that $\Lambda_n \to \Lambda$ in certain sense and Λ_n satisfies the assumptions from Theorem 3.1 and Λ is a probability measure on $L^1(\Omega) \times L^1(\Omega)$ such that

$$\Lambda\{\varrho \geq 0\} = 1,\ \Lambda\left\{0 < \underline{\varrho} \leq \int_\Omega \varrho \,\mathrm{d}x \leq \overline{\varrho} < \infty\right\} = 1 \tag{23}$$

and

$$\int_{L^1(\Omega)\times L^1(\Omega)} \left| \int_\Omega \left[\frac{1}{2}\frac{(\varrho\mathbf{u})_0^2}{\varrho_0} + P(\varrho)\right] \mathrm{d}x \right|^r \mathrm{d}\Lambda \leq C \tag{24}$$

for some $r \geq 4$. In particular, let ϱ_0 and $\mathbf{u}_0$ be $L^1(\Omega) \times L^1(\Omega)$-valued random variable whose law is given by Λ. Then, we construct $\varrho_{0,n}$ and $\mathbf{u}_{0,n}$ as the initial conditions satisfying

$$(\varrho_{0,n}, \mathbf{u}_{0,n}) \in C^{2+\nu}(\Omega) \times X_n, \qquad \frac{1}{\varrho_{0,n}} \in L^\infty(\Omega),$$

and the following convergencies hold true:

$$\begin{aligned} \varrho_{0,n} &\rightharpoonup \varrho_0 \text{ in } L^P(\Omega), \\ \varrho_{0,n}\mathbf{u}_{0,n} &\rightharpoonup (\varrho\mathbf{u})_0 \text{ in } L^G(\Omega), \\ \int_\Omega \left[\frac{1}{2}\varrho_{0,n}\mathbf{u}_{0,n}^2 + P(\varrho_{0,n})\right] \mathrm{d}x &\to \int_\Omega \left[\frac{1}{2}\frac{(\varrho\mathbf{u})_0^2}{\varrho_0} + P(\varrho_0)\right] \mathrm{d}x \end{aligned}$$

$\mathbb{P}$-almost surely where G is some N-function. Consequently, we take Λ_n to be the law of $\varrho_{0,n}$ and $\mathbf{u}_{0,n}$.

Throughout this section, we work with the sequence $(\varrho_n, \mathbf{u}_n, \mathbb{S}_n)$ of solutions constructed in the previous section and emanating from $\varrho_{0,n}$ and $\mathbf{u}_{0,n}$ whose laws are given by Λ_n, and we let $n \to \infty$. In order to reach this goal, we use the advantage of stochastic Young measures (as stated in Lemma 2.2 and Lemma 2.3 in [8]). In particular, we use (9) in order to deduce the following bounds,

$$\begin{aligned} \mathbb{E}\left(\left|\int_Q F^*(\mathbb{S}_n)\,\mathrm{d}x\mathrm{d}t\right|^r\right) &\le C,\\ \mathbb{E}\left(\|P(\varrho_n)\|^r_{L^\infty(L^1(\Omega))}\right) &\le C,\\ \mathbb{E}\left(\left|\int_Q F(Du_n)\,\mathrm{d}x\mathrm{d}t\right|^r\right) &\le C, \end{aligned} \tag{25}$$

where C depends on initial data, right-hand side, and Ω.

Lemma 4.1 *Let F satisfy* (5). *Assume $\mathbf{u}_n \restriction_{\partial\Omega}= 0$ and let* $(25)_3$ *hold. Then, there is an N-function $g_0 : [0,\infty) \to [0,\infty)$ such that*

$$\mathbb{E}\left(\left|\int_Q g_0(|\mathbf{u}_n|)\,\mathrm{d}x\mathrm{d}t\right|^r\right) \le C.$$

Proof Let $\{\mathbf{u}_{n,1}\}$ be the sequence of the first coordinates of $\mathbf{u}_n$. For every $x = (x_1, x_2, x_3) \in \partial\Omega$, let $x_b = (x_{1b}, x_2, x_3)$, $x_{1b} < x_1$ be the point of the boundary such that all points with coordinates (s, x_2, x_3), $s \in (x_{1b}, x_1)$ belong to Ω. Assume without loss of generality that Ω is small enough that $x_{1b} < x_1 < x_{1b}+1$ (otherwise, there will appear some additional constant in the following computation), and for the purpose of this proof, assume that $\partial_1\mathbf{u}_{n,1}(s, x_2, x_3) = 0$ whenever $s \notin (x_{1b}, x_1)$ (this just means that we artificially extend $\mathbf{u}_n$ by a constant outside of the relevant interval).

Let g_1 be an (arbitrary) N-function. Then the Jensen inequality yields

$$\begin{aligned} \int_\Omega g_1(|\mathbf{u}_{n,1}|)\,\mathrm{d}x &\le \int_\Omega g_1\left(\int_{x_{1b}}^{x_1} |\partial_1\mathbf{u}_{n,1}|\,\mathrm{d}s\right)\mathrm{d}x_1\mathrm{d}x_2\mathrm{d}x_3\\ &\le \int_\Omega g_1\left(\int_{x_{1b}}^{x_{1b}+1} |\partial_1\mathbf{u}_{n,1}|\,\mathrm{d}s\right)\mathrm{d}x_1\mathrm{d}x_2\mathrm{d}x_3\\ \le \int_\Omega\left(\int_{x_{1b}}^{x_{1b}+1} g_1(|\partial_1\mathbf{u}_{n,1}|)\,\mathrm{d}s\right)\mathrm{d}x_2\mathrm{d}x_3 &\le \int_\Omega g_1(|D\mathbf{u}_n|)\,\mathrm{d}x\\ &\le \int_\Omega F(D\mathbf{u}_n)\,\mathrm{d}x \end{aligned} \tag{26}$$

assuming $g_1 = g$. This computation can be repeated in order to get g_2 and g_3 for the remaining variables. Furthermore, there exists a constant $c_{2,\infty} \in (0, \infty)$ such that any N-function g_0 satisfies

$$\int_\Omega g_0(|\mathbf{u}_n|)\,\mathrm{d}x \le \int_\Omega g_0(c_{2,\infty} \max\{|\mathbf{u}_{n,1}|, |\mathbf{u}_{n,2}|, |\mathbf{u}_{n,3}|\})\,\mathrm{d}x$$
$$\le \int_\Omega g_0(c_{2,\infty}|\mathbf{u}_{n,1}|)\,\mathrm{d}x + \int_\Omega g_0(c_{2,\infty}|\mathbf{u}_{n,2}|)\,\mathrm{d}x + \int_\Omega g_0(c_{2,\infty}|\mathbf{u}_{n,3}|)\,\mathrm{d}x,$$

and the choice $g_0(t) = \max\left\{g_1\left(\frac{t}{c_{2,\infty}}\right), g_2\left(\frac{t}{c_{2,\infty}}\right), g_3\left(\frac{t}{c_{2,\infty}}\right)\right\}$ together with (26) yields

$$\int_\Omega g_0(|\mathbf{u}_n|)\,\mathrm{d}x \le \int_\Omega g_1(|\mathbf{u}_{n,1}|)\,\mathrm{d}x + \int_\Omega g_2(|\mathbf{u}_{n,2}|)\,\mathrm{d}x + \int_\Omega g_3(|\mathbf{u}_{n,3}|)\,\mathrm{d}x$$
$$\le 3 \int_\Omega F(D\mathbf{u}_n)\,\mathrm{d}x,$$

which completes the proof. □

We state the path space $\mathcal{X}$ as

$$\mathcal{X} = \mathcal{X}_{\varrho 0} \times \mathcal{X}_{(\varrho\mathbf{u})_0} \times \mathcal{X}_{\varrho\mathbf{u}\otimes\mathbf{u}} \times \mathcal{X}_{p(\varrho)} \times \mathcal{X}_E \times \mathcal{X}_\nu \mathcal{X}_W$$
$$= L^P(\Omega, w) \times L^G(\Omega, w) \times L^\infty(0, T; \mathcal{M}(\Omega), w^*)$$
$$\times L^\infty(0, T; \mathcal{M}(\Omega), w^*) \times L^\infty(0, T; \mathcal{M}(\Omega), w^*) \times L^\infty((0, T) \times \Omega; \mathbb{P}(\mathbb{R}^{22}),$$
$$\times C([0, T]; \mathfrak{U}_0)w^*),$$

and we work with the sequence $(\varrho_{0,n}, (\varrho_{0,n}\mathbf{u}_{0,n}), \varrho_n\mathbf{u}_n \otimes \mathbf{u}_n, p(\varrho_n), E_n, \times W_n \nu_n) \in \mathcal{X}$ where

$$\nu_n = \delta_{\varrho_n} \times \delta_{\mathbf{u}_n} \times \delta_{\mathbb{S}_n} \times \delta_{D\mathbf{u}_n}$$

and E_n is the abbreviation for

$$E_n = \frac{1}{2}\varrho_n|\mathbf{u}_n|^2 + P(\varrho_n).$$

Clearly, the boundedness and convergences of $\varrho_{0,n}$ and $\varrho_n\mathbf{u}_{0,n}$ imply the tightness of their laws in $\mathcal{X}_{\varrho 0}$ and $\mathcal{X}_{(\varrho\mathbf{u})_0}$.

Next, the sequence ν_n is tight in $\mathcal{X}_\nu$. Indeed, this follows as the set

$$B_R := \Big\{\nu \in L^\infty((0,T)\times\Omega;\mathbb{P}(\mathbb{R}^{22}),w^*),\\ \int_0^T\int_\Omega\int_{\mathbb{R}^{22}}[|r|+|\mathbf{w}|+|S|+|D|]\,\mathrm{d}\nu\,\mathrm{d}x\mathrm{d}t \le R\Big\}$$

is compact (see [12, Proposition 2.8.5]). Here, $(r,\mathbf{w},S,D)\in([0,\infty)\times\mathbb{R}^3\times\mathbb{R}^{3\times3}_{\rm sym}\times\mathbb{R}^{3\times3}_{\rm sym})$ are dummy variables for the density, velocity, and stress tensor. Next, we have due to the Chebyshev and Young inequalities

$$\begin{aligned}\mathcal{L}(\nu_n)(B_R^C) &= \mathbb{P}\left(\int_0^T\int_\Omega\int_{\mathbb{R}^{22}}[|r|+|\mathbf{w}|+|S|+|D|]\,\mathrm{d}\nu_n\,\mathrm{d}x\mathrm{d}t \ge R\right)\\ &\le \mathbb{P}\left(\int_0^T\int_\Omega\big(P(\varrho_n)+g(\mathbf{u}_n)+F^*(\mathbb{S}_n)+F(D\mathbf{u}_n)\big)\,\mathrm{d}x\mathrm{d}t \ge R-C\right)\\ &\le \frac{C}{R-C}\end{aligned}$$

which gives the desired tightness.

Finally, the bounded sets in $L^\infty(0,T,\mathcal{M}(\Omega))$ are relatively compact in weak star topology, and, therefore, the a priori bounds directly yield the desired compactness of the laws of $\varrho_n\mathbf{u}_n\otimes\mathbf{u}_n$, $p(\varrho_n)$, and E_n.

As a matter of fact, we deduce the existence of the sequence of representations $(\tilde{\varrho}_0, (\widetilde{\varrho\mathbf{u}})_0, (\widetilde{\varrho_n\mathbf{u}_n\otimes\mathbf{u}_n}), \widetilde{p(\varrho_n)}, \tilde{E}_n, \tilde{\nu}_n)$ which converges in the appropriate topologies to $(\tilde{\varrho}_0, (\widetilde{\varrho\mathbf{u}})_0, (\widetilde{\varrho\mathbf{u}\otimes\mathbf{u}}), \widetilde{p(\varrho)}, \tilde{E}, \tilde{\nu})$. As both P and F are Young functions, we use Theorem 2.3 and Lemma 4.1 in order to deduce that the parametrized measure $\nu_{\omega,t,x}$ supported on $[0,\infty)\times\mathbb{R}^3\times\mathbb{R}^{3\times3}_{\rm sym}\times\mathbb{R}^{3\times3}_{\rm sym}$ satisfies

$$\begin{aligned}\tilde{\varrho}_n(t,x) &\rightharpoonup \langle\nu_{\omega,t,x}, r\rangle =: \varrho(t,x),\\ \tilde{\mathbf{u}}_n(t,x) &\rightharpoonup \langle\nu_{\omega,t,x}, \mathbf{w}\rangle =: \mathbf{u}(t,x),\\ \tilde{\mathbb{S}}_n(t,x) &\rightharpoonup \langle\nu_{\omega,t,x}, S\rangle =: \mathbb{S}(t,x),\\ D\tilde{u}_n &\rightharpoonup \langle\nu_{\omega,t,x}, D\rangle =: \overline{Du},\end{aligned} \tag{27}$$

where $\tilde{\varrho}_n$, $\tilde{\mathbf{u}}_n$, and $\tilde{\mathbb{S}}_n$ are suitable quantities defined on the standard probability space which coincide with ϱ_n, $\mathbf{u}_n$, and $\mathbb{S}_n$ in law.

The limit in the energy inequality yields

$$\mathbb{E}\left[\left|\sup_{\tau\in[0,T]}\int_\Omega\left\langle \nu_{\omega,t,x}, \frac{1}{2}r|\mathbf{w}|^2 + P(r)\right\rangle \mathrm{d}x + \sup_{\tau\in[0,T]}\int_\Omega \mathrm{d}\mathcal{D}_t\right|^r \right.$$
$$\left. + \left|\int_0^T\int_\Omega \left\langle \nu_{\omega,t,x}, F^*(S) + F(D)\right\rangle \mathrm{d}x\mathrm{d}t\right|^r\right]$$
$$\leq C\mathbb{E}\left[\left|\int_\Omega \frac{1}{2}\varrho_0|\mathbf{u}_0|^2 + P(\varrho_0)\,\mathrm{d}x\right|^r + 1\right],$$

where $\mathcal{D}_t$ is the defect measure arising as

$$\mathcal{D}_t = \lim\left(\frac{1}{2}\varrho_n|\mathbf{u}_n|^2 + P(\varrho_n)\right) - \left\langle \nu_{\omega,t,x}, \frac{1}{2}r|\mathbf{w}|^2 + P(r)\right\rangle,$$

where the limit is understood in the weak sense.

Consequently,

$$\widetilde{\varrho_n\mathbf{u}_n\otimes\mathbf{u}_n} \rightharpoonup \langle \nu, r\mathbf{w}\otimes\mathbf{w}\rangle + \Theta_t,$$
$$\widetilde{p(\varrho_n)} \rightharpoonup \langle \nu, p(r)\rangle + \Lambda_t$$

for almost all $t \in [0, T]$. The total variations of measures Θ_t and Λ_t are dominated by the energy inequality—this is the claim of [8, Lemma 2.3]. In particular, we have

$$\int_\Omega \mathrm{d}|\Theta_t| + \int_\Omega \mathrm{d}|\Lambda_t| \leq \int_\Omega \mathrm{d}\mathcal{D}_t \tag{28}$$

As a matter of fact, the Young measure ν satisfies the continuity equation

$$\int_\Omega \varrho(T)\varphi(T) - \varrho_0\varphi(0)\,\mathrm{d}x - \int_0^T\int_\Omega \varrho\partial_t\varphi\,\mathrm{d}x\mathrm{d}t - \int_0^T\int_\Omega \langle \nu, r\mathbf{w}\rangle\nabla\varphi\,\mathrm{d}x\mathrm{d}t = 0 \tag{29}$$

for $\varphi \in C_c^\infty(\overline{Q_T})$ $\mathbb{P}$-a.s.

The momentum equation takes the form

$$-\int_0^T \partial_t\Phi\int_\Omega \langle \nu, r\mathbf{w}\rangle\cdot\boldsymbol{\varphi}\,\mathrm{d}x\mathrm{d}t - \Phi(0)\int_\Omega \varrho_0\mathbf{u}_0\cdot\boldsymbol{\varphi}(0)\,\mathrm{d}x$$
$$= \int_0^T \Phi\int_\Omega \langle \nu, r\mathbf{w}\otimes\mathbf{w}\rangle : \nabla\boldsymbol{\varphi} + \langle \nu, p(r)\rangle\,\mathrm{div}\,\boldsymbol{\varphi}\,\mathrm{d}x\mathrm{d}t$$

$$+\int_0^T \Phi \int_\Omega \nabla \boldsymbol{\varphi} : \mathrm{d}\Theta_t \mathrm{d}t + \int_0^T \Phi \int_\Omega \operatorname{div} \boldsymbol{\varphi} \mathrm{d}\Lambda_t \mathrm{d}t$$

$$-\int_0^T \Phi \int_\Omega \mathbb{S} : \nabla \boldsymbol{\varphi} \, \mathrm{d}x \mathrm{d}t + \int_\Omega \int_0^T \Phi \boldsymbol{\varphi} \, \mathrm{d}M^1(t) \mathrm{d}x \tag{30}$$

where $\Phi \in C_c^\infty([0,T))$, $\boldsymbol{\varphi} \in C_c^\infty(\Omega)$ and where M^1 is a $W^{-r,2}(\Omega)$-valued martingale satisfying

$$\langle\!\langle f(t), M^1 \rangle\!\rangle = \sum_{i,j} \left(\sum_{k=1}^\infty \langle \langle \nu_{s,x}^\omega, r\mathbf{F}_k(r, \mathbf{w}) \rangle, g_i \rangle \langle \mathbb{D}^s f(e_k), g_j \rangle \, \mathrm{d}s \right) g_i \otimes g_j, \tag{31}$$

where $\{g_i\} \subset W^{-r,2}(\Omega)$ is the orthonormal basis for some $r > 0$ and f is an arbitrary random process f adapted to $(\mathfrak{F}_t)_{t \geq 0}$ satisfying

$$\mathrm{d}f = D^d f \, \mathrm{d}t + \mathbb{D}^s f \, \mathrm{d}W,$$

$$f \in C([0,T], W^{1,q}(\Omega) \cap C(\Omega)), \; \mathbb{E}\left[\sup_{t \in [0,T]} \|f\|_{W^{1,q}}^2 \right]^q < \infty,$$

$\mathbb{P}$-almost surely for all $q \in [1, \infty)$,

and

$$D^d f \in L^q(O; L^q(0,T; W^{1,q}(\Omega))), \; \mathbb{D}^s f \in L^2(O; L^2(0,T; L_2(\mathfrak{U}; L^2(\Omega)))),$$

$$\left(\sum_{k=1}^\infty |\mathbb{D}^s f(e_k)|^q \right)^{1/q} \in L^q(O; L^q(0,T; L^q(\Omega))).$$

Indeed, let M_n be the martingale

$$M_n = \int_0^t \varrho_n \Pi_n \left[\mathbb{F}(\varrho_n, \mathbf{u}_n) \right] \mathrm{d}W,$$

and let f be as above. Then, it holds that

$$\langle\!\langle f, M_n(t) \rangle\!\rangle = \sum_{i,j=1}^\infty \left(\sum_{k=1}^\infty \int_0^t \langle \varrho_n \Pi_n \left[\mathbf{F}_k(\varrho_n, \mathbf{u}_n) \right], g_i \rangle \langle \mathbb{D}_t^s f(e_k), g_j \rangle \, \mathrm{d}s \right) g_i \otimes g_j.$$

Properties of cross variation yield that

$$f(t) M_n(t) - \langle\!\langle f(t), M_n(t) \rangle\!\rangle$$

is a martingale, and, therefore,

$$\mathbb{E}\Big[L_\tau(\Phi_n)\Big(\langle M_n(t), g_i\rangle\langle f(t), g_j\rangle - \langle M_n(\tau), g_i\rangle\langle f(\tau), g_j\rangle - \sum_{k=1}^{\infty}\int_\tau^t \langle \varrho_n \Pi_n\left[\mathbf{F}_k(\varrho_n, \mathbf{u}_n)\right], g_i\rangle\langle \mathbb{D}_t^s f(e_k), g_j\rangle \,\mathrm{d}s\Big)\Big] = 0$$

whenever $0 \le \tau \le t \le T$. Here, L_τ is any bounded continuous functional on $\Psi_n = (\varrho_0, \mathbf{u}_0, \varrho_n\mathbf{u}_n \otimes \mathbf{u}_n, p(\varrho_n), E_n, \nu_n, W)$ depending only on the values of Ψ_n restricted to $[0, \tau]$. Due to the convergencies of ϱ_n and $\mathbf{u}_n$, we get that

$$L_\tau(\Psi_n) \to L_\tau(\Psi)\ \mathbb{P}\text{-almost surely.}$$

Recall the assumption on $\mathbf{F}$—especially recall that $\mathbf{F} = 0$ whenever ϱ is large. Therefore,

$$\varrho_n \Pi_n[\mathbf{F}_k(\varrho_n, \mathbf{u}_n)] \in L^\infty((0, T) \times \Omega)$$

uniformly with respect to n, and Theorem 2.3 yields that

$$\varrho_n \Pi_n\left[\mathbf{F}_k(\varrho_n, \mathbf{u}_n)\right] \rightharpoonup \langle \nu, \mathbf{G}_k(r, \mathbf{w})\rangle \ \text{ in } L^q((0, T) \times \Omega)$$

for all $q \in (1, \infty)$. The Sobolev embedding theorem then gives $L^q \subset W^{-m,2}$ for any $m \in \mathbb{N}$ assuming q is sufficiently large and thus

$$\varrho_n \Pi_n\left[\mathbf{F}_k(\varrho_n, \mathbf{u}_n)\right] \rightharpoonup \langle \nu, \mathbf{G}_k(r, \mathbf{w})\rangle \ \text{ in } L^2(0, T; W^{-m,2}(\Omega))$$

for some $m \in \mathbb{N}$. Consequently,

$$\mathbb{E}\Big[L_\tau(\Phi)\Big(\langle M^1(t), g_i\rangle\langle f(t), g_j\rangle - \langle M^1(\tau), g_i\rangle\langle f(\tau), g_j\rangle - \sum_{k=1}^{\infty}\int_\tau^t \langle\langle \nu, \mathbf{G}_k(r, \mathbf{w})\rangle, g_i\rangle\langle \mathbb{D}_t^s f(e_k), g_j\rangle \,\mathrm{d}s\Big)\Big] = 0$$

which gives that

$$M^1(t) f(t) - \sum_{i,j=1}^{\infty}\left(\sum_{k=1}^{\infty}\langle\langle \nu, r\mathbf{F}_k(r, \mathbf{w})\rangle, g_i\rangle \left\langle \mathbb{D}_t^s f(e_k), g_j\right\rangle \mathrm{d}s\right) g_i \otimes g_j$$

is a martingale. This implies the precise form of the cross variation for M^1 and f and, consequently, also the desired result (31).

In order to deduce the energy inequality, we recall that F and F^* are convex and therefore sequentially weakly lower semicontinuous. Thus, we deduce from (9) that

$$\begin{aligned}\Big[\int_\Omega \Big\langle \nu_{\omega,\tau,\cdot}, \frac{1}{2} r|\mathbf{w}|^2(\tau,\cdot) + P(r)\Big\rangle \,\mathrm{d}x + \int_\Omega \mathrm{d}\mathcal{D}_\tau\Big]_{\tau=0}^{\tau=t} \\ + \int_0^t \int_\Omega \big\langle \nu, F(S) + F^*(D)\big\rangle \,\mathrm{d}x\mathrm{d}\tau \\ \le \frac{1}{2}\int_0^t \int_\Omega \sum_{k=1}^\infty \langle \nu_{t,x}^\omega, r^{-1}|\mathbf{G}_k(r, r\mathbf{w})|^2\rangle \,\mathrm{d}x\mathrm{d}\tau + \int_0^t \mathrm{d}M^2(\tau) \qquad (32)\end{aligned}$$

where $\mathcal{D}_\tau$ is the defect measure and M^2 is the martingale arising as the limit of $\int_0^t \int_\Omega \varrho_n \Pi_n[\mathbb{F}(\varrho_n, \mathbf{u}_n)] \cdot \mathbf{u}_n \,\mathrm{d}x\mathrm{d}W$ which is shown below. In particular, we want to show that

$$\mathbb{E}[M^2(t)|\mathfrak{F}_s] = M^2(s)$$

whenever $0 \le s \le t \le T$. As above, we rather show that

$$\mathbb{E}\Big[L_s(\Psi)(M^2(t) - M^2(s))\Big] = 0$$

where $\Psi = (\varrho_0, \mathbf{u}_0, \varrho\mathbf{u} \otimes \mathbf{u}, p(\varrho), E, \nu, W)$ and L_s is any bounded continuous functional that depends only on the values of Ψ restricted to $[0, s]$ on the path space $\mathcal{X}$. However, we know that

$$\mathbb{E}\Big[L_s(\Psi_n)\int_s^t \int_\Omega \varrho_n \Pi_n[\mathbb{F}(\varrho_n, \mathbf{u}_n)] \cdot \mathbf{u}_n \,\mathrm{d}x\mathrm{d}W\Big] = 0$$

where $\Psi_n = (\varrho_0, \mathbf{u}_0, \varrho_n\mathbf{u}_n \otimes \mathbf{u}_n, p(\varrho_n), E_n, \nu_n, W)$. The convergencies in the appropriate topology of $\mathcal{X}$ yield that $\Psi_n \to \Psi$ $\mathbb{P}$-almost surely. This, together with the Vitali convergence theorem, allows us to pass to the limit in n, and we get the demanded martingale property of the limit M^2.

The above considerations yield the correct notion of a measure-valued martingale solution.

Definition 4.1 Let Λ be a Borel probability measure on $L^\gamma(\Omega) \times L^{2\gamma/(\gamma+1)}(\Omega)$. A trio $[(\mathcal{O}, \mathfrak{F}, (\mathfrak{F}_t)_{t\ge 0}, \mathbb{P}), \nu_{t,x}^\omega, W]$, where $(\mathcal{O}, \mathfrak{F}, (\mathfrak{F}_t)_{t\ge 0}, \mathbb{P})$ is a stochastic basis with a complete right-continuous filtration, W is $(\mathfrak{F}_t)$-cylindrical Wiener process, and $\nu^{w}_{t,x}$ is a family of probabilistic measures on $([0,\infty), \mathbb{R}^3, \mathbb{R}^{3\times3}, \mathbb{R}^{3\times3})$, is called *measure-valued solution* to (1) if[2] the following hold:

- The density ϱ satisfies

$$\mathbb{E}\left[\left(\operatorname{esssup}_{t\in(0,T)} \int_\Omega P(\varrho)\,\mathrm{d}x\right)^r\right] < \infty$$

  for all $r \in (1,\infty)$.
- There exists an N-function g such that

$$\mathbb{E}\left[\left(\int_0^T \int_\Omega g(|\mathbf{u}|)\,\mathrm{d}x\mathrm{d}t\right)^r\right] < \infty$$

  for all $r \in (1,\infty)$.
- The shear rate $D\mathbf{u}$ and the stress tensor $\mathbb{S}$ satisfy

$$\mathbb{E}\left[\left(\int_0^T \int_\Omega F(D\mathbf{u})\,\mathrm{d}x\mathrm{d}t\right)^r + \left(\int_0^T \int_\Omega F^*(\mathbb{S})\,\mathrm{d}x\mathrm{d}t\right)^r\right] < \infty$$

  for all $r \in (1,\infty)$.
- The continuity equation (29) is satisfied for all $\varphi \in C_c^\infty([0,T) \times \overline{\Omega})$ $\mathbb{P}$-almost surely.
- There is a $W^{-m,2}(\Omega)$-valued square integrable continuous martingale M^1 fulfilling (31) and measures Θ_t and Λ_t such that the momentum equation (30) is satisfied $\mathbb{P}$-almost surely and for all $\Phi \in C_c^\infty([0,T))$ and $\varphi \in C_c^\infty(\Omega)$.
- There exists a defect measure $\mathcal{D}$ and a martingale M^2 such that the total variations of the measures Θ_t and Λ_t satisfy the estimate (28) together with the energy inequality (32) for almost all $t \in (0,T)$ $\mathbb{P}$-almost surely.

The existence of such solution was shown above. However, our findings are clearly summarized in the following theorem:

Theorem 4.1 *Let the domain Ω be of class $C^{2+\nu}$ for some $\nu > 0$ and the right-hand side G satisfy* (4), *and let the law of the initial data Λ satisfy* (23) *and* (24). *Next, let $\mathbb{S} = \mathbb{S}(D\mathbf{u})$ be given as a subdifferential of a convex function F satisfying* (5).

Then, there exists a measure-valued martingale solution to (1) *as stated in the above definition.*

Acknowledgments Václav Mácha would like to acknowledge the support of RVO:67985840 and of Praemium Academiæ of Š. Nečasová.

[2] We use the abbreviation stated in (27).

References

1. Málek, J., Rajagopal, K.R.: Compressible generalized Newtonian fluids. Z. Angew. Math. Phys. **61**(6), 1097–1110 (2010). DOI https://doi.org/10.1007/s00033-010-0061-8
2. Mamontov, A.E.: On the global solvability of the multidimensional Navier-Stokes equations of a nonlinearly viscous fluid. I. Sibirsk. Mat. Zh. **40**(2), 408–420, iii (1999)
3. Mamontov, A.E.: On the global solvability of the multidimensional Navier-Stokes equations of a nonlinearly viscous fluid. II. Sibirsk. Mat. Zh. **40**(3), 635–649, iii (1999). DOI https://doi.org/10.1007/BF02679762
4. Feireisl, E., Liao, X., Málek, J.: Global weak solutions to a class of non-Newtonian compressible fluids. Math. Methods Appl. Sci. **38**(16), 3482–3494 (2015). DOI https://doi.org/10.1002/mma.3432
5. Zhikov, V.V., Pastukhova, S.E.: On the solvability of the Navier-Stokes system for a compressible non-Newtonian fluid. Dokl. Akad. Nauk **427**(3), 303–307 (2009). DOI https://doi.org/10.1134/S1064562409040164
6. Abbatiello, A., Feireisl, E., Novotný, A.: Generalized solutions to models of compressible viscous fluids. Discrete Contin. Dyn. Syst. **41**(1), 1–28 (2021). DOI https://doi.org/10.3934/dcds.2020345
7. Basarić, D.: Existence of dissipative (and weak) solutions for models of general compressible viscous fluids with linear pressure. J. Math. Fluid Mech. **24**(2), 22 (2022). DOI https://doi.org/10.1007/s00021-022-00688-1. Id/No 56
8. Hofmanová, M., Koley, U., Sarkar, U.: Measure-valued solutions to the stochastic compressible Euler equations and incompressible limits. Commun. Partial Differ. Equations **47**(9), 1907–1943 (2022). DOI https://doi.org/10.1080/03605302.2022.2101002
9. Hofmanová, M., Zhu, R., Zhu, X.: On ill- and well-posedness of dissipative martingale solutions to stochastic 3d Euler equations. Commun. Pure Appl. Math. **75**(11), 2446–2510 (2022). DOI https://doi.org/10.1002/cpa.22023
10. Lions, P.L.: Mathematical topics in fluid mechanics. Vol. 2, *Oxford Lecture Series in Mathematics and its Applications*, vol. 10. The Clarendon Press, Oxford University Press, New York (1998). Compressible models, Oxford Science Publications
11. Feireisl, E., Novotný, A., Petzeltová, H.: On the existence of globally defined weak solutions to the Navier-Stokes equations. J. Math. Fluid Mech. **3**(4), 358–392 (2001). DOI https://doi.org/10.1007/PL00000976
12. Breit, D., Feireisl, E., Hofmanová, M.: Stochastically forced compressible fluid flows, *De Gruyter Series in Applied and Numerical Mathematics*, vol. 3. De Gruyter, Berlin (2018)
13. Breit, D., Hofmanová, M.: Stochastic Navier-Stokes equations for compressible fluids. Indiana Univ. Math. J. **65**(4), 1183–1250 (2016). DOI https://doi.org/10.1512/iumj.2016.65.5832
14. Adams, R.A., Fournier, J.J.F.: Sobolev spaces, *Pure and Applied Mathematics (Amsterdam)*, vol. 140, second edn. Elsevier/Academic Press, Amsterdam (2003)
15. Bogachev, V.I.: Measure theory. Vol. I, II. Springer-Verlag, Berlin (2007). DOI https://doi.org/10.1007/978-3-540-34514-5
16. Pedregal, P.: Parametrized measures and variational principles, *Progress in Nonlinear Differential Equations and their Applications*, vol. 30. Birkhäuser Verlag, Basel (1997). DOI https://doi.org/10.1007/978-3-0348-8886-8
17. Debussche, A., Vovelle, J.: Scalar conservation laws with stochastic forcing. J. Funct. Anal. **259**(4), 1014–1042 (2010). DOI https://doi.org/10.1016/j.jfa.2010.02.016
18. Jakubowski, A.: The almost sure Skorokhod representation for subsequences in nonmetric spaces. Teor. Veroyatnost. i Primenen. **42**(1), 209–216 (1997). DOI https://doi.org/10.1137/S0040585X97976052
19. Feireisl, E., Gwiazda, P., Świerczewska Gwiazda, A., Wiedemann, E.: Dissipative measure-valued solutions to the compressible Navier-Stokes system. Calc. Var. Partial Differential Equations **55**(6), Art. 141, 20 (2016). DOI https://doi.org/10.1007/s00526-016-1089-1

20. Florescu, L.C., Godet-Thobie, C.: Young measures and compactness in measure spaces. De Gruyter, Berlin (2012). DOI https://doi.org/10.1515/9783110280517
21. Málek, J., Nečas, J., Rokyta, M., Růžička, M.: Weak and measure-valued solutions to evolutionary PDEs, *Applied Mathematics and Mathematical Computation*, vol. 13. Chapman & Hall, London (1996). DOI https://doi.org/10.1007/978-1-4899-6824-1
22. Novotný, A., Straškraba, I.: Introduction to the mathematical theory of compressible flow, *Oxford Lecture Series in Mathematics and its Applications*, vol. 27. Oxford University Press, Oxford (2004)

Weak-Strong Uniqueness for a Compressible Fluid-Rigid Body Interaction Problem with General Inflow-Outflow Boundary Data

Šárka Nečasová and Ana Radošević

Abstract The chapter deals with the problem of the motion of a rigid body in a domain filled by the compressible fluid. We consider the nonhomogeneous boundary condition of the velocity and inflow boundary condition for the density. We study the weak-strong uniqueness property for such problem. The novelty lies in the fact we are considering the general inflow-outflow boundary data which were not considered before in the context of fluid-structure interaction of weak-strong uniqueness.

1 Introduction

In this chapter, we examine the weak-strong uniqueness property for the motion of a rigid body within a viscous, compressible fluid, considering general inflow-outflow boundary conditions. We consider $\Omega \subset \mathbb{R}^3$ to be a given bounded domain with a sufficiently regular boundary and let $S_0 \subset \Omega$ be sufficiently regular part of the domain Ω with a strictly positive distance from $\partial\Omega$. In the setting of a rigid body moving within a fluid, S_0 represents the part of the domain initially occupied by the rigid body, and the corresponding fluid domain at the initial time $t = 0$ is given by $\Omega_F = \Omega \setminus \overline{S_0}$. Furthermore, for $T > 0$, we introduce the following time-space regions,

$$Q_T = (0, T) \times \Omega, \qquad Q_T^F = \bigcup_{t\in(0,T)} \{t\} \times F(t),$$

$$Q_T^S = \bigcup_{t\in(0,T)} \{t\} \times S(t), \qquad Q_T^{\partial S} = \bigcup_{t\in(0,T)} \{t\} \times \partial S(t),$$

Š. Nečasová (✉)
Institute of Mathematics, Czech Academy of Sciences, Prague, Czech Republic
e-mail: matus@math.cas.cz

A. Radošević
Faculty of Economics and Business, University of Zagreb, Zagreb, Croatia
e-mail: aradosevic@efzg.hr

T. Bodnár et al. (eds.), *Stochastics in Fluids*, Lecture Notes in Mathematical Fluid Mechanics, https://doi.org/10.1007/978-3-032-02299-8_4

where $\Omega_F(t) = \Omega \setminus \overline{S(t)}$ represents the fluid region at time t, while $S(t)$ denotes the portion of the domain occupied by the rigid body at time t.

Regarding the inhomogeneous boundary conditions, $\mathbf{u}_B$ represents the prescribed velocity on the boundary of Ω, and we define the inflow and outflow parts of the boundary as follows,

$$\Gamma_{in} = \{x \in \partial\Omega \mid \mathbf{u}_B \cdot \mathbf{n} < 0\}, \qquad \Gamma_{out} = \{x \in \partial\Omega \mid \mathbf{u}_B \cdot \mathbf{n} \geq 0\},$$

where $\mathbf{n}$ is the outer normal to $\Omega_F(t)$. On Γ_{in}, we prescribe time-independent fluid density ϱ_B.

The fluid flow is described by the compressible Navier-Stokes equations

$$\begin{aligned}
&\left.\begin{aligned}
&\partial_t \varrho_F + \mathrm{div}_x(\varrho_F \mathbf{u}_F) = 0,\\
&\partial_t(\varrho_F \mathbf{u}_F) + \mathrm{div}_x(\varrho_F \mathbf{u}_F \otimes \mathbf{u}_F) + \nabla p(\varrho_F) - \mathrm{div}_x S(\boldsymbol{u}_F) = 0,
\end{aligned}\right\} \text{ in } Q_T^F,\\
&\mathbf{u}_F = \mathbf{u}_S, \quad \text{on } Q_T^{\partial S},\\
&\mathbf{u}_F = \mathbf{u}_B \quad \text{on } (0, T) \times \partial\Omega\\
&\varrho_F = \varrho_B \quad \text{on } (0, T) \times \Gamma_{in}\\
&\varrho_F(0) = \varrho_0 \quad \text{and} \quad (\varrho_F \mathbf{u}_F)(0) = \mathbf{m}_0 \text{ in } \Omega_F,
\end{aligned} \tag{1}$$

where $\mathbf{u}_F$ and ϱ_F denote the velocity and the density of the fluid and $\mathbf{u}_S$ is the velocity of the rigid body. For the Newtonian fluid, the viscous part of the stress tensor is given by

$$\mathbb{S}(\mathbf{u}_F) = 2\mu\mathbb{D}(\mathbf{u}_F) + \lambda \mathrm{div}_x \mathbf{u}_F \mathbb{i}, \qquad \mathbb{D}(\mathbf{u}_F) = \frac{1}{2}\left(\nabla \mathbf{u}_F + (\mathbf{u}_F)^T\right)$$

with fluid viscosities μ and λ such that $\mu > 0$, $\mu + \lambda \geq 0$. We assume that the fluid pressure is given by the equation of state

$$p(\varrho_F) = a\varrho_F^\gamma, \quad a > 0,\ \gamma > \frac{3}{2},$$

and the pressure potential is given by $P(\varrho_f) = a\dfrac{\varrho_f^\gamma}{\gamma - 1}$.

For the rigid body, we use the following standard notation:

- $m = \int_{S(t)} \varrho_S(t, \boldsymbol{x})\, \mathrm{d}x$ represents the total mass of the rigid body.
- $\mathbf{q}(t) = \frac{1}{m} \int_{S(t)} \boldsymbol{x} \varrho_S(t, \boldsymbol{x})\, \mathrm{d}x$ denotes the center of mass.
- $\mathbf{a}(t) = \frac{\mathrm{d}}{\mathrm{d}t}\mathbf{q}(t)$ is the velocity of the center of mass.
- $\mathbb{J}(t) = \int_{S(t)} \varrho_s(t, \boldsymbol{x})\Big(|\boldsymbol{x} - \mathbf{q}(t)|^2 \,\mathbb{i} - (\boldsymbol{x} - \mathbf{q}(t))^T (\boldsymbol{x} - \mathbf{q}(t))\Big)\, \mathrm{d}x$ represents the inertia tensor of the rigid body.

The velocity of the rigid body is given by

$$\mathbf{u}_S(t,\boldsymbol{x}) = \mathbf{a}(t) + \boldsymbol{\omega}(t) \times (\boldsymbol{x} - \mathbf{q}(t)), \qquad (t,\boldsymbol{x}) \in Q_T^S. \tag{2}$$

The fluid equations are coupled with the following balance laws:

$$\begin{aligned}
&\left.\begin{aligned}
m\frac{\mathrm{d}^2}{\mathrm{d}t^2}\mathbf{q} &= -\int_{\partial S(t)} (\mathbb{S}(\varrho_F,\mathbf{u}_F) - p(\varrho_F)I)\,\mathbf{n}\,\mathrm{d}\boldsymbol{\gamma}(\boldsymbol{x}),\\
\frac{\mathrm{d}}{\mathrm{d}t}(\mathbb{J}\boldsymbol{\omega}) &= -\int_{\partial S(t)} (\boldsymbol{x} - \mathbf{q}(t)) \times (\mathbb{S}(\varrho_F,\mathbf{u}_F) - p(\varrho_F)I)\,\mathbf{n}\,\mathrm{d}\boldsymbol{\gamma}(\boldsymbol{x})
\end{aligned}\right\} \text{ in } (0,T),\\
&\varrho_S(0) = \varrho_0 \text{ in } S_0\\
&\mathbf{q}(0) = \mathbf{q}_0, \quad \mathbf{a}(0) = \mathbf{a}_0, \quad \boldsymbol{\omega}(0) = \boldsymbol{\omega}_0
\end{aligned} \tag{3}$$

1.1 Weak Solution

In the following, we will use global density and velocity field defined in the following way:

$$\varrho = \begin{cases} \varrho_F & \text{in } Q_T^F \\ \varrho_S & \text{in } Q_T^S \end{cases} \quad \text{and} \quad \mathbf{u} = \begin{cases} \mathbf{u}_F & \text{in } Q_T^F \\ \mathbf{u}_S & \text{in } Q_T^S \end{cases} \tag{4}$$

To derive the corresponding weak formulation, it is necessary to suitably extend the boundary velocity into the interior of the domain. This is achieved through the following lemma; see Axmann et al. [1].

Lemma 1.1 *Let $\Omega \subset \mathbb{R}^3$ be a bounded domain with Lipschitz boundary and $\mathbf{u}_B \in W^{1,\infty}(\partial\Omega;\mathbb{R}^3)$. Then, there exists $h_0 > 0$ such that for any $h \in (0,h_0)$ there is an extension vector field $\mathbf{u}_\infty \in W^{1,\infty}(\mathbb{R}^3) \cap C_c(\mathbb{R}^3)$, such that*

$$\begin{aligned}
\mathbf{u}_\infty|_{\partial\Omega} &= \mathbf{u}_B,\\
\operatorname{div}_x \mathbf{u}_\infty &\geq 0 \text{ a.e. in } U_h = \{\boldsymbol{x} \in \mathbb{R}^3 : \operatorname{dist}(\boldsymbol{x},\partial\Omega) < h\},\\
\mathbf{u}_\infty &= 0 \text{ on } \mathbb{R}^3 \setminus U_{2h}.
\end{aligned}$$

Definition 1.1 We say that triple $(\varrho, \mathbf{u}, \eta)$ is a weak solution to problems (1)–(3) on $[0,T]$ if $\mathbf{u}_\infty$ is an extension of the boundary data $\mathbf{u}_B$ from Lemma 1.1 and

1. The functions ϱ and $\mathbf{u}$ belong to regularity classes

$$\begin{aligned}
&\varrho \geq 0 \text{ a.e. in } Q_T, \quad \varrho \in L^\infty\big(0,T;L^\gamma(\Omega)\big)\\
&\mathbf{u} - \mathbf{u}_\infty \in L^2\big(0,T;W_0^{1,2}(\Omega)\big), \quad \varrho\mathbf{u} \in C_{\text{weak}}([0,T];L^1(\Omega))
\end{aligned}$$

and the velocity field $\mathbf{u}$ is *compatible* with parametrized isometries $\eta[t]$ describing the evolution of $S(t)$, such that the mapping $t \mapsto \eta[t](\mathbf{y})$ is absolutely continuous on $[0, T]$ and

$$S(t) = \eta[t](S_0), \quad \eta[t](\mathbf{y}) = \mathbf{q}(t) + \mathbb{O}(t)\mathbf{y}, \qquad \mathbb{O} \in SO(3), \ t \in [0, T]$$

$$\Big(\frac{\mathrm{d}}{\mathrm{d}t}\eta[t]\Big)\eta[t]^{-1}(\boldsymbol{x}) = \mathbf{u}(t, \boldsymbol{x}), \ \boldsymbol{x} \in \overline{S}(t) \text{ and a.e. } t \in (0, T)$$

2. Functions $(\varrho, \mathbf{u})$ satisfy in the sense of distributions the continuity equation

$$\begin{aligned}
&\int_\Omega (\varrho\psi)(\tau)\,\mathrm{d}x - \int_\Omega \varrho_0\psi(0)\,\mathrm{d}x \\
&\quad = \int_0^\tau \int_\Omega \big(\varrho\partial_t\psi + \varrho\mathbf{u}\cdot\nabla\psi\big)\,\mathrm{d}x\,\mathrm{d}t - \int_0^\tau \int_{\Gamma_{in}} \varrho_B\mathbf{u}_B\cdot\mathbf{n}\psi\,\mathrm{d}\boldsymbol{\gamma}(\boldsymbol{x})\,\mathrm{d}t \\
&\qquad \forall \tau \in [0, T],\ \forall\psi \in C^1_c\big([0, T] \times (\Omega \cup \Gamma_{in})\big)
\end{aligned} \tag{5}$$

as well as its renormalized version

$$\begin{aligned}
&\int_\Omega (b(\varrho)\psi)(\tau)\ \mathrm{d}x - \int_\Omega b(\varrho_0)\psi(0)\ \mathrm{d}x \\
&\quad = \int_0^\tau \int_\Omega \Big(b(\varrho)\partial_t\psi + b(\varrho)\mathbf{u}\cdot\nabla\psi - \psi\big(b'(\varrho)\varrho - b(\varrho)\big)\mathrm{div}_x\mathbf{u}\Big)\,\mathrm{d}x\,\mathrm{d}t \\
&\qquad - \int_0^\tau \int_{\Gamma_{in}} b(\varrho_B)\mathbf{u}_B\cdot\mathbf{n}\psi\mathrm{d}\boldsymbol{\gamma}(\boldsymbol{x})\,\mathrm{d}t \\
&\forall \tau \in [0, T],\ \forall\psi \in C^1_c\big([0, T] \times (\Omega \cup \Gamma_{in})\big), \forall b \in C^1\big([0, \infty)\big) \text{ with} \\
&\qquad b' \in C^1_c\big([0, \infty)\big)
\end{aligned} \tag{6}$$

3. The weak variational formulation of the momentum equation holds true

$$\begin{aligned}
&\int_\Omega (\varrho\mathbf{u}\cdot\boldsymbol{\varphi})(\tau)\ \mathrm{d}x - \int_\Omega \mathbf{m}_0\cdot\boldsymbol{\varphi}(0)\ \mathrm{d}x \\
&\quad = \int_0^\tau \int_\Omega \Big(\varrho\mathbf{u}\cdot\partial_t\boldsymbol{\varphi} + (\varrho\mathbf{u}\otimes\mathbf{u}) : \mathbb{D}(\boldsymbol{\varphi}) + p(\varrho)\mathrm{div}_x\boldsymbol{\varphi} - \mathbb{S}(\mathbf{u}) : \mathbb{D}(\boldsymbol{\varphi})\Big)\,\mathrm{d}x\,\mathrm{d}t \\
&\qquad \forall\tau \in [0, T], \\
&\forall\boldsymbol{\varphi} \in \mathcal{R}(Q_S) = \big\{\boldsymbol{\varphi} \in C^\infty_c([0, T]\times\Omega)\ :\ \mathbb{D}(\boldsymbol{\varphi}) = 0 \text{ on some neighborhood of } Q_S\big\}
\end{aligned} \tag{7}$$

4. We have the energy inequality

$$\int_\Omega \Big(\frac{1}{2}\varrho|\mathbf{u}-\mathbf{u}_\infty|^2 + P(\varrho)\Big)(\tau)\,\mathrm{d}x + \int_0^\tau\int_\Omega \mathbb{S}(\mathbf{u}-\mathbf{u}_\infty) : \mathbb{D}(\mathbf{u}-\mathbf{u}_\infty)\,\mathrm{d}x\,\mathrm{d}t$$
$$\leq \int_\Omega \Big(\frac{1}{2}\frac{|(\varrho\mathbf{u})_0-\varrho_0\mathbf{u}_\infty|^2}{\varrho_0} + P(\varrho_0)\Big)\,\mathrm{d}x - \int_0^\tau\int_\Omega \varrho\mathbf{u}\cdot\nabla\mathbf{u}_\infty\cdot(\mathbf{u}-\mathbf{u}_\infty)\,\mathrm{d}x\,\mathrm{d}t$$
$$-\int_0^\tau\int_\Omega p(\varrho)\mathrm{div}_x\mathbf{u}_\infty\,\mathrm{d}x\,\mathrm{d}t - \int_0^\tau\int_\Omega \mathbb{S}(\mathbf{u}_\infty) : \mathbb{D}(\mathbf{u}-\mathbf{u}_\infty)\,\mathrm{d}x\,\mathrm{d}t$$
$$-\int_0^\tau\int_{\Gamma_{in}} P(\varrho_B)\mathbf{u}_B\cdot\mathbf{n}\,\gamma(x)\,\mathrm{d}t, \tag{8}$$

for a.e. $\tau \in (0, T)$

The existence of weak solutions was shown by Axmann et al. [1], and we shall state this result.

Theorem 1.1 *Suppose that* $\mathrm{dist}(S_0, \partial\Omega) > h > 0$ *and*

$$(\varrho\mathbf{u})_0 \in L^2(\Omega), \quad \varrho_B \in C(\partial\Omega), \quad \mathbf{u}_B \in C^2(\partial\Omega),$$

$$\varrho_0 \geq 0, \quad \int_\Omega \varrho_0\,\mathrm{d}x > 0, \quad \varrho_B \geq \underline{\varrho}_B > 0,$$

$$\int_\Omega \left(\frac{1}{2}\rho_0|\mathbf{u}_0|^2 + P(\varrho_0)\right)\,\mathrm{d}x < \infty.$$

Then, for $\mathbf{u}_\infty$*, an extension of the boundary data* $\mathbf{u}_B$ *from Lemma 1.1, there exists a time* $T > 0$ *and a weak solution in the sense of Definition 1.1 such that* $\mathrm{dist}(S(t), \partial\Omega) \geq h$*, for all* $t \in [0, T]$.

We will consider the following strong solution. We will say that $(\mathbf{u}, \varrho, \mathbf{q}, \boldsymbol{\omega})$ is a strong solution to the problems (1)–(3) if it has the following regularity class $\mathcal{W}$

$$\mathbf{u}_F \in L^2(0,T;H^3)\cap C(0,T;H^3)\cap H^1(0,T;H^2)\cap C^1(0,T;H^2)\cap H^2(0,T;L^2),$$
$$\mathbf{u}_F \in L^2(0,T;H^4)\cap C(0,T;H^3)\cap C^1(0,T;H^1)\cap H^2(0,T;L^2),$$
$$\mathbf{a} = \frac{\mathrm{d}}{\mathrm{d}t}\mathbf{q},\ \boldsymbol{\omega} \in H^2(0,T)\cap C^1([0,T]).$$

and satisfies (1)–(3) in the classical sense.

Remark 1.1 Let us mention that there is no result concerning the strong solution with inflow boundary condition for the density and inhomogeneous Dirichlet condition for the velocity field.

The goal is to prove the following result.

Theorem 1.2 *Let* $(\varrho_1, \mathbf{u}_1, \eta_1)$ *be a weak solution to* (1)–(3) *in the sense of Definition 1.1 such that* $\mathrm{dist}(S_1(t), \partial\Omega) \geq h > 0$, *for all* $t \in [0, T_1]$. *Suppose that* $(\mathbf{u}_2, \varrho_2, \mathbf{q}_2, \boldsymbol{\omega}_2)$ *is a strong solution to* (1)–(3) *defined on* $(0, T_2)$ *such that* $\mathrm{dist}(S_2(t), \partial\Omega) \geq h > 0 \quad \forall t \in [0, T_2]$. *Then,* $S_1(t) = S_2(t)$ *and*

$$(\varrho_1, \mathbf{u}_1) = (\varrho_2, \mathbf{u}_2) \tag{9}$$

on time interval $[0, T]$, *where* $T = \min\{T_1, T_2\}$.

1.2 Brief Overview of Known Results

We will provide insights into recent developments in the theory of compressible fluids, as well as the challenges related to fluid-structure interaction.

Before we start with compressible case, let us mention that first result of the global-in-time solutions for large data in the context of nonlinear models of fluid mechanics goes back to the seminal work of Leray for incompressible Navier-Stokes (NS) equations (now called *weak solutions*) in the 1930s. The compressible case was intensively studied much later, and the first global-in-time solutions, but only for data near equilibrium, were constructed by Matsumura and Nishida in the 1980s (*strong solutions*) [37]. In the article [48], the authors established the local-in-time existence of strong solutions in the presence of inflow and outflow of the fluid through the boundary. Moreover, in the same work, they also gave the proof of global in time existence for small data in the absence of the inflow. The theory of *weak solutions* to the compressible Navier-Stokes equation equipped with homogeneous boundary data is starting from the pioneering works of Lions [35, 36] more than a decade later. After that, Feireisl and his coworkers in [23] extended his results to a more general case of pressure laws and were also able to treat the full Navier-Stokes-Fourier (NSF) system [17, 24]. P.-L. Lions proved in [35, 36] the global existence of renormalized weak solution with bounded energy for an isentropic fluid (i.e., $p(\rho) = \rho^\gamma$) with the adiabatic constant $\gamma > 3d/(d+2)$, where d is the space dimension. After that by E. Feireisl *et al.*, the result was generalized to cover the range $\gamma > 3/2$ in dimension 3 and $\gamma > 1$ in dimension 2 in [23].

Let us mention different approach presented in the recent work of Bresch and Jabin [5] where the authors introduce a completely new method to obtain compactness of the density. The well-posedness issues of the compressible Navier-Stokes equations for critical regularity data can be found in the work of R. Danchin [12]. For further references, see [20, 41].

For the study of the problem under the general boundary data, we can start with initial results for some special cases [27, 40]; after that, the existence of weak solutions for isentropic case [8, 9], as well as for the full Navier-Stokes-Fourier system [7, 21], was investigated. Let us also mention the general nonzero inflow-outflow problem; see [32]. For more informations see also [2, 15, 22, 25, 38, 42].

The mathematical investigation of systems describing the motion of a rigid body in a viscous incompressible fluid is nowadays well developed. The proof of existence of weak solutions until a first collision can be found in several papers; see [11, 13, 28, 31, 45]. Later, the possibility of collisions in the case of a weak solution was included; see [14, 18, 44]. For the case where the electromagnetic field is present in fluid, see [3].

Let us also mention results on strong solutions, see, for example, [26, 47, 49].

Concerning the compressible case with rigid structure only, a few results are available. The existence of **strong** solutions in the L^2-framework for small data up to a collision was shown in [4, 43]. The existence of strong solutions in the L^p setting based on $\mathcal{R}$-bounded operators was applied in the barotropic case [30] and in the full system [29].

The existence of a *weak* solution with the Dirichlet boundary conditions, also up to a collision but without smallness assumptions, was shown in [14]. Generalization of this result allowing collisions was given in [16, 46]. The weak-strong uniqueness of a compressible fluid with a rigid body can be found in [33]. The existence of weak solutions in the case of the Navier boundary conditions was established in [39]. Moreover, for interaction of compressible heat-conducting fluid and rigid body, see [6, 29].

2 Relative Energy Inequality

For a weak solution $(\varrho, \mathbf{u})$ and a pair of test functions $(r, \mathbf{U})$ where r is defined on Q_T^F and $\mathbf{U}$ is defined on Q_T, we define the relative energy

$$\begin{aligned}\mathcal{E}([\varrho, \mathbf{u}]|[r, \mathbf{U}])(\tau) = &\int_\Omega \frac{1}{2}\varrho|\mathbf{u} - \mathbf{U}|^2(\tau, \cdot)\,\mathrm{d}x \\ &+ \int_{\Omega_F(t)} \left(P(\varrho) - P'(r)(\varrho - r) - P(r)\right),\end{aligned} \tag{10}$$

where

$$P'(r)r - P(r) = p(r). \tag{11}$$

The first aim is to derive the relative entropy inequality.

Lemma 2.1 *Let* $(\varrho, \mathbf{u}, \eta)$ *be a weak solution to* (1)–(3) *in the sense of Definition 1.1 such that* $\operatorname{dist}(S(t), \partial\Omega) \geq h > 0$, *for all* $t \in [0, T_1]$.

Then, $(\varrho, \mathbf{u}, \eta)$ *satisfies the following relative energy inequality*

$$\mathcal{E}([\varrho, \mathbf{u}]|[r, \mathbf{U}])(\tau) + \int_0^\tau \int_\Omega \mathbb{S}(\mathbf{U} - \mathbf{u}) : \mathbb{D}(\mathbf{U} - \mathbf{u}) \, \mathrm{d}x \, \mathrm{d}t$$
$$\leq \mathcal{E}([\varrho, \mathbf{u}]|[r, \mathbf{U}])(0) + \int_0^\tau \mathcal{R}(\varrho, \mathbf{u}, r, \mathbf{U})(t) \, \mathrm{d}x \, \mathrm{d}t \qquad (12)$$

for a.a. $\tau \in (0, T)$ *and any pair* (r, U) *such that* $r \in C^\infty, r > 0$, $U \in C^\infty$, $\mathbf{U} = \mathbf{u} = \mathbf{u}_B$ *on* $\partial\Omega$, *and* $r = \varrho = \varrho_B$ *on* Γ_{in} *and* $\mathbb{D}(U) = 0$ *in the neighborhood of* $S(t)$.

The remainder $\mathcal{R}$ *has a form*

$$\mathcal{R}(\varrho, \mathbf{u}, r, \mathbf{U})(t) = \int_\Omega \varrho \, (\partial_t \mathbf{U} + \mathbf{u} \cdot \nabla \mathbf{U}) \cdot (\mathbf{U} - \mathbf{u}) \, \mathrm{d}x$$
$$+ \int_{\Omega_F(t)} (\mathbb{S}(\mathbf{U}) : \mathbb{D}(\mathbf{U} - \mathbf{u}) + (p(r) - p(\varrho))\mathrm{div}_x \mathbf{U}) \, \mathrm{d}x$$
$$+ \int_{\Omega_F(t)} \big((r\mathbf{U} - \varrho\mathbf{u}) \cdot \nabla P'(r) + (r - \varrho)\partial_t P'(r)\big) \, \mathrm{d}x$$
$$+ \int_{\partial S(t)} p(r)(\mathbf{u} - \mathbf{U}) \cdot \mathbf{n} \, \mathrm{d}\boldsymbol{\gamma}(\boldsymbol{x}).$$

Remark 2.1 Let us mention that pressure $p(\rho)$ is constant on the rigid part.

Proof We begin by testing the momentum equation (7) with the test function $\boldsymbol{\varphi} = \mathbf{U} - \mathbf{u}_\infty$

$$\int_\Omega (\varrho\mathbf{u} \cdot (\mathbf{U} - \mathbf{u}_\infty))(\tau, \cdot) \, \mathrm{d}\boldsymbol{x} - \int_\Omega (\varrho\mathbf{u} \cdot (\mathbf{U} - \mathbf{u}_\infty))(0, \cdot) \, \mathrm{d}\boldsymbol{x}$$
$$= \int_0^\tau \int_\Omega (\varrho\mathbf{u} \cdot \partial_t \mathbf{U} + \varrho\mathbf{u} \otimes \mathbf{u} : \mathbb{D}(\mathbf{U} - \mathbf{u}_\infty)$$
$$+ p(\varrho)\mathrm{div}_x(\mathbf{U} - \mathbf{u}_\infty) - \mathbb{S}(\mathbf{u}) : \mathbb{D}(\mathbf{U} - \mathbf{u}_\infty))\mathrm{d}\boldsymbol{x} \, \mathrm{d}t. \qquad (13)$$

Next, we test the continuity equation (5) with $\psi = \frac{1}{2}(|\mathbf{U}|^2 - |\mathbf{u}_\infty|^2)$

$$\int_\Omega \frac{1}{2}\varrho \left(|\mathbf{U}|^2 - |\mathbf{u}_\infty|^2\right)(\tau, \cdot) \, \mathrm{d}\boldsymbol{x} - \int_\Omega \frac{1}{2}\varrho \left(|\mathbf{U}|^2 - |\mathbf{u}_\infty|^2\right)(0, \cdot) \, \mathrm{d}\boldsymbol{x}$$
$$= \int_0^\tau \int_\Omega \frac{1}{2}\left(\varrho\partial_t|\mathbf{U}|^2 + \varrho\mathbf{u} \cdot \nabla|\mathbf{U}|^2 - \varrho\mathbf{u} \cdot \nabla|\mathbf{u}_\infty|^2\right) \mathrm{d}\boldsymbol{x} \, \mathrm{d}t, \qquad (14)$$

since $\psi = 0$ on $\partial\Omega$.

The weak solution $(\varrho, \mathbf{u})$ satisfies the energy inequality

$$
\begin{aligned}
&\int_\Omega \Big(\frac{1}{2}\varrho|\mathbf{u}-\mathbf{u}_\infty|^2 + P(\varrho)\Big)(\tau)\ \mathrm{d}x + \int_0^\tau \int_\Omega \mathbb{S}(\mathbf{u}-\mathbf{u}_\infty) : \mathbb{D}(\mathbf{u}-\mathbf{u}_\infty)\ \mathrm{d}x\,\mathrm{d}t \\
&\le \int_\Omega \Big(\frac{1}{2}\frac{|(\varrho\mathbf{u})_0 - \varrho_0\mathbf{u}_\infty|^2}{\varrho_0} + P(\varrho_0)\Big)\ \mathrm{d}x - \int_0^\tau \int_\Omega \varrho\mathbf{u}\cdot\nabla\mathbf{u}_\infty\cdot(\mathbf{u}-\mathbf{u}_\infty)\ \mathrm{d}x\,\mathrm{d}t \\
&\qquad - \int_0^\tau \int_\Omega p(\varrho)\mathrm{div}_x\,\mathbf{u}_\infty\ \mathrm{d}x\,\mathrm{d}t - \int_0^\tau \int_\Omega \mathbb{S}(\mathbf{u}_\infty) : \mathbb{D}(\mathbf{u}-\mathbf{u}_\infty)\ \mathrm{d}x\,\mathrm{d}t \\
&\qquad\qquad - \int_0^\tau \int_{\Gamma_{in}} P(\varrho_B)\mathbf{u}_B\cdot\mathbf{n}\,\boldsymbol{\gamma}(\boldsymbol{x})\,\mathrm{d}t. \qquad (15)
\end{aligned}
$$

Adding (15) and (14) and subtracting (13), we obtain

$$
\begin{aligned}
&\int_\Omega \Big(\frac{1}{2}\varrho|\mathbf{U}-\mathbf{u}|^2 + P(\varrho)\Big)(\tau)\ \mathrm{d}x + \int_0^\tau \int_\Omega \mathbb{S}(\mathbf{U}-\mathbf{u}) : \mathbb{D}(\mathbf{U}-\mathbf{u})\ \mathrm{d}x\,\mathrm{d}t \\
&\le \int_\Omega \Big(\frac{1}{2}\frac{|(\varrho\mathbf{u})_0 - \varrho_0\mathbf{U}(0,\cdot)|^2}{\varrho_0} + P(\varrho_0)\Big)\ \mathrm{d}x + \int_0^\tau \int_\Omega \varrho\,(\partial_t\mathbf{U} + \mathbf{u}\cdot\nabla\mathbf{U})\cdot(\mathbf{U}-\mathbf{u})\ \mathrm{d}x\,\mathrm{d}t \\
&\qquad - \int_0^\tau \int_\Omega p(\varrho)\mathrm{div}_x\mathbf{U}\ \mathrm{d}x\,\mathrm{d}t + \int_0^\tau \int_\Omega \mathbb{S}(\mathbf{U}) : \mathbb{D}(\mathbf{U}-\mathbf{u})\ \mathrm{d}x\,\mathrm{d}t \\
&\qquad\qquad - \int_0^\tau \int_{\Gamma_{in}} P(\varrho_B)\mathbf{u}_B\cdot\mathbf{n}\,\boldsymbol{\gamma}(\boldsymbol{x})\,\mathrm{d}t, \qquad (16)
\end{aligned}
$$

where we have simplified terms as follows:

$$
\frac{1}{2}\varrho|\mathbf{u}-\mathbf{u}_\infty|^2 + \frac{1}{2}\varrho\left(|\mathbf{U}|^2 - |\mathbf{u}_\infty|^2\right) - \varrho\mathbf{u}\cdot(\mathbf{U}-\mathbf{u}_\infty) = \frac{1}{2}\varrho|\mathbf{U}-\mathbf{u}|^2,
$$

$$
\begin{aligned}
&\frac{1}{2}\frac{|(\varrho\mathbf{u})_0 - \varrho_0\mathbf{u}_\infty|^2}{\varrho_0} + \frac{1}{2}\varrho\left(|\mathbf{U}|^2 - |\mathbf{u}_\infty|^2\right)(0,\cdot) - (\varrho\mathbf{u})_0\cdot(\mathbf{U}-\mathbf{u}_\infty)(0,\cdot) \\
&\qquad = \frac{1}{2}\frac{|(\varrho\mathbf{u})_0 - \varrho_0\mathbf{U}(0,\cdot)|^2}{\varrho_0},
\end{aligned}
$$

$$
\begin{aligned}
&\mathbb{S}(\mathbf{u}-\mathbf{u}_\infty) : \mathbb{D}(\mathbf{u}-\mathbf{u}_\infty) + \mathbb{S}(\mathbf{u}_\infty) : \mathbb{D}(\mathbf{u}-\mathbf{u}_\infty) - \mathbb{S}(\mathbf{u}) : \mathbb{D}(\mathbf{U}-\mathbf{u}_\infty) \\
&\qquad = \mathbb{S}(\mathbf{u}-\mathbf{U}) : \mathbb{D}(\mathbf{u}-\mathbf{U}) + \mathbb{S}(\mathbf{U}) : \mathbb{D}(\mathbf{u}-\mathbf{U}),
\end{aligned}
$$

$$
\begin{aligned}
&-\varrho\mathbf{u}\cdot\nabla\mathbf{u}_\infty\cdot(\mathbf{u}-\mathbf{u}_\infty) + \frac{1}{2}(\varrho\partial_t|\mathbf{U}|^2 + \varrho\mathbf{u}\cdot\nabla|\mathbf{U}|^2 - \varrho\mathbf{u}\cdot\nabla|\mathbf{u}_\infty|^2) \\
&\qquad - (\varrho\mathbf{u}\cdot\partial_t\mathbf{U} + \varrho\mathbf{u}\otimes\mathbf{u} : \mathbb{D}(\mathbf{U}-\mathbf{u}_\infty)) = \varrho\,(\partial_t\mathbf{U} + \mathbf{u}\cdot\nabla\mathbf{U})\cdot(\mathbf{U}-\mathbf{u}),
\end{aligned}
$$

$$
-p(\varrho)\mathrm{div}_x\mathbf{u}_\infty - p(\varrho)\mathrm{div}_x(\mathbf{U}-\mathbf{u}_\infty) = -p(\varrho)\mathrm{div}_x\mathbf{U}
$$

Now, we observe the term with the pressure potential. On a solid domain, we have

$$\int_{S(t)} P(\varrho)(\tau)\ \mathrm{d}x = \int_{S_0} P(\varrho_0)(\tau)\ \mathrm{d}x. \tag{17}$$

On the fluid part, we proceed by multiplying the continuity equation $(1)_1$ by $\psi = P'(r)$ and integrate over Q_F. By using the Reynolds transport theorem, we obtain

$$\begin{aligned}&\int_{\Omega_F(t)} (\varrho P'(r))(\tau)\,\mathrm{d}x - \int_{\Omega_F(0)} \varrho_0 P'(r)(0)\,\mathrm{d}x\\ &= \int_0^\tau \int_{\Omega_F(t)} \big(\varrho \partial_t P'(r) + \varrho\mathbf{u}\cdot\nabla P'(r)\big)\,\mathrm{d}x\,\mathrm{d}t - \int_0^\tau \int_{\Gamma_{in}} \varrho_B \mathbf{u}_B\cdot\mathbf{n}P'(\varrho_B)\ .\mathrm{d}\boldsymbol{\gamma}(\boldsymbol{x})\,\mathrm{d}t\end{aligned} \tag{18}$$

The Reynolds transport theorem also gives us

$$\begin{aligned}&\int_{\Omega_F(t)} p(r)(\tau)\,\mathrm{d}x - \int_{\Omega_F(0)} p(r)(0)\,\mathrm{d}x\\ &\qquad = \int_0^\tau \int_{\Omega_F(t)} \partial_t p(r)\,\mathrm{d}x\,\mathrm{d}t + \int_0^\tau \int_{\partial S(t)} p(r)\mathbf{u}\cdot\mathbf{n}\,\mathrm{d}\boldsymbol{\gamma}(\boldsymbol{x})\,\mathrm{d}t\end{aligned} \tag{19}$$

Since

$$P'(r)r - P(r) = p(r),$$

we have

$$\mathrm{div}_x(p(r)\mathbf{U}) = \nabla P'(r)r\cdot\mathbf{U} + p(r)\mathrm{div}_x\mathbf{U},$$

and therefore

$$\begin{aligned}&\int_0^\tau \int_{\Omega_F(t)} \big(r\mathbf{U}\cdot\nabla P'(r) + p(r)\mathrm{div}_x\mathbf{U}\big)\,\mathrm{d}x\,\mathrm{d}t\\ &\qquad = \int_0^\tau \int_{\partial\Omega_F(t)} p(r)\mathbf{U}\cdot\mathbf{n}\,\mathrm{d}\boldsymbol{\gamma}(\boldsymbol{x})\,\mathrm{d}t\\ &\qquad = \int_0^\tau \int_{\partial S(t)} p(r)\mathbf{U}\cdot\mathbf{n}\,\mathrm{d}\boldsymbol{\gamma}(\boldsymbol{x})\,\mathrm{d}t + \int_0^\tau \int_{\Gamma_{in}} p(\varrho_B)\mathbf{u}_B\cdot\mathbf{n}\,\mathrm{d}\boldsymbol{\gamma}(\boldsymbol{x})\,\mathrm{d}t.\end{aligned} \tag{20}$$

By adding (20) and (19) and subtracting (18), we obtain

$$\int_{\Omega_F(t)} \big(-P'(r)(\varrho - r) - P(r)\big)(\tau)\,\mathrm{d}x = \int_{\Omega_F(0)} \big(-P'(r)(\varrho_0 - r) - P(r)\big)(0)\,\mathrm{d}x$$
$$+ \int_0^\tau \int_{\Omega_F(t)} \big((r\mathbf{U} - \varrho\mathbf{u}) \cdot \nabla P'(r) + p(r)\mathrm{div}_x\mathbf{U} + (r - \varrho)\partial_t P'(r)\big)\,\mathrm{d}x\,\mathrm{d}t$$
$$- \int_0^\tau \int_{\partial S(t)} p(r)(\mathbf{U} - \mathbf{u}) \cdot \mathbf{n}\,\mathrm{d}\boldsymbol{\gamma}(\boldsymbol{x})\,\mathrm{d}t + \int_0^\tau \int_{\Gamma_{in}} P(\varrho_B)\mathbf{u}_B \cdot \mathbf{n}\,\mathrm{d}\boldsymbol{\gamma}(\boldsymbol{x})\,\mathrm{d}t. \tag{21}$$

This follows since

$$r\mathbf{U} \cdot \nabla P'(r) + p(r)\mathrm{div}_x\mathbf{U} + \partial_t p(r) - \big(\varrho\partial_t P'(r) + \varrho\mathbf{u} \cdot \nabla P'(r)\big)$$
$$= (r\mathbf{U} - \varrho\mathbf{u}) \cdot \nabla P'(r) + p(r)\mathrm{div}_x\mathbf{U} + (r - \varrho)\partial_t P'(r),$$

$$p(r) - \varrho P'(r) = P'(r)r - P(r) - \varrho P'(r) = -P'(r)(\varrho - r) - P(r),$$

$$p(\varrho_B) - \varrho_B P'(\varrho_B) = P'(\varrho_B)\varrho_B - P(\varrho_B) - \varrho_B P'(\varrho_B) = -P(\varrho_B).$$

Finally, by adding (16) and (21), we obtain

$$\int_\Omega \frac{1}{2}\left(\varrho|\mathbf{U} - \mathbf{u}|^2\right)(\tau)\,\mathrm{d}x + \int_{\Omega_F(\tau)} \big(P(\varrho) - P'(r)(\varrho - r) - P(r)\big)(\tau)\,\mathrm{d}x$$
$$+ \int_0^\tau \int_\Omega \mathbb{S}(\mathbf{U} - \mathbf{u}) : \mathbb{D}(\mathbf{U} - \mathbf{u})\,\mathrm{d}x\,\mathrm{d}t$$
$$\leq \int_\Omega \frac{1}{2}\frac{|(\varrho\mathbf{u})_0 - \varrho_0\mathbf{U}(0,\cdot)|^2}{\varrho_0}\,\mathrm{d}x + \int_{\Omega_F(0)} \big(P(\varrho_0) - P'(r)(\varrho_0 - r) - P(r)\big)(0)\,\mathrm{d}x$$
$$+ \int_0^\tau \int_\Omega \varrho\,(\partial_t\mathbf{U} + \mathbf{u} \cdot \nabla\mathbf{U}) \cdot (\mathbf{U} - \mathbf{u})\,\mathrm{d}x\,\mathrm{d}t + \int_0^\tau \int_\Omega \mathbb{S}(\mathbf{U}) : \mathbb{D}(\mathbf{U} - \mathbf{u})\,\mathrm{d}x\,\mathrm{d}t$$
$$+ \int_0^\tau \int_\Omega (p(r) - p(\varrho))\mathrm{div}_x\mathbf{U}\,\mathrm{d}x\,\mathrm{d}t$$
$$+ \int_0^\tau \int_{\Omega_F(t)} \big((r\mathbf{U} - \varrho\mathbf{u}) \cdot \nabla P'(r) + (r - \varrho)\partial_t P'(r)\big)\,\mathrm{d}x\,\mathrm{d}t$$
$$- \int_0^\tau \int_{\partial S(t)} p(r)(\mathbf{U} - \mathbf{u}) \cdot \mathbf{n}\,\mathrm{d}\boldsymbol{\gamma}(\boldsymbol{x})\,\mathrm{d}t.$$

□

3 Change of Coordinates

We cannot as in the fixed domain compare weak and strong solutions directly since the domain of weak solutions does not coincide with the domain of a strong solution, and we have to transform the strong solution to the physical domain of weak solution. Firstly, such approach was done in the incompressible framework by Chemetov et al. [10] and later in compressible framework by Kreml et al. [33].

Let $(\varrho_1, \mathbf{u}_1, \eta_1)$ be a weak solution to (1)–(3) and $(\varrho_2, \mathbf{u}_2, \mathbf{q}_2, \boldsymbol{\omega}_2)$ be a strong solution with the regularity $\mathcal{W}$.

We denote by

- $S_1(t)$ and $\Omega^1_F(t)$ the domains occupied by solid and fluid at time t for the weak solution
- $S_2(t)$ and $\Omega^2_F(t)$ fluid and solid domains for the strong solution
- $\mathbf{a}_2 = \frac{\mathrm{d}}{\mathrm{d}t}\mathbf{q}_2$

We begin by transforming the strong solution into the domain of the weak solution. To this end, we introduce cut-off functions with compact support, defined as

$$\zeta_i(t, \boldsymbol{x}) = \begin{cases} 1, \text{ in a neighborhood of } S_i(t), \\ 0, \text{ in a neighborhood of } \partial\Omega, \end{cases}$$

for $i = 1, 2$, ensuring that they localize the motion near solid body while vanishing near the boundary of the domain.

We extend the rigid body velocity $\mathbf{u}^i_S$ to the whole $(0, T) \times \mathbb{R}^3$ by

$$\mathbf{u}^i_S(t, \boldsymbol{x}) = \mathbf{a}_i(t) + \boldsymbol{\omega}_i(t) \times (\boldsymbol{x} - \mathbf{q}_i(t)), \quad \text{for } \boldsymbol{x} \in \mathbb{R}^3,$$

and set

$$\Lambda_i(t, \boldsymbol{x}) = \zeta_i(t, \boldsymbol{x})\mathbf{u}^i_S(t, \boldsymbol{x}).$$

By construction, Λ_i coincides with a rigid motion in a neighborhood of $S_i(t)$ and vanishes near the boundary $\partial\Omega$.

We then define the transformation maps $\mathbf{X}_i$ as solutions to the following system of ordinary differential equations:

$$\frac{\mathrm{d}}{\mathrm{d}t}\mathbf{X}_i(t, \mathbf{y}) = \Lambda_i(t, \mathbf{X}_i(t, \mathbf{y})) \quad \forall \mathbf{y} \in \mathbb{R}^3, \ t \in (0, T),$$

$$\mathbf{X}_i(0, \mathbf{y}) = \mathbf{y}$$

Next, we introduce $\mathbf{Y}_i(t,\cdot) = \mathbf{X}_i^{-1}(t,\cdot)$ and define mappings

$$\widetilde{\mathbf{X}}_1(t,\boldsymbol{x}) = \mathbf{X}_1(t,\mathbf{Y}_2(t,\boldsymbol{x})), \tag{22}$$

$$\widetilde{\mathbf{X}}_2(t,\boldsymbol{x}) = \mathbf{X}_2(t,\mathbf{Y}_1(t,\boldsymbol{x})). \tag{23}$$

Near the rigid bodies, we have

$$\widetilde{\mathbf{X}}_1(t,\boldsymbol{x}) = \mathbf{q}_1(t) + \mathbb{O}_1(t)\mathbb{O}_2^T(t)(\boldsymbol{x} - \mathbf{q}_2(t)) \quad \text{in the neighborhood of } S_2(t) \text{ and} \tag{24}$$

$$\widetilde{\mathbf{X}}_2(t,\boldsymbol{x}) = \mathbf{q}_2(t) + \mathbb{O}_2(t)\mathbb{O}_1^T(t)(\boldsymbol{x} - \mathbf{q}_1(t)) \quad \text{in the neighborhood of } S_1(t). \tag{25}$$

For simplicity of notation, we denote

$$\widetilde{\mathbb{O}}(t) = \mathbb{O}_2(t)\mathbb{O}_1^T(t). \tag{26}$$

We are now in a position to define the transformed strong solution $(r^s, \mathbf{U}^s)$ by

$$r^s(t,\boldsymbol{x}) = \varrho_2(t,\widetilde{\mathbf{X}}_2(t,\boldsymbol{x})) \qquad \text{for all } (t,\boldsymbol{x}) \in Q_F^1, \tag{27}$$

$$\mathbf{U}^s(t,\boldsymbol{x}) = \nabla\widetilde{\mathbf{X}}_1(t,\widetilde{\mathbf{X}}_2(t,\boldsymbol{x}))\mathbf{u}_2(t,\widetilde{\mathbf{X}}_2(t,\boldsymbol{x})) \quad \text{for all } (t,\boldsymbol{x}) \in Q_T. \tag{28}$$

In particular, for $\boldsymbol{x} \in S_1(t)$, the transformed velocity field $\mathbf{U}^s$ satisfies

$$\mathbf{U}^s(t,\boldsymbol{x}) = \mathbf{a}^s(t) + \boldsymbol{\omega}^s(t) \times (\boldsymbol{x} - \mathbf{q}_1(t)) \tag{29}$$

where the transformed translational and angular velocities are given by

$$\mathbf{a}^s(t) = \widetilde{\mathbb{O}}^T(t)\mathbf{a}_2(t), \qquad \boldsymbol{\omega}^s(t) = \widetilde{\mathbb{O}}^T(t)\boldsymbol{\omega}_2(t). \tag{30}$$

Moreover, in a neighborhood of the boundary $\partial\Omega$, we have

$$\mathbf{U}^s(t,\boldsymbol{x}) = \mathbf{u}_2(t,\boldsymbol{x}) \quad \text{in the neighborhood of } \partial\Omega, \tag{31}$$

$$r^s(t,\boldsymbol{x}) = \varrho_2(t,\boldsymbol{x}) \quad \text{in the neighborhood of } \partial\Omega. \tag{32}$$

To compare the weak and transformed strong solutions, we need precise estimates of the transformation mappings $\widetilde{\mathbf{X}}_i$. The following lemma, proven in Kreml et al. [33], provides bounds on these mappings and their derivatives, which will be essential for controlling the differences between the solutions.

Lemma 3.1

(i) It holds

$$\widetilde{\mathbb{O}}^T(t)\frac{\mathrm{d}\widetilde{\mathbb{O}}}{\mathrm{d}t}(t)\boldsymbol{x} = (\boldsymbol{\omega}^s - \boldsymbol{\omega}_1)(t) \times \boldsymbol{x}. \tag{33}$$

(ii) The following estimates hold for $\boldsymbol{x} \in \mathrm{d}S_1(t)$, $t \in (0, T)$

$$|\widetilde{\mathbf{X}}_2(t, \boldsymbol{x}) - \boldsymbol{x}| \le C\left(\|\mathbf{a}_1 - \mathbf{a}^s\|_{L^2(0,t)} + \|\boldsymbol{\omega}_1 - \boldsymbol{\omega}^s\|_{L^2(0,t)}\right), \tag{34}$$

$$|\partial_t\widetilde{\mathbf{X}}_2(t, \boldsymbol{x})| \le C\left(|\mathbf{a}_1 - \mathbf{a}^s|(t) + |\boldsymbol{\omega}_1 - \boldsymbol{\omega}^s|(t)\right). \tag{35}$$

(iii) The following estimates hold for $t \in (0, T)$

$$\|\widetilde{\mathbf{X}}_2(t, \cdot) - \mathrm{id}\|_{W^{3,\infty}(\Omega_F^1(t))} \le C\left(\|\mathbf{a}_1 - \mathbf{a}^s\|_{L^2(0,t)} + \|\boldsymbol{\omega}_1 - \boldsymbol{\omega}^s\|_{L^2(0,t)}\right), \tag{36}$$

$$\|\partial_t\widetilde{\mathbf{X}}_2(t, \cdot)\|_{W^{1,\infty}(\Omega_F^1(t))} \le C\left(|\mathbf{a}_1 - \mathbf{a}^s|(t) + |\boldsymbol{\omega}_1 - \boldsymbol{\omega}^s|(t)\right). \tag{37}$$

The transformed strong solution satisfies the following system of equations in the fluid part of the domain Ω

$$\begin{cases} \partial_t r^s + \mathrm{div}_x(r^s\mathbf{U}^s) = G(r^s, \mathbf{U}^s) & \text{in} \quad Q_F^1, \\ \partial_t(r^s\mathbf{U}^s) + \mathrm{div}_x(r^s\mathbf{U}^s \otimes \mathbf{U}^s) - \mathrm{div}_x\mathbb{S}(\nabla\mathbf{U}^s) + \nabla p(r^s) = \mathbf{F}(r^s, \mathbf{U}^s) & \text{in} \quad Q_F^1, \\ \mathbf{U}^s = 0 & \text{on } (0, T) \times \partial\Omega, \\ r^s(0, \cdot) = \varrho_0 & \text{in} \quad \Omega \setminus S_0, \\ \mathbf{U}^s(0, \cdot) = \mathbf{u}_0 & \text{in} \quad \Omega \setminus S_0. \end{cases} \tag{38}$$

To express the terms on the right-hand sides of the transformed continuity and momentum equations, we introduce the following notation, employing the summation convention:

$$\mathbb{H} = (\nabla\widetilde{\mathbf{X}}_2)^{-1} = \nabla\widetilde{\mathbf{X}}_1, \tag{39}$$

$$\mathbb{G} = \mathbb{H}\mathbb{H}^T, \tag{40}$$

$$\Gamma^i_{\alpha\beta} = \partial_l(\widetilde{\mathbf{X}}_1)_i\partial_{\alpha\beta}(\widetilde{\mathbf{X}}_2)_l \tag{41}$$

Using these expressions, we obtain the following:

$$G(r^s, \mathbf{U}^s) = \mathbb{H}_{jk}\partial_t(\widetilde{\mathbf{X}}_2)_k\partial_j r^s, \tag{42}$$

$$\mathbf{F}_i(r^s, \mathbf{U}^s) = -r^s\mathbb{H}_{ij}\partial_t\partial_\beta(\widetilde{\mathbf{X}}_2)_j\mathbf{U}^s_\beta + r^s\Gamma^i_{\alpha\beta}\mathbb{H}_{\alpha j}(\partial_t\widetilde{\mathbf{X}}_2)_j\mathbf{U}^s_\beta + \mathbb{H}_{\alpha j}(\partial_t\widetilde{\mathbf{X}}_2)_j\partial_\alpha(r^s\mathbf{U}^s_i) \tag{43}$$

$$- r^s \Gamma^i_{\alpha\beta} \mathbf{U}^s_\alpha \mathbf{U}^s_\beta + (\mu + \lambda)(\mathbb{G} - \mathrm{i})_{i\alpha} \partial_\alpha \mathrm{div}_x \mathbf{U}^s - (\mathbb{G} - \mathrm{i})_{i\alpha} \partial_\alpha p(r^s)$$
$$+ \mu((\mathbb{G} - \mathbb{I})_{\alpha\beta} \partial_{\alpha\beta} \mathbf{U}^s_i + \Delta(\widetilde{\mathbf{X}}_1)_\beta \partial_\beta \mathbf{U}^s_i + 2\Gamma^i_{\alpha\gamma} \mathbb{G}_{\alpha\beta} \partial_\beta \mathbf{U}^s_\gamma)$$
$$+ \mu(\Gamma^i_{\alpha\gamma} \Delta(\widetilde{\mathbf{X}}_1)_\alpha \mathbf{U}^s_\gamma + \mathbb{H}_{ij} \mathbb{G}_{\alpha\beta} \partial_{\alpha\beta\gamma} (\widetilde{\mathbf{X}}_2)_j \mathbf{U}^s_\gamma)$$

When we express the equations for the rigid body, we get

$$m \frac{\mathrm{d}\mathbf{a}^s}{\mathrm{d}t} = -m(\boldsymbol{\omega}^s - \boldsymbol{\omega}_1) \times \mathbf{a}^s - \int_{\partial \mathcal{B}_1(t)} \left(\mathbb{S}(\nabla \mathbf{U}^s) - p(r^s)\mathbb{I} \right) \mathbf{n} \, \mathrm{d}S \quad \text{in } (0, T), \tag{44}$$

$$\mathbb{J}_1 \frac{\mathrm{d}\boldsymbol{\omega}^s}{\mathrm{d}t} = \mathbb{J}_1 \boldsymbol{\omega}^s \times \boldsymbol{\omega}^s - \mathbb{J}_1 \left((\boldsymbol{\omega}^s - \boldsymbol{\omega}_1) \times \boldsymbol{\omega}^s \right) \tag{45}$$
$$- \int_{\partial \mathcal{B}_1(t)} (\boldsymbol{x} - \mathbf{q}_1(t)) \times \left(\mathbb{S}(\nabla \mathbf{U}^s) - p(r^s)\mathbb{I} \right) \mathbf{n} \, \mathrm{d}S \quad \text{in } (0, T),$$

$$\mathbf{a}^s(0) = \mathbf{a}_0, \qquad \boldsymbol{\omega}^s(0) = \boldsymbol{\omega}_0, \tag{46}$$

where we used $\mathbb{J}_1 = \widetilde{\mathbb{O}}^T \mathbb{J}_2 \widetilde{\mathbb{O}}$.

4 Weak-Strong Uniqueness

The aim of this section is to show the weak-strong uniqueness result stated in Theorem 1.2. To do it, we would like to apply the relative energy inequality (12) with $r = r^s$ and $\mathbf{U} = \mathbf{U}^s$. We follow the result of Kreml et al. [33] and Chemetov et al. [10].

Since in the energy inequality $\mathbf{U}$ must satisfy

$$\mathbb{D}(\mathbf{U}) = 0 \text{ in a neighborhood of } S_1(t) \tag{47}$$

and $\mathbf{U}^s$ is a rigid motion only on $S_1(t)$, we need to introduce some approximation of the test function.

For given $\varepsilon > 0$, we introduce a smooth cut-off function $\xi^\varepsilon : Q_T \to Q_T$ such that $\xi^\varepsilon(t, \boldsymbol{x}) \in [0, 1]$ and

$$\xi^\varepsilon(t, \boldsymbol{x}) = \begin{cases} 1, & \boldsymbol{x} \in [S_1(t)]_\varepsilon, \\ 0, & \boldsymbol{x} \in \Omega \setminus [S_1(t)]_{2\varepsilon}. \end{cases} \tag{48}$$

We define

$$\mathbf{U}^s_\varepsilon = \xi_\varepsilon \mathbf{U}^s_S + (1 - \xi_\varepsilon) \mathbf{U}^s_F, \tag{49}$$

and we have

$$\mathbb{D}(\mathbf{U}^s_\varepsilon) = 0 \text{ on } [S_1(t)]_\varepsilon, \quad \mathbf{U}^s_\varepsilon = \mathbf{U}^s \text{ on } S_1(t) \cup (\Omega \setminus [S_1(t)]_{2\varepsilon}), \tag{50}$$

where we denote by X_ε the ε-neighborhood of the set X.

We now apply the relative energy inequality (12) with $(\varrho, \mathbf{u}) = (\varrho_1, \mathbf{u}_1)$ and $(r, \mathbf{U}) = (r^s, \mathbf{U}^s_\varepsilon)$. In what follows, we will use $(\varrho, \mathbf{u})$ instead of $(\varrho_1, \mathbf{u}_1)$, $(r, \mathbf{U}_\varepsilon)$ instead of $(r^s, \mathbf{U}^s_\varepsilon)$, and $S(t)$, $\Omega_F(t)$ instead of $S_1(t)$, $\Omega^1_F(t)$. We further introduce the sets

$$\Gamma^\varepsilon(t) = [S(t)]_{2\varepsilon} \setminus S(t), \quad \Omega^\varepsilon_F(t) = \Omega \setminus [S(t)]_{2\varepsilon}. \tag{51}$$

Let us examine the first term of the relative energy inequality (12). Noting that

$$\int_\Omega \frac{1}{2}\varrho \left|\mathbf{u} - \mathbf{U}_\varepsilon\right|^2 (\tau, \cdot)\,\mathrm{d}x$$
$$= \int_\Omega \frac{1}{2}\varrho \left(|\mathbf{u} - \mathbf{U}|^2 + 2(\mathbf{u} - \mathbf{U}) \cdot (\mathbf{U} - \mathbf{U}_\varepsilon) + |\mathbf{U} - \mathbf{U}_\varepsilon|^2\right) (\tau, \cdot)\,\mathrm{d}x$$

and observing that $\mathbf{U} \neq \mathbf{U}_\varepsilon$ only in the set Γ^ε, it is easy to verify that

$$\lim_{\varepsilon \to 0} \int_{\Gamma^\varepsilon(\tau)} \left(\varrho(\mathbf{u} - \mathbf{U}) \cdot (\mathbf{U} - \mathbf{U}_\varepsilon) + \frac{1}{2}\varrho \left|\mathbf{U} - \mathbf{U}_\varepsilon\right|^2\right) (\tau, \cdot)\,\mathrm{d}x = 0.$$

This implies that

$$\lim_{\varepsilon \to 0} \int_\Omega \frac{1}{2}\varrho \left|\mathbf{u} - \mathbf{U}_\varepsilon\right|^2 (\tau, \cdot)\,\mathrm{d}x = \int_\Omega \frac{1}{2}\varrho \left|\mathbf{u} - \mathbf{U}\right|^2 (\tau, \cdot)\,\mathrm{d}x. \tag{52}$$

Therefore, after taking the limit $\varepsilon \to 0$, we proceed to estimate the left-hand side of the relative energy inequality (12), producing the following lower bound:

$$\mathcal{E}([\varrho, \mathbf{u}]|[r, \mathbf{U}])(\tau) + \int_0^\tau \int_\Omega \mathbb{S}(\mathbf{U} - \mathbf{u}) : \mathbb{D}(\mathbf{U} - \mathbf{u})\,\mathrm{d}x\,\mathrm{d}t$$
$$\geq \mathcal{E}([\varrho, \mathbf{u}]|[r, \mathbf{U}])(\tau) + c \left\|\mathbb{S}(\mathbf{U} - \mathbf{u})\right\|^2_{L^2(Q_F)}$$

For the right-hand side of the relative energy inequality, it is not difficult to see that

$$\lim_{\varepsilon \to 0} \mathcal{E}\Big([\varrho_0, \mathbf{u}_0]|[r(0, \cdot), \mathbf{U}_\varepsilon(0, \cdot)]\Big) = 0$$

and

$$\lim_{\varepsilon\to 0}\int_0^\tau \mathcal{R}(\varrho,\mathbf{u},r,\mathbf{U}_\varepsilon)\,\mathrm{d}t = \int_0^\tau \mathcal{R}(\varrho,\mathbf{u},r,\mathbf{U})\,\mathrm{d}t.$$

Now, passing to the limit $\varepsilon \to 0$ in (12), we obtain

$$\mathcal{E}\Big([\varrho,\mathbf{u}]|[r,\mathbf{U}]\Big)(\tau) + c\,\|\mathbb{S}(\mathbf{U}-\mathbf{u})\|^2_{L^2(Q_F)} \leq \int_0^\tau \mathcal{R}(\varrho,\mathbf{u},r,\mathbf{U})(t)\,\mathrm{d}t. \tag{53}$$

Therefore, it remains to estimate the remainder term

$$\begin{aligned}\mathcal{R}(\varrho,\mathbf{u},r,\mathbf{U})(t) &= \int_{\Omega_F(t)} \varrho(\partial_t\mathbf{U}+\mathbf{u}\cdot\nabla\mathbf{U})\cdot(\mathbf{U}-\mathbf{u})\,\mathrm{d}x \\ &+ \int_{S(t)} \varrho(\partial_t\mathbf{U}+\mathbf{u}\cdot\nabla\mathbf{U})\cdot(\mathbf{U}-\mathbf{u})\,\mathrm{d}x \\ &+ \int_{\Omega_F(t)} \Big(\mathbb{S}(\mathbf{U}):(\nabla\mathbf{U}-\nabla\mathbf{u}) + \mathrm{div}_x\mathbf{U}(p(r)-p(\varrho)) + (r-\varrho)\partial_t P'(r)\Big)\,\mathrm{d}x \\ &+ \int_{\Omega_F(t)} (r\mathbf{U}-\varrho\mathbf{u})\cdot\nabla P'(r)\,\mathrm{d}x + \int_{\partial S(t)} p(r)(\mathbf{u}-\mathbf{U})\cdot\mathbf{n}\,\mathrm{d}S \\ &= \mathcal{I}_1^F(t) + \mathcal{I}_1^B(t) + \sum_{i=2}^{6}\mathcal{I}_i(t).\end{aligned} \tag{54}$$

We start with the term $\mathcal{I}_1^F$. Since

$$\mathbf{u}\cdot\nabla_x\mathbf{U}\cdot(\mathbf{U}-\mathbf{u}) = \mathbf{U}\cdot\nabla_x\mathbf{U}\cdot(\mathbf{U}-\mathbf{u}) + (\mathbf{u}-\mathbf{U})\cdot\nabla_x\mathbf{U}\cdot(\mathbf{U}-\mathbf{u}) \tag{55}$$

and $\nabla_x\mathbf{U}$ is bounded, we get

$$\left|\int_0^\tau\int_{\Omega_F(t)} \varrho(\mathbf{u}-\mathbf{U})\cdot\nabla_x\mathbf{U}\cdot(\mathbf{U}-\mathbf{u})\,\mathrm{d}x\,\mathrm{d}t\right| \leq \int_0^\tau h_1(t)\mathcal{E}([\varrho,\mathbf{u}]|[r,\mathbf{U}])\,\mathrm{d}t \tag{56}$$

for some nonnegative $h_1 \in L^1(0,T)$. Likewise, throughout the rest of this section, $h_k(t)$ denotes a nonnegative integrable function.

By combining the equations $(38)_1$ and $(38)_2$ for $(r,\mathbf{U})$, we obtain

$$\partial_t\mathbf{U}+\mathbf{U}\cdot\nabla\mathbf{U} = \frac{1}{r}\left(\mathrm{div}_x\mathbb{S}(\mathbf{U}) - \nabla p(r) + \mathbf{F}(r,\mathbf{U}) - G(r,\mathbf{U})\mathbf{U}\right),$$

and by (55) and (56), it remains to estimate

$$
\begin{aligned}
&\int_0^\tau \int_{\Omega_F(t)} \varrho(\partial_t \mathbf{U} + \mathbf{U} \cdot \nabla \mathbf{U}) \cdot (\mathbf{U} - \mathbf{u})\,\mathrm{d}x\,\mathrm{d}t \\
&= \int_0^\tau \int_{\Omega_F(t)} \frac{\varrho}{r} (\operatorname{div}_x \mathbb{S}(\mathbf{U}) - \nabla p(r) + \mathbf{F}(r, \mathbf{U}) - G(r, \mathbf{U})\mathbf{U}) \cdot (\mathbf{U} - \mathbf{u})\,\mathrm{d}x\,\mathrm{d}t \\
&= \sum_{i=1}^{4} \mathcal{J}_i.
\end{aligned} \tag{57}
$$

First, we split $\mathcal{J}_1$ into two parts using

$$
\frac{\varrho}{r} = 1 + \frac{\varrho - r}{r}.
$$

Then, since $\mathbf{U} - \mathbf{u} = \mathbf{u}_B - \mathbf{u}_B = 0$ on $\partial\Omega$, we have

$$
\begin{aligned}
&\int_0^\tau \int_{\Omega_F(t)} \operatorname{div}_x \mathbb{S}(\mathbf{U}) \cdot (\mathbf{U} - \mathbf{u})\,\mathrm{d}x\,\mathrm{d}t + \int_0^\tau \mathcal{I}_2(t)\,\mathrm{d}t \\
&\int_0^\tau \int_{\Omega_F(t)} \operatorname{div}_x \mathbb{S}(\mathbf{U}) \cdot (\mathbf{U} - \mathbf{u}) + \mathbb{S}(\mathbf{U}) : \nabla(\mathbf{U} - \mathbf{u})\,\mathrm{d}x\,\mathrm{d}t \\
&= \int_0^\tau \int_{\Omega_F(t)} \operatorname{div}_x (\mathbb{S}(\mathbf{U})(\mathbf{U} - \mathbf{u}))\,\mathrm{d}x\,\mathrm{d}t = \int_0^\tau \int_{\partial S(t)} \mathbb{S}(\nabla_x \mathbf{U})\mathbf{n} \cdot (\mathbf{U} - \mathbf{u})\,\mathrm{d}S\,\mathrm{d}t \\
&= \mathcal{J}_5.
\end{aligned}
$$

In order to estimate

$$
\int_0^\tau \int_{\Omega_F(t)} \frac{\varrho - r}{r} \operatorname{div}_x \mathbb{S}(\mathbf{U}) \cdot (\mathbf{U} - \mathbf{u})\,\mathrm{d}x\,\mathrm{d}t, \tag{58}
$$

we employ a similar procedure as in [19], with further details provided in [34, Section 6.3]. The integral in (58) is decomposed into three parts, and separate estimates are made for the cases where ϱ is small, comparable to r, or large. Additionally, the estimate in the final case differs depending on whether $\gamma < 2$ or $\gamma \geq 2$. This procedure is also applied later in the proof when estimating the terms involving $\mathbf{F}(r, \mathbf{U})$ and $G(\varrho, \mathbf{U})$, as discussed in [33]. We now proceed to obtain

$$
\begin{aligned}
&\left| \int_0^\tau \int_{\Omega_F(t)} \frac{\varrho - r}{r} \operatorname{div}_x \mathbb{S}(\mathbf{U}) \cdot (\mathbf{U} - \mathbf{u})\,\mathrm{d}x\,\mathrm{d}t \right| \\
&\leq \delta \, \| \mathbb{S}(\mathbf{U} - \mathbf{u}) \|_{L^2(Q_F)}^2 + C(\delta) \int_0^\tau h_2(t) \mathcal{E}([\varrho, \mathbf{u}] | [r, \mathbf{U}])\,\mathrm{d}t.
\end{aligned} \tag{59}
$$

Next, we combine the terms $\mathcal{I}_4$, $\mathcal{I}_5$, and $\mathcal{J}_2$ to obtain

$$\int_0^\tau \int_{\Omega_F(t)} (r-\varrho)\partial_t P'(r) + (r\mathbf{U} - \varrho\mathbf{u}) \cdot \nabla_x P'(r) - \frac{\varrho}{r}\nabla_x p(r) \cdot (\mathbf{U} - \mathbf{u})\, \mathrm{d}x\, \mathrm{d}t \tag{60}$$

$$= \int_0^\tau \int_{\Omega_F(t)} (r-\varrho)\left(\partial_t P'(r) + \mathbf{U} \cdot \nabla P'(r)\right) \mathrm{d}x\, \mathrm{d}t.$$

We apply the continuity equation $(38)_1$, satisfied by the strong solution $(r, \mathbf{U})$, to derive

$$\begin{aligned} \partial_t P'(r) + \mathbf{U} \cdot \nabla_x P'(r) &= P''(r)\left(\partial_t r + \mathbf{U} \cdot \nabla_x r\right) \\ &= P''(r)(G(r, \mathbf{U}) - r\mathrm{div}_x \mathbf{U}) = -p'(r)\mathrm{div}_x \mathbf{U} + P''(r)G(r, \mathbf{U}). \end{aligned} \tag{61}$$

Substituting (61) into (60), we find

$$\begin{aligned} &\mathcal{J}_2 + \int_0^\tau [\mathcal{I}_4(t) + \mathcal{I}_5(t)]\, \mathrm{d}t \\ &= \int_0^\tau \int_{\Omega_F(t)} -p'(r)(r-\varrho)\mathrm{div}_x \mathbf{U} + P''(r)(r-\varrho)G(r, \mathbf{U})\, \mathrm{d}x\, \mathrm{d}t, \end{aligned} \tag{62}$$

and we then combine the first integral in (62) with $\mathcal{I}_3$, which yields

$$\begin{aligned} &\left| \int_0^\tau \int_{\Omega_F(t)} \mathrm{div}_x \mathbf{U}\left(p(\varrho) - p(r) - p'(r)(\varrho - r)\right) \mathrm{d}x\, \mathrm{d}t \right| \\ &\leq \int_0^\tau h_3(t)\mathcal{E}([\varrho, \mathbf{u}] | [r, \mathbf{U}])\, \mathrm{d}t. \end{aligned}$$

By combining the above estimates, we arrive at

$$\begin{aligned} \int_0^\tau \left[\mathcal{I}_1^F(t) + \sum_{i=2}^5 \mathcal{I}_i(t)\right] \mathrm{d}t \leq & \delta \, \|\mathbb{S}(\mathbf{U} - \mathbf{u})\|^2_{L^2(Q_F)} \\ & + C(\delta) \int_0^\tau h_4(t)\mathcal{E}([\varrho, \mathbf{u}] | [r, \mathbf{U}])\, \mathrm{d}t \\ & + \int_0^\tau \int_{\Omega_F(t)} P''(r)(r-\varrho)G(r, \mathbf{U})\, \mathrm{d}x\, \mathrm{d}t \\ & + \sum_{i=3}^5 \mathcal{J}_i. \end{aligned} \tag{63}$$

Now, we aim to combine $\mathcal{I}_1^B$ with the boundary integrals $\mathcal{I}_6$ and $\mathcal{J}_5$. We begin by applying (55) once more to estimate the quadratic term in $(\mathbf{u}-\mathbf{U})$ from $\mathcal{I}_1^B$, using a similar approach to that in (56). This leads to

$$\begin{aligned}\int_0^\tau \mathcal{I}_1^B(t)\,\mathrm{d}t &= \int_0^\tau \int_{S(t)} \varrho(\partial_t \mathbf{U} + \mathbf{U}\cdot\nabla_x \mathbf{U})\cdot(\mathbf{U}-\mathbf{u})\,\mathrm{d}x\,\mathrm{d}t \qquad (64)\\ &= -\int_0^\tau \int_{\partial S(t)} (\mathbb{S}(\mathbf{U}) - p(r)\mathbb{I})\mathbf{n}\cdot(\mathbf{U}-\mathbf{u})\,\mathrm{d}S\,\mathrm{d}t\\ &- \int_0^\tau m((\boldsymbol{\omega}^s - \boldsymbol{\omega}_1)\times \mathbf{a}^s)\cdot(\mathbf{a}^s - \mathbf{a}_1) - \mathbb{J}_1((\boldsymbol{\omega}^s-\boldsymbol{\omega}_1)\times\boldsymbol{\omega}^s)\cdot(\boldsymbol{\omega}^s-\boldsymbol{\omega}_1)\,\mathrm{d}t,\end{aligned}$$

where the second equality is derived by applying the equations satisfied by the transformed rigid velocity, given in (44)–(45), following the procedure outlined in Kreml et al. [33, Appendix]. One can easily verify that

$$\begin{aligned}&\left|\int_0^\tau m((\boldsymbol{\omega}^s - \boldsymbol{\omega}_1)\times \mathbf{a}^s)\cdot(\mathbf{a}^s - \mathbf{a}_1) - \mathbb{J}_1((\boldsymbol{\omega}^s-\boldsymbol{\omega}_1)\times\boldsymbol{\omega}^s)\cdot(\boldsymbol{\omega}^s-\boldsymbol{\omega}_1)\,\mathrm{d}t\right|\\ &\quad\le \int_0^\tau h_4(t)\mathcal{E}([\varrho,\mathbf{u}]|[r,\mathbf{U}])\,\mathrm{d}t,\end{aligned}$$

and thus, (64) implies

$$\left|\int_0^\tau [\mathcal{I}_1^B(t) + \mathcal{I}_6(t)]\,\mathrm{d}t + \mathcal{J}_5\right| \le \int_0^\tau h_4(t)\mathcal{E}([\varrho,\mathbf{u}]|[r,\mathbf{U}])\,\mathrm{d}t. \qquad (65)$$

Finally, we estimate the integrals over the fluid part involving $\mathbf{F}(r,\mathbf{U})$ and $G(r,\mathbf{U})$ from (57) and (63), specifically

$$\begin{aligned}&\int_0^\tau \int_{\Omega_F(t)} \frac{\varrho}{r}\mathbf{F}(r,\mathbf{U})\cdot(\mathbf{U}-\mathbf{u}) - \frac{\varrho}{r}G(r,\mathbf{U})\mathbf{U}\cdot(\mathbf{U}-\mathbf{u})\\ &\quad + P''(r)(r-\varrho)G(r,\mathbf{U})\,\mathrm{d}x\,\mathrm{d}t. \qquad (66)\end{aligned}$$

We divide the integrals into three regions: one where ϱ is close to r, one where ϱ is small, and one where ϱ is large. The regions are defined as follows,

$$\begin{aligned}S_1(t) &= \left\{\boldsymbol{x}\in Q_F:\quad \frac{r_-}{2} < \varrho(t,\boldsymbol{x}) < 2r_+\right\},\\ S_2(t) &= \left\{\boldsymbol{x}\in Q_F:\quad \varrho(t,\boldsymbol{x}) \le \frac{r_-}{2}\right\},\\ S_3(t) &= \{\boldsymbol{x}\in Q_F:\quad \varrho(t,\boldsymbol{x}) \ge 2r_+\},\end{aligned}$$

where $r_- = \inf_{Q_F} r(t, \boldsymbol{x})$ and $r_+ = \sup_{Q_F} r(t, \boldsymbol{x})$. Recall that

$$
\begin{aligned}
P(\varrho) - P'(r)(\varrho - r) - P(r) &\geq c(r)(\varrho - r)^2 \quad \text{for } \frac{r_-}{2} < \varrho < 2r_+, \\
&\geq c(r)(1 + \varrho^\gamma) \quad \text{otherwise.}
\end{aligned}
\tag{67}
$$

We begin with the last term of (66). Using (42), Lemma 3.1, and (67), we estimate

$$
\begin{aligned}
&\left| \int_{S_1(t)} P''(r)(r - \varrho) G(r, \mathbf{U}) \, \mathrm{d}x \right| \leq C \| r - \varrho \|_{L^2(S_1(t))} \| \partial_t \widetilde{\mathbf{X}}_2 \|_{L^\infty(\Omega_F(t))} \\
&\quad \leq C(\| r - \varrho \|^2_{L^2(S_1(t))} + |\mathbf{a}^s - \mathbf{a}_1|^2 + |\boldsymbol{\omega}^s - \boldsymbol{\omega}_1|^2) \leq C\mathcal{E}([\varrho, \mathbf{u}]|[r, \mathbf{U}])(t),
\end{aligned}
\tag{68}
$$

$$
\begin{aligned}
&\left| \int_{S_2(t)} P''(r)(r - \varrho) G(r, \mathbf{U}) \, \mathrm{d}x \right| \leq C \| 1 \|_{L^2(S_2(t))} \| \partial_t \widetilde{\mathbf{X}}_2 \|_{L^\infty(\Omega_F(t))} \\
&\quad \leq C(\| 1 \|^2_{L^2(S_2(t))} + |\mathbf{a}^s - \mathbf{a}_1|^2 + |\boldsymbol{\omega}^s - \boldsymbol{\omega}_1|^2) \leq C\mathcal{E}([\varrho, \mathbf{u}]|[r, \mathbf{U}])(t),
\end{aligned}
\tag{69}
$$

and for large ϱ, we treat two cases: if $\gamma \leq 2$,

$$
\begin{aligned}
&\left| \int_{S_3(t)} P''(r)(r - \varrho) G(r, \mathbf{U}) \, \mathrm{d}x \right| \leq C \| \varrho \|_{L^\gamma(S_3(t))} \| \partial_t \widetilde{\mathbf{X}}_2 \|_{L^\infty(\Omega_F(t))} \\
&\quad \leq C(\| \varrho \|^2_{L^\gamma(S_3(t))} + |\mathbf{a}^s - \mathbf{a}_1|^2 + |\boldsymbol{\omega}^s - \boldsymbol{\omega}_1|^2) \\
&\quad \leq C \left(\mathcal{E}([\varrho, \mathbf{u}]|[r, \mathbf{U}])^{2/\gamma}(t) + \mathcal{E}([\varrho, \mathbf{u}]|[r, \mathbf{U}])(t) \right) \\
&\quad \leq C\mathcal{E}([\varrho, \mathbf{u}]|[r, \mathbf{U}])(t),
\end{aligned}
\tag{70}
$$

and if $\gamma > 2$,

$$
\begin{aligned}
&\left| \int_{S_3(t)} P''(r)(r - \varrho) G(r, \mathbf{U}) \, \mathrm{d}x \right| \leq C \| \varrho^{\gamma/2} \|_{L^2(S_3(t))} \| \partial_t \widetilde{\mathbf{X}}_2 \|_{L^\infty(\Omega_F(t))} \\
&\quad \leq C(\| \varrho \|^\gamma_{L^\gamma(S_3(t))} + |\mathbf{a}^s - \mathbf{a}_1|^2 + |\boldsymbol{\omega}^s - \boldsymbol{\omega}_1|^2) \leq C\mathcal{E}([\varrho, \mathbf{u}]|[r, \mathbf{U}])(t).
\end{aligned}
\tag{71}
$$

Thus, we conclude:

$$
\left| \int_{\Omega_F(t)} P''(r)(r - \varrho) G(r, \mathbf{U}) \, \mathrm{d}x \right| \leq C\mathcal{E}([\varrho, \mathbf{u}]|[r, \mathbf{U}])(t) \tag{72}
$$

Next, we estimate the second term of (66). To do this, we split it into two parts, similar to the approach used for $\mathcal{J}_1$, using the identity

$$
\frac{\varrho}{r} = 1 + \frac{\varrho - r}{r}.
$$

Then, we estimate

$$\begin{aligned}\left|\int_{S_1(t)} G(r,\mathbf{U})\mathbf{U}\cdot(\mathbf{U}-\mathbf{u})\,\mathrm{d}x\right| &\leq C\|\mathbf{U}-\mathbf{u}\|_{L^6(S_1(t))}\|\partial_t\widetilde{\mathbf{X}}_2\|_{L^\infty(\Omega_F(t))}\\ &\leq \delta\|\mathbb{S}(\mathbf{U}-\mathbf{u})\|^2_{L^2(S_1(t))}+C(\delta)(|\mathbf{a}^s-\mathbf{a}_1|^2+|\boldsymbol{\omega}^s-\omega_1|^2)\\ &\leq \delta\|\mathbb{S}(\mathbf{U}-\mathbf{u})\|^2_{L^2(\Omega_F(t))}+C\mathcal{E}([\varrho,\mathbf{u}]|[r,\mathbf{U}])(t),\end{aligned}$$

$$\begin{aligned}&\left|\int_{S_1(t)} \frac{\varrho-r}{r}G(r,\mathbf{U})\mathbf{U}\cdot(\mathbf{U}-\mathbf{u})\,\mathrm{d}x\right|\\ &\leq C\|\mathbf{U}-\mathbf{u}\|_{L^6(S_1(t))}\|\varrho-r\|_{L^2(S_1(t))}\|\partial_t\widetilde{\mathbf{X}}_2\|_{L^\infty(\Omega_F(t))}\\ &\leq \delta\|\mathbb{S}(\mathbf{U}-\mathbf{u})\|^2_{L^2(S_1(t))}+C(\delta)\|\varrho-r\|^2_{L^2(S_1(t))}(|\mathbf{a}^s-\mathbf{a}_1|^2+|\boldsymbol{\omega}^s-\omega_1|^2)\\ &\leq \delta\|\mathbb{S}(\mathbf{U}-\mathbf{u})\|^2_{L^2(\Omega_F(t))}+C\mathcal{E}([\varrho,\mathbf{u}]|[r,\mathbf{U}])(t).\end{aligned}$$

The estimates for small ϱ and large ϱ follow similarly as in (69), (70), and (71), yielding

$$\int_{\Omega_F(t)} \frac{\varrho}{r}G(r,\mathbf{U})\mathbf{U}\cdot(\mathbf{U}-\mathbf{u})\,\mathrm{d}x \leq \delta\|\mathbb{S}(\mathbf{U}-\mathbf{u})\|^2_{L^2(\Omega_F(t))}+C\mathcal{E}([\varrho,\mathbf{u}]|[r,\mathbf{U}])(t). \tag{73}$$

Finally, we estimate the first term in (66), following a similar approach to that used for the second term. The primary objective is to establish an estimate for $\|\mathbf{F}(r,\mathbf{U})\|_{L^\infty(\Omega_F(t))}$. To achieve this, we apply formulas (39)–(43) along with Lemma 3.1. We obtain

$$\left|\int_{\Omega_F(t)} \frac{\varrho}{r}F(r,\mathbf{U})\cdot(\mathbf{U}-\mathbf{u})\,\mathrm{d}x\right| \leq \delta\|\mathbb{S}(\mathbf{U}-\mathbf{u})\|^2_{L^2(\Omega_F(t))}+C\mathcal{E}([\varrho,\mathbf{u}]|[r,\mathbf{U}])(t). \tag{74}$$

By combining the estimates (63), (65), (72), (73), and (74), we end up with

$$\left|\int_0^\tau \mathcal{R}(\varrho,\mathbf{u},r,\mathbf{U})(t)\right| \leq \delta\,\|\mathbb{S}(\mathbf{U}-\mathbf{u})\|^2_{L^2(Q_F)}+C(\delta)\int_0^\tau h_5(t)\mathcal{E}([\varrho,\mathbf{u}]|[r,\mathbf{U}])(t)\,\mathrm{d}t.$$

Using this estimate along with (53), we arrive at the inequality for sufficiently small $\delta>0$

$$\mathcal{E}([\varrho,\mathbf{u}]|[r,\mathbf{U}])(\tau) \leq \int_0^\tau h(t)\mathcal{E}([\varrho,\mathbf{u}]|[r,\mathbf{U}])(t)\,\mathrm{d}t,$$

where $h \in L^1(0, T)$ is a nonnegative function. Applying the Gronwall lemma, we can conclude that

$$\mathcal{E}([\varrho, \mathbf{u}]|[r, \mathbf{U}])(\tau) = 0 \tag{75}$$

for almost every $\tau \in (0, T)$, which, in particular, implies the equalities $\mathbf{a}_1 = \mathbf{a}^s$, $\boldsymbol{\omega}_1 = \boldsymbol{\omega}^s$, $\varrho = r$, and $\mathbf{u} = \mathbf{U}$.

In order to finish the proof, we need to show that $\mathbf{a}_1 = \mathbf{a}_2$, $\boldsymbol{\omega}_1 = \boldsymbol{\omega}_2$, $S_1(t) = S_2(t)$, and $(\varrho_{1F}, \mathbf{u}_{1F}) = (\varrho_{2F}, \mathbf{u}_{2F})$. It follows directly from the proof similarly as in [33].

Acknowledgments The work of Š.N. has been supported by Praemium Academiæ of Š. Nečasová, and A.R. has been supported by the Croatian Science Foundation under project number IP-2022-10-29. Finally, the Institute of Mathematics, CAS, is supported by RVO:67985840.

References

1. Š. Axmann,Š. Nečasová, A. Radoševič, Rigid body in compressible flow with general inflow-outflow boundary data. Z. Angew. Math. Phys. 76 (2025), no. 2, Paper No. 46, 28 pp.
2. H. Al Baba, A. Ghosh, B. Muha, Š. Nečasová, L^p-strong solution to fluid-rigid body interaction system with Navier slip boundary condition, Journal of Elliptic and Parabolic Equations, 7, 2, 439–489, (2021).
3. B. Benešová, Š. Nečasová, J. Scherz, A. Schlömerkemper, Fluid-rigid body interaction in an incompressible electrically conducting fluid, SIAM J. Math. Anal., 55, 2, 929–965, (2023).
4. M. Boulakia, S. Guerrero, A regularity result for a solid-fluid system associated to the compressible Navier-Stokes equations, Ann. Inst. H. Poincaré Anal. Non Linéaire, 26, (3), (2009), 777—813.
5. D. Bresch, P. E. Jabin, Global existence of weak solutions for compressible Navier-Stokes equations: thermodynamically unstable pressure and anisotropic viscous stress tensor, Ann. of Math., 2, 188, 577–684, (2018).
6. J. Březina, Selected mathematical problems in the thermodynamics of viscous compressible fluids, Charles University, (2008).
7. N. Chaudhuri, E. Feireisl, Navier–Stokes–Fourier system with Dirichlet boundary conditions, Applicable Analysis, 101, 12, 4076–4094, (2022).
8. HJ Choe, A. Novotný, M. Yang, Compressible Navier–Stokes system with hard sphere pressure law and general inflow–outflow boundary conditions, Journal of Differential Equations, 266, 6, 3066–3099, (2019).
9. T. Chang, B. J. Jin, A. Novotný, Compressible Navier–Stokes System with General Inflow-Outflow Boundary Data, SIAM Journal on Mathematical Analysis, 51, 2, 1238–1278, (2019).
10. N. V. Chemetov, Š. Nečasová, B. Muha, Weak-strong uniqueness for fluid-rigid body interaction problem with slip boundary condition. J. Math. Phys. 60 (2019), no. 1, 011505, 13 pp.
11. C. Conca, Carlos, J. San Martín, M. Tucsnak, Motion of a rigid body in a viscous fluid, Comptes Rendus de l'Académie des Sciences-Series I-Mathematics, 328, 6, 473–478, (1999).
12. R. Danchin, Global existence in critical spaces for compressible Navier-Stokes equations, Invent. Math., 141, 579–614, (2000).
13. B. Desjardins, M. Esteban, Existence of weak solutions for the motion of rigid bodies in a viscous fluid, Archive for Rational Mechanics and Analysis, 146, 1, 59–71, (1999).

14. B. Desjardins, M. J. Esteban, On weak solutions for fluid-rigid structure interaction: Compressible and incompressible model, Commun. Partial Differential Equations, (2000), 25, 1399–1413.
15. R.J. DiPerna, P.-L. Lions, Ordinary differential equations, transport theory and Sobolev spaces, Inventiones mathematical, 98, 511–547, (1989).
16. E. Feireisl, On the motion of rigid bodies in a viscous compressible fluid, Archive for rational mechanics and analysis, 167, 4, 281–308, (2003).
17. E. Feireisl, A. Novotný: *Singular limits in thermodynamics of viscous fluids.* Birkhäuser, Basel, (2009).
18. E. Feireisl, On the motion of rigid bodies in a viscous incompressible fluid, Journal of Evolution Equations, (2003), 3, 419–441.
19. E. Feireisl, B. J. Jin, A. Novotný. Relative entropies, suitable weak solutions, and weak-strong uniqueness for the compressible Navier-Stokes system, J. Math. Fluid Mech. 14 (2012), no. 4, 717–730.
20. E. Feireisl, T. G. Karper, M. Pokorný, Mathematical theory of compressible viscous fluids, Springer, (2016)
21. E. Feireisl, A. Novotný, Navier–Stokes–Fourier system with general boundary conditions, Communications in Mathematical Physics, 386, 975–1010, (2021).
22. E. Feireisl, A. Novotný, Mathematics of open fluid systems, (2022), Springer
23. E. Feireisl, A. Novotný, H. Petzeltová, On the existence of globally defined weak solutions to the Navier–Stokes equations, Journal of Mathematical Fluid Mechanics, 3, 4, 358–392, (2001).
24. E. Feireisl: *Dynamics of viscous compressible fluids.* Oxford University Press, Oxford, (2004).
25. G. P. Galdi, On the steady self-propelled motion of a body in a viscous incompressible fluid, Archive for rational mechanics and analysis, 148, 1, 53–88, 1999.
26. M. Geissert, K. Götze, M. Hieber, L^p-theory for strong solutions to fluid-rigid body interaction in Newtonian and generalized Newtonian fluids, Trans. Amer. Math. Soc., (2013), 365, 3, 1393–1439.
27. V. Girinon, Navier–Stokes equations with nonhomogeneous boundary conditions in a bounded three-dimensional domain, Journal of Mathematical Fluid Mechanics, 13, 3, 309–339, (2011).
28. M. D. Gunzburger, H. C. Lee, A. Seregin, Global existence of weak solutions for viscous incompressible flow around a moving rigid body in three dimensions, J. Math. Fluid Mech., 2000, 2, 219–266.
29. B. H. Haak, D. Maity, T. Takahashi, M. Tucsnak, Mathematical analysis of the motion of a rigid body in a compressible Navier–Stokes–Fourier fluid, Mathematische Nachrichten, 292, 9, 1972–2017, (2019),
30. M. Hieber, M. Murata, The L^p-approach to the fluid-rigid body interaction problem for compressible fluids, Evolution Equations & Control Theory, 4, 1, (2015)
31. K.H. Hoffmann, V. N. Starovoitov, On a motion of a solid body in a viscous fluid. Two-dimensional case, Adv. Math. Sci. Appl., 9, (1999), 2, 633–648.
32. S. Kračmar, Y-S. Kwon, Š. Nečasová, A. Novotný, Weak solutions for a bifluid model for a mixture of two compressible noninteracting fluids with general boundary data, SIAM J. Math. Anal., 54, (2022), 1, 818–871.
33. O. Kreml, Š. Nečasová, T. Piasecki, Weak-strong uniqueness for the compressible fluid-rigid body interaction, Journal of Differential Equations, 268, 8, 4756–4785, (2020).
34. O. Kreml, Š. Nečasová, T. Piasecki. Local existence of strong solutions and weak-strong uniqueness for the compressible Navier-Stokes system on moving domains, Proceedings of the Royal Society of Edinburgh Section A: Mathematics 150.5 (2020): 2255–2300.
35. P.-L. Lions, Compacité des solutions des équations de Navier-Stokes compressibles isentropiques, Comptes rendus de l'Académie des sciences. Série 1, Mathématique, 317, 1, 115–120, (1993).
36. P.-L. Lions, Mathematical topics in fluid mechanics. Vol. 2, Oxford Lecture Series in Mathematics and its Applications, 10,The Clarendon Press, Oxford University Press, New York, 1998.

37. A. Matsumura, T. Nishida, T. The initial value problem for the equations of motion of viscous and heat-conductive gases, J. Math. Kyoto Univ., 20, 67–104, (1980).
38. B. Muha, Š. Nečasová, A. Radošević, A uniqueness result for 3D incompressible fluid-rigid body interaction problem, Journal of Mathematical Fluid Mechanics, 23, 1–39, (2021).
39. Š. Nečasová, M. Ramaswamy, Mythily, A. Roy, A. Schlömerkemper, Motion of a rigid body in a compressible fluid with Navier-slip boundary condition Journal of Differential Equations, 338, 256–320, (2022).
40. S. Novo, Compressible Navier–Stokes model with inflow-outflow boundary conditions, Journal of Mathematical Fluid Mechanics, 7, 485–514, (2005).
41. A. Novotný, I. Straškraba, Introduction to the mathematical theory of compressible flow, 27, 2004, Oxford.
42. T. Piasecki, On an inhomogeneous slip-inflow boundary value problem for a steady flow of a viscous compressible fluid in a cylindrical domain, Journal of Differential Equations, 248, 8, 2171–2198, 2010.
43. A. Roy, T. Takahashi, Stabilization of a rigid body moving in a compressible viscous fluid, Journal of Evolution Equations, 1424–3202, (2020).
44. J.A. San Martín, V. Starovoitov, M. Tucsnak, Global Weak Solutions for the Two-Dimensional Motion of Several Rigid Bodies in an Incompressible Viscous Fluid, Archive for Rational Mechanics and analysis, 161, 2, 113–147, (2002).
45. D. Serre, Chute libre d'un solide dans un fluide visqueux incompressible. Existence, Jap. J. Appl. Math., 1987, 4, 99–110.
46. J. Scherz, Fluid structure interaction in viscous compressible fluids, Julius-Maximilians-Universität Würzburg, (2019).
47. T. Takahashi, Analysis of strong solutions for the equations modeling the motion of a rigid-fluid system in a bounded domain, Adv. Differential Equations, 2003, 8, 12, 1499–1532.
48. A. Valli, W. M. Zajaczkowski, Navier-Stokes equations for compressible fluids: global existence and qualitative properties of the solutions in the general case, Comm. Math. Phys., 103, 259–296, (1986).
49. C. Wang, Strong solutions for the fluid-solid systems in a 2-D domain, Asymptot. Anal., (2014), 89, 3–4, 263–306.

Calculation of Pressure and Forces in Stochastic Vortex Methods

Andrzej F. Nowakowski

Abstract This chapter discusses pressure calculations and aerodynamic forces in external flows. The dynamic variables are recovered from the kinematic velocity-vorticity formulation of the Navier-Stokes equations. Both velocity and vorticity fields may result from vortex methods. In the example presented, the kinematic variables are generated from the stochastic vortex blob simulation of the viscous incompressible flow. The vorticity field is a discontinuous function. This contribution demonstrates how to find pressure as a solution to the variational problem which forms a weak version of the momentum equation. It does not require any explicit boundary conditions for pressure, and no derivatives of vorticity have to be calculated in this approach. The correctness of the calculation of the pressure field and consequently of the aerodynamic forces is verified by an alternative method of calculating the aerodynamic forces which directly applies formulas involving the velocity and vorticity fields.

1 Introduction

Vortex methods are a class of Lagrangian numerical methods that could be applied to solve the incompressible Navier-Stokes equations in their velocity-vorticity formulation. Vortex methods have been used in many complex flow applications, and results demonstrate quantitatively correct results. They provide efficient procedures for the simulation of convection-dominated flows as a result of their negligible dispersion error. The significant key milestones and a discussion of challenging issues related to the simulation of the flow problem using vortex methods were presented in the review papers [1–3] and monographs [4, 5]. These methods simulate the advection-diffusion process of vorticity transport. Vorticity is created by imposing a boundary condition on the velocity field. In vortex methods, the space-time evolution of the vorticity field is simulated by a large number of

A. F. Nowakowski (✉)
School of Mechanical, Aerospace and Civil Engineering, University of Sheffield, Sheffield, UK
e-mail: a.f.nowakowski@sheffield.ac.uk

T. Bodnár et al. (eds.), *Stochastics in Fluids*, Lecture Notes in Mathematical Fluid Mechanics, https://doi.org/10.1007/978-3-032-02299-8_5

interacting discrete vorticity carriers. The characteristic feature of vortex methods is the elimination of pressure field from the equations describing the motion. At the same time, we know that pressure is the most important flow parameter from the point of view of applications of numerical methods. Knowing this parameter, one can calculate the forces acting on the bodies. Determining the resultant of forces acting on an immersed body is one of the most fundamental problems of fluid mechanics. Determination of the pressure field and aerodynamic forces for simple flow configurations allows us to validate a given numerical method with experimental results and thus to evaluate its usefulness in the simulation of practical problems. This contribution presents a method that allows determining the pressure field for known velocity and vorticity fields. It is assumed that velocity and vorticity fields have been determined beforehand. A feature of these fields is their low smoothness.

2 Velocity-Vorticity Formulation of Navier-Stokes Equations

The incompressible flow in the domain Ω is governed by the Navier-Stokes equations:

$$\frac{\partial \mathbf{u}}{\partial t} + (\mathbf{u} \cdot \nabla)\mathbf{u} = -\frac{1}{\rho}\nabla p + \nu\nabla^2\mathbf{u} + \mathbf{f_e}, \tag{1}$$

$$\nabla \cdot \mathbf{u} = 0, \quad \text{for} \quad \mathbf{x} \in \Omega, \tag{2}$$

$$\mathbf{u} = \mathbf{b}, \quad \text{for} \quad \mathbf{x} \in \Gamma \tag{3}$$

The momentum equation (1) relates the velocity vector field $\mathbf{u} = (u, v, w)^T$ and the pressure field p. The given vector field $\mathbf{f_e}$ represents the external body force field, ρ density, and ν kinematic viscosity. The initial velocity field $\mathbf{u}_0$ is assumed to be divergent-free $\nabla \cdot \mathbf{u}_0 = 0$, and its value at the surface Γ of a rigid body is $\mathbf{b}$. Pressure, which is one of the unknown functions, does not appear in the form of a time derivative. This is an expression of the non-evolutionary nature of the continuity equation. Boundary or initial conditions are not formulated explicitly for pressure field either.

The momentum equation (1) of the Navier-Stokes equations for incompressible and viscous flows can be replaced by the vorticity-velocity formulation. The equation for the evolution of the vorticity $\boldsymbol{\omega} = \nabla \times \mathbf{u}$ is obtained by taking the curl of the momentum equation (1)

$$\frac{D\boldsymbol{\omega}}{Dt} = (\boldsymbol{\omega} \cdot \nabla)\mathbf{u} - \boldsymbol{\omega}(\nabla \cdot \mathbf{u}) + \nu\nabla^2\boldsymbol{\omega} + \nabla \times \mathbf{f_e}. \tag{4}$$

In the two-dimensional case $\boldsymbol{\omega} = (0, 0, \partial_x v - \partial_y u)^T$, the vorticity vector has a direction perpendicular to the plane of motion and can be treated as a scalar $\zeta =$

$\partial_x v - \partial_y u$. Equation (4) is then simplified to the form

$$\frac{\partial \zeta}{\partial t} + (\mathbf{u} \cdot \nabla)\zeta - \nu \nabla^2 \zeta = \nabla \times \mathbf{f_e}. \tag{5}$$

The motion of a viscous fluid is therefore described as a continuous process of convection and diffusion of the vorticity field. In order to close the problem (5), a method for determining the velocity from the vorticity has to be provided. This is possible by introducing a stream function. Its relation to the velocity has the form $\mathbf{u} = \nabla \times \boldsymbol{\psi}$. The stream function is related to the vorticity by the Poisson equation $\nabla^2 \boldsymbol{\psi} = -\zeta$.

Equation (5) is formally linear with respect to vorticity and has boundary conditions imposed implicitly as a function of velocity. Equation (5), however, has the advantage that it allows for a purely kinematic description of the equation of motion, eliminating somehow the troublesome quantity of pressure. In the following chapter, the implementation of the 2D viscous vortex method will constitute the basis for the method to recover the pressure field from (5).

3 Stochastic Vortex Method

The velocity-vorticity formulation (5) allows the vortex method to restrict the computational domain to points with nonzero vorticity. This feature of vortex methods enables the vorticity field to be represented as a finite set of interacting vortices or vortex particles. The induced velocity field in combination with the Lagrangian description allows to eliminate the use, or at least minimize the importance, of computational grids. The discretization of the computational domain is not performed on a grid but on a set of particles. The vorticity of each particle is distributed on a blob, which in 2D is represented by a disk with a constant vorticity distribution and a compact support of diameter d. Velocity and vorticity are coupled to each other with the Biot-Savart law. Equation (5) is supplemented with appropriate boundary and initial conditions, which are expressed in terms of velocity: the no-slip condition at the obstacle boundary and uniform flow at infinity (the flow domain is assumed to be infinite). The flow is accelerated from the rest. The stochastic vortex method was described in [6]. The method is based on a formal similarity between Eq. (5) and the Planck-Fokker equation describing the evolution of a stochastic Wiener process. The motion of a viscous fluid can therefore be modeled by following a family of elementary vortex particles referred to as vortex blobs. Each of them follows the stochastic Ito equation:

$$d\mathbf{x} = \mathbf{V}dt + \sqrt{2\nu}\, d\mathbf{W} \tag{6}$$

$d\mathbf{x}$ represents an infinitesimal displacement of a representative vortex blob, and $\mathbf{V}$ is the deterministic velocity field combining the potential background flow and the vortex-to-vortex interactions. Such a velocity decomposition allows the formulation of a boundary integral equation defining the boundary value of vorticity, and thus the intensity of vortex blobs formed at the boundaries. The last term on the right-hand side of (6) represents the so-called random walk which models viscous diffusion with $d\mathbf{W}$ being an infinitesimal increment of a non-anticipating Wiener process. A viscous splitting algorithm is used in which advective and diffusion effects are handled separately and sequentially. The fast multipole method by Greengard and Rokhlin [7] was applied to compute the induction of the vortex blobs.

4 Variational Formulation of Pressure Problem

Pressure does not explicitly appear in the kinematic formulation of stochastic vortex method. In order to be able to determine the forces, one can try to first obtain the pressure field and then the forces and moments.

The Navier-Stokes equation (1) can be written in a reduced form,

$$\nabla q = \mathbf{F}, \tag{7}$$

where

$$q = \frac{u^2}{2} + p \qquad\qquad u^2 = \mathbf{u} \cdot \mathbf{u}$$

$$\mathbf{F} = \frac{\partial \mathbf{u}}{\partial t} + \mathbf{u} \times \boldsymbol{\omega} - \nu \nabla \times \boldsymbol{\omega} + \mathbf{f_e}$$

The major steps to obtain the reduced form (7) are presented in [8]. The left-hand side q can be interpreted as the total pressure. If q is found by solving (7), the pressure field could be calculated from the formula $p = q - u^2/2$. In the formulation (7), it is assumed that the vector field F is known as it is calculated from the known velocity and vorticity fields, and the quantity sought is a scalar field q. A necessary condition for the existence of a solution to the Eq. (7) is $\nabla \times \mathbf{F} = O$ which is equivalent to 2D vorticity transport Eq. (5). This means that the vector field $\mathbf{F}$ is correctly determined only if the velocity and vorticity fields satisfy (5). Satisfying this postulate exactly may not be possible as a result of the numerical representation of $\mathbf{F}$. Consequently, the problem remains ill-posed. In addition, the vortex method may not provide sufficient vorticity field smoothness, which is necessary for the calculation $\nabla \times \boldsymbol{\omega}$. In a multiconnected domain (this is always the case of flow around an external body or system of bodies), the necessary condition $\nabla \times \mathbf{F} = O$ may not be a sufficient condition for the existence of the solution. One can then apply some measures for the filtration of the potential error [8] while using the variational formulation of (7).

The variational formulation assumes that the flow domain $\Omega \subset \mathcal{R}^N$ is open, bounded, and connected and its boundary Γ is at least Lipschitz continuous. For each vector function $\mathbf{F} \in [L^2(\Omega)]^2$, there exist scalar functions $q, \Phi \in H^1(\Omega)$ such that

$$\mathbf{F} = \nabla q + \nabla \times \Phi \tag{8}$$

$$\mathbf{n} \cdot (\mathbf{F} - \nabla q) = 0. \tag{9}$$

∇q and $\nabla \times \Phi$ are orthogonal in $H^1(\Omega)$, and $\mathbf{n}$ is generalized normal component on Γ. If these relations are satisfied, the scalar field q will be a solution to the following variational formulation:

$$\int_\Omega \nabla q \cdot \nabla \mu \, d\Omega = \int_\Omega \mathbf{F} \cdot \nabla \mu \, d\Omega \quad \forall \mu \in H^1(\Omega) \tag{10}$$

For each arbitrary test function μ from $H^1(\Omega)$, vectors $\nabla\mu$ and $\nabla \times \Phi$ are orthogonal:

$$\int_\Omega \nabla \mu \cdot \nabla \times \Phi \, d\Omega = -\int_\Omega \mu \nabla(\nabla \times \Phi) \, d\Omega + \int_\Gamma \mu \mathbf{n} \cdot \nabla \times \Phi \, d\Gamma = 0 \tag{11}$$

The solution of (10) may be regarded as a least square solution of the ill-posed problem in case

$$\nabla \times \mathbf{F} \neq 0.$$

The variational formulation (10) filters out the rotational part of the vector field $\mathbf{F}$. If our considerations are restricted to 2D incompressible and viscous flow problem, the right-hand side of (10) simplifies to four integrals,

$$\int_\Omega \nabla q \cdot \nabla \mu \, d\Omega = -\oint_\Gamma \mu \frac{\partial \mathbf{u}}{\partial t} \cdot \mathbf{n} \, d\Gamma \; + \tag{12}$$

$$+ \nu \oint_\Gamma \zeta \frac{\partial \mu}{\partial \sigma} \, d\Gamma \; + \tag{13}$$

$$+ \int_\Omega \zeta \cdot (\nabla \mu \times \mathbf{u}) \, d\Omega \; + \tag{14}$$

$$+ \int_\Omega \mathbf{f_e} \cdot \nabla \mu \, d\Omega, \tag{15}$$

where σ denotes the direction tangent to the boundary Γ. Some of the area integrals have been transformed into integrals over the boundary of the area, which reduces the necessary calculations. The formulation of the problem of determining the pressure in the form (13, 14) allows avoiding differentiation of the vorticity field. There is also no need to directly define boundary conditions. A very important advantage of the variational formulation is the possibility of determining the pressure in a fragment of the flow region without determining it in the remaining part of the region. This possibility results from the fact that the region Ω and the region boundary Γ can be chosen arbitrarily.

5 Numerical Implementation

The variational problem can be solved using the finite element method. The analyzed flow region is divided into a finite number of elements of characteristic size h. The piecewise linear approximation of the fields $\mathbf{u}$ and q is denoted by $\mathbf{u}_h$ and q_h. In the stochastic vortex method, the vorticity field is created by a finite number of blobs of constant vorticity. The diameter of a single blob is much smaller than the dimensions of the computational area and smaller than the size of the computational grid.

The finite-dimensional linear problem is

$$\begin{aligned}\int_\Omega \nabla q_h \cdot \nabla \mu_h \, d\Omega = & -\oint_\Gamma \mu_h \frac{\partial \mathbf{u}_h}{\partial t} \cdot \mathbf{n} \, d\Gamma + \\ & + \nu \oint_\Gamma \zeta \frac{\partial \mu_h}{\partial \sigma} \, d\Gamma + \\ & + \int_\Omega \zeta \cdot (\nabla \mu_h \times \mathbf{u}_h) \, d\Omega + \\ & + \int_\Omega \mathbf{f_e} \cdot \nabla \mu_h \, d\Omega \qquad \forall \mu \in V_h .\end{aligned}$$

6 Aerodynamic Forces

The calculation of aerodynamic forces in the case when the pressure field and the vorticity field are known comes down to the numerical implementation of the formula

$$\mathbf{Force} = \oint_\Gamma \boldsymbol{\tau} \, d\Gamma - \oint_\Gamma p \, \mathbf{n} \, d\Gamma = \oint_\Gamma (2\mu \mathbf{D}(\mathbf{u}) - p \, \mathbf{n}) \, d\Gamma, \tag{16}$$

where $\mathbf{D}(\mathbf{u})$ is

$$\mathbf{D}(\mathbf{u}) = \frac{1}{2} \left[\nabla \mathbf{u} + (\nabla \mathbf{u})^T \right].$$

On the surface of a stationary body, the formula (16) can be transformed into

$$\mathbf{Force} = \oint_{\Gamma} (2\mu \mathbf{D}(\mathbf{u}) - p\,\mathbf{n})\, d\Gamma = -\oint_{\Gamma} (p\,\mathbf{n} + \nu \mathbf{n} \times \boldsymbol{\omega})\, d\Gamma. \tag{17}$$

This last formulation can be implemented after calculating the pressure for a known vorticity field.

7 Computation of Pressure Field Around a Two-Dimensional Circular Cylinder

For the 2D test case, a cylinder with circular cross section is considered. The velocity and vorticity fields in consecutive moments of time are generated using a numerical implementation of the stochastic vortex method developed by Szumbarski and Wald [6]. The example velocity field is presented in Fig. 1.

A numerical implementation of the method developed by Szumbarski and Wald [6] is used, which guarantees obtaining a physically interpretable pressure field. Example viscous flow simulations were performed for different Reynolds numbers and a cylinder of radius $R = 1$. The Reynolds number Re is based on the value of the velocity at infinity $U_\infty = 1$ and the cylinder diameter $2R$

$$Re = \frac{U_\infty 2R}{\nu} = \frac{2}{\nu}.$$

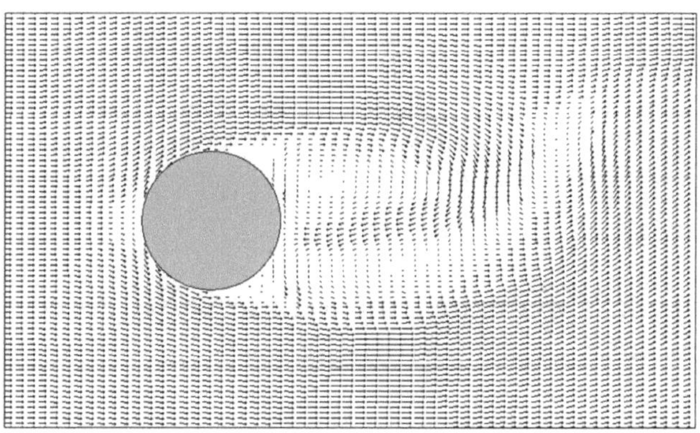

Fig. 1 The velocity vector field $\mathbf{u}$ at the nondimensional time $T = 100$, $Re = 186000$. The inner boundary corresponds to the cylinder surface; the outer boundary is fictitious

Fig. 2 The map of pressure coefficient c_p at the nondimensional time $T = 100$, $Re = 186000$. The inner boundary corresponds to the cylinder surface; the outer boundary is fictitious

The pressure coefficient c_p is defined by the formula

$$c_p = \frac{p - p_\infty}{\rho_\infty {u_\infty}^2/2} = 2(p - p_\infty), \qquad (\rho_\infty = 1).$$

A computational grid made of $160 \times 50 \times 2$ triangles was generated for pressure calculations. The mesh has a rectangular topology and is spread over 160 nodes on the cylinder surface and 50 nodes in the radial direction. The maximum number of vorticity carriers (blobs of diameter $\delta = 0.034$) was about 100,000. Figure 2 shows the main advantage of the variational formulation (10), which is the ability to determine the pressure in a part of the flow region without determining it in the rest of the region. Figure 2 shows the lift coefficient for $Re = 186000$ and for the moment $T = 100$ ($T = Ut/R$) for which the flow is fully developed. The characteristic quantities such as drag and lift coefficients are computed in order to compare them with other methods.

8 Quartapelle-Napolitano Approach to Verify Force Calculations

Both forces and moments acting on closed contours can be calculated based on the velocity and vorticity fields known in the flow region. In the approach proposed by Quartapelle and Napolitano [9], there is no need to solve the pressure problem. The classical way of determining forces by integrating the pressure and shear stresses over the contour surface has been replaced by a general formula based on information about the velocity and vorticity fields. This approach is particularly attractive in methods in which pressure has been eliminated from the equations describing the motion. The basis for deriving general formulas for the force is the projection of the Navier-Stokes equation onto the subspace of harmonic functions. Let $\nabla^2\eta = 0$. By multiplying the momentum Eq. (1) by $\nabla\eta$ and omitting $\mathbf{f_e}$ term, one obtains

$$-\oint_\Gamma p\nabla\eta\cdot\mathbf{n}\,d\Gamma = \oint_\Gamma \mathbf{n}\cdot\frac{\partial\mathbf{u}}{\partial t}\eta\,d\Gamma + \int_\Omega (\mathbf{u}\cdot\nabla)\mathbf{u}\cdot\nabla\eta\,d\Omega + \nu\oint_\Gamma \mathbf{n}\times\boldsymbol{\omega}\cdot\nabla\eta\,d\Gamma.$$

Now, let us consider an arbitrary solid body placed in the flow. Let Γ_1 be the surface of this body and Γ_2 any surface enclosing this body and bounding the flow region Ω. The resultant force acting on the body is given by (17). The normal vector $\mathbf{x}$ in the formulation (17) will be directed outward from the computational region and into the material body.

It is assumed that the harmonic function η_x satisfies the following Neumann boundary conditions

$$\mathbf{n}\cdot\nabla\eta_x|_{\Gamma_1} = -\mathbf{n}\cdot\mathbf{x}, \tag{18}$$

$$\mathbf{n}\cdot\nabla\eta_x|_{\Gamma_2} = 0. \tag{19}$$

where $\mathbf{x}$ is a unit vector in the direction of the abscissa axis of the Cartesian coordinate system. Consequently,

$$\oint_\Gamma p\nabla\eta\cdot\mathbf{n}\,d\Gamma = -\oint_\Gamma p\mathbf{x}\cdot\mathbf{n}\,d\Gamma_1, \tag{20}$$

and the force formula (17) projected in the $\mathbf{x}$ direction is

$$F_x = \oint_{\Gamma_1+\Gamma_2} \mathbf{n}\cdot\frac{\partial\mathbf{u}}{\partial t}\eta_x\,d\Gamma + \int_\Omega (\mathbf{u}\cdot\nabla)\mathbf{u}\cdot\nabla\eta_x\,d\Omega + \nu\oint_{\Gamma_1+\Gamma_2} \mathbf{n}\times\boldsymbol{\omega}\cdot\nabla\eta_x\,d\Gamma + \nu\mathbf{x}\oint_{\Gamma_1} \mathbf{n}\times\boldsymbol{\omega}\,d\Gamma. \tag{21}$$

The equation can be further simplified by taking advantage of the fact that on the material boundary the velocity is zero and on the external boundary the relation (19) holds.

The final form of the formula for the force (17) in the $\mathbf{x}$ direction will be

$$
\begin{aligned}
F_x = & \frac{d}{dt} \oint_{\Gamma_2} \mathbf{n} \cdot \mathbf{u}\, \eta_x \, d\Gamma - \int_{\Omega} \mathbf{u} \times \boldsymbol{\omega} \cdot \nabla \eta_x \, d\Omega \\
& + \nu \oint_{\Gamma_1} \mathbf{n} \times \boldsymbol{\omega} (\nabla \eta_x + \mathbf{x}) \, d\Gamma - \nu \oint_{\Gamma_2} \mathbf{n} \times \boldsymbol{\omega} \cdot \nabla \eta_x \, d\Gamma
\end{aligned} \tag{22}
$$

and, respectively, in the $\mathbf{y}$ direction

$$
\begin{aligned}
F_y = & \frac{d}{dt} \oint_{\Gamma_2} \mathbf{n} \cdot \mathbf{u}\, \eta_y \, d\Gamma - \int_{\Omega} \mathbf{u} \times \boldsymbol{\omega} \cdot \nabla \eta_y \, d\Omega \\
& + \nu \oint_{\Gamma_1} \mathbf{n} \times \boldsymbol{\omega} (\nabla \eta_y + \mathbf{y}) \, d\Gamma - \nu \oint_{\Gamma_2} \mathbf{n} \times \boldsymbol{\omega} \cdot \nabla \eta_y \, d\Gamma .
\end{aligned} \tag{23}
$$

The harmonic function η_y was selected analogously to (18) and (19), and the unit vector $\mathbf{y}$ has the direction of the ordinate axes of the Cartesian coordinate system.

The obtained formulas allow determination of the forces acting on the body without referring to the information about the pressure field.

In Sect. 7, the aerodynamic forces were determined after solving the pressure problem. They were calculated based on the pressure and vorticity values at the solid boundary according to formula (17). This approach may suggest that due to using only the boundary values of pressure and vorticity the solution is burdened with a certain error. However, it turns out that calculating the forces according to formulas (22) and (23) gives similar results (Figs. 3 and 4). Since the method [9] does not require knowledge of the pressure field, it can be concluded that the method of determining the pressure field is correct.

9 Conclusions

The expressions of recovering the pressure field and body forces from velocity and vorticity data were presented. In a large variety of fluid dynamic problems, it is often impossible to directly measure or calculate the instantaneous aerodynamic or hydrodynamic forces. The presented variational formulation of pressure problem has the following advantages:

- It avoids differentiation of the vorticity field, which is of significant importance for Lagrangian methods characterized by a low smoothness of the vorticity field.
- It does not require direct determination of boundary conditions and therefore allows calculating the pressure in a fragment of the flow region, without determining it in the rest of the region.

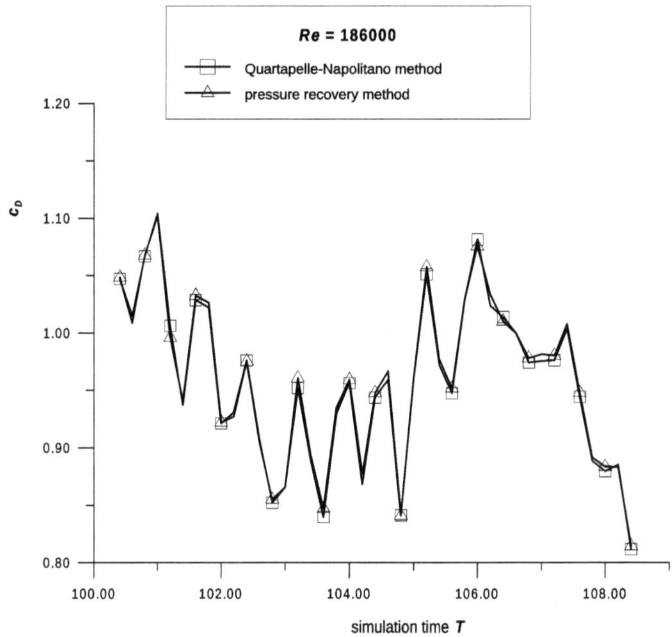

Fig. 3 Comparison of the drag coefficient calculated using the presented method recovering pressure from velocity and vorticity fields with the drag coefficient calculated directly from kinematic quantities

- It can be applied to determine the pressure for single-connected and multi-connected regions.
- When applied to the ill-posed problem of $\nabla q = \mathbf{F}$, it allows to filter out the numerical error. Since the vector $\mathbf{F}$ is the result of numerical calculations, the necessary condition for the existence of the solution ($\nabla \times \mathbf{F} = O$) is generally not met.

The approach was used to verify the simulation of unsteady flow obtained by the stochastic vortex method. For the verification purposes, the forces acting on the body were also calculated without referring to the information about the pressure field.

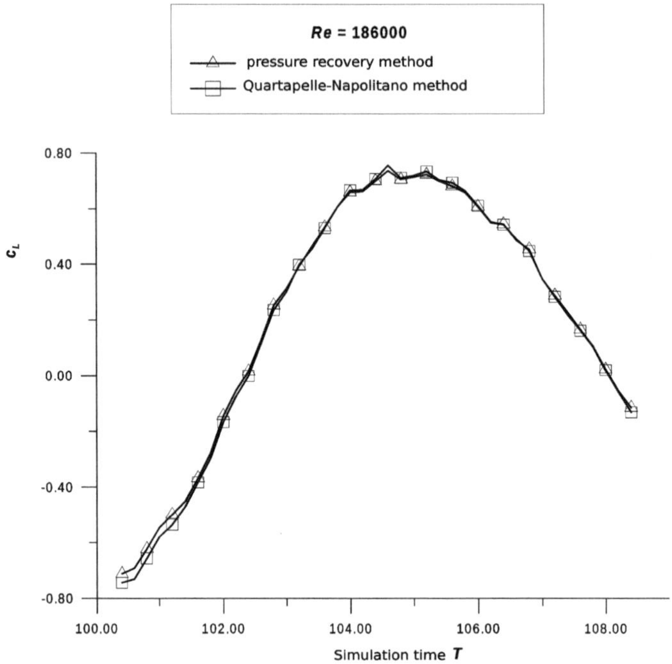

Fig. 4 Comparison of the lift coefficient calculated using the presented method recovering pressure from velocity and vorticity fields with the lift coefficient calculated directly from kinematic quantities

Acknowledgments This work was initiated and mainly carried out during the author's stay at the Institute of Aeronautics and Applied Mechanics of Warsaw University of Technology.

References

1. Mimeau C, Mortazavi I (2021) A review of vortex methods and their applications: From creation to recent advances. Fluids, https://doi.org/10.3390/fluids6020068
2. Winckelmans G (2018) Vortex methods. In: Stein E, de Borst R, Hughes TJR (ed) Encyclopedia of computational mechanics, 2nd edn. Wiley
3. Yokota R, Obi S (2011) Vortex methods for the simulation of turbulent flows: Review. Journal of Fluid Science and Technology, https://doi.org/10.1299/jfst.6.14
4. Cottet G-H, Koumoutsakos P (2000) Vortex methods – theory and practice. Cambridge University Press, Cambridge
5. Gustafson KE, Sethian JA (1991) Vortex methods and vortex motion. SIAM, Philadelphia
6. Szumbarski J, Wald P (1996) The stochastic vortex blob simulation of an unsteady viscous flow in a multiconnected domain. European series in applied and industrial mathematics. 1: 153–167
7. Greengard L, Rokhlin V (1987) A Fast Algorithm for Particle Interaction. Journal of Computational Physics, https://doi.org/10.1016/0021-9991(87)90140-9

8. Nowakowski A, Rokicki J, Styczek A (1996) The pressure problem in the stochastic vortex blob method European series in applied and industrial mathematics 1: 125–134
9. Quartapelle L, Napolitano M (1983) Force and moment in incompressible flows. AIAA Journal, https://doi.org/10.2514/3.8171

On the Steady Motion of a Navier-Stokes Flow Across a Sieve with Prescribed Pressure Drop in a Finite Pipe

Gianmarco Sperone

Abstract The steady motion of a viscous incompressible fluid through a sieve (i.e., a wall perforated with a large number of small holes), in a pipe of finite length, is modeled through the Navier-Stokes equations under mixed boundary conditions involving the Bernoulli pressure and the tangential velocity on the inlet and outlet of the tube, while the pressure drop is prescribed along the pipe. Applying the classical energy method in homogenization theory, we study the asymptotic behavior of the solutions to this system, without any restriction on the magnitude of the data, as the diameters of the perforations vanish. Regardless of the initial scaling and distribution of the holes, we show that the sieve asymptotically becomes a wall, meaning that the effective equations are two, independent, stationary Navier-Stokes systems with a no-slip boundary condition on the wall. In the absence of external forces, we prove, furthermore, that the fluid motion becomes quiescent in the homogenization limit.

1 Introduction and Presentation of the Problem

In his celebrated paper from 1976, concerning uniqueness questions in the theory of viscous flow [22], Heywood established the problem of determining the steady motion of a viscous incompressible fluid in a three-dimensional *aperture domain*, that is, in an unbounded region (with a non-compact boundary) of the following type,[1]

$$\Omega_H \doteq \left\{(x, y, z) \in \mathbb{R}^3 \mid z \neq 0 \quad \text{or} \quad (x, y) \in S\right\},$$

[1] In [22, Section 6] Heywood showed that, for such aperture domain Ω_H, the completion, in the Dirichlet norm, of the space of compactly supported solenoidal vector fields in $C^\infty(\Omega_H)$ *does not* coincide with the set of divergence-free vector fields in the Sobolev space $W_0^{1,2}(\Omega_H)$.

G. Sperone (✉)
Facultad de Matemáticas, Pontificia Universidad Católica de Chile, Santiago, Chile
e-mail: gianmarco.sperone@uc.cl

T. Bodnár et al. (eds.), *Stochastics in Fluids*, Lecture Notes in Mathematical Fluid Mechanics, https://doi.org/10.1007/978-3-032-02299-8_6

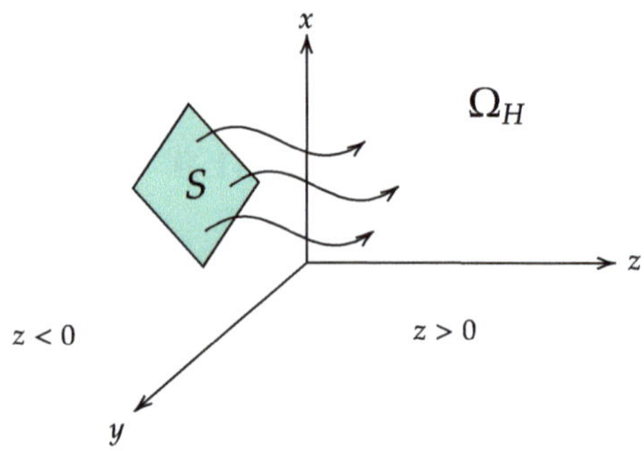

Fig. 1 Representation of an *aperture domain* in the sense of Heywood

where $S \subset \mathbb{R}^3$ is a bounded and connected subset of the XY-plane having a Lipschitz boundary; in other words, the fluid is allowed to flow between the half-space $z < 0$ and the half-space $z > 0$ only through a *perforation* located on the XY-plane, defined by the set S; see Fig. 1 for a simple illustration.

Among several other issues, Heywood discovered [22, Section 6] (see also [23]) that the hydrodynamic equations governing the motion of the *jet* can be correctly formulated by prescribing, as auxiliary conditions, either the *transverse flux* through S or the *pressure drop* across S (further properties of such steady jets can be found in the works of Borchers, Galdi, and Pileckas [5, 6]). Later on, the use of these physical quantities as admissible boundary conditions for the Stokes or Navier-Stokes equations (in both steady and unsteady regimes) in bounded domains comprised the main subject of various investigations, of numerical and analytical nature, such as [1, 13, 14, 24, 26, 33, 39]. This raises a first natural question:

How does the motion of the jet change if we allow it to transit through several apertures, driven by a pressure difference across the wall containing the holes?

Furthermore,

Can we describe the asymptotic motion (percolation) of the pressure-driven jet when, simultaneously, the number of apertures goes to infinity but their diameters vanish?

Naturally, these two inquiries should be tackled in the framework of homogenization theory concerning *sieve problems*, where a partial differential equation is studied in, say, a three-dimensional domain whose intersection with a given plane contains a large number of tiny perforations. The objective is then to describe the limiting behavior of the solutions to such PDE as the sieve or grill becomes finer and finer, yielding an *effective equation* in the homogeneous medium. Starting from the seminal work of Marchenko and Khruslov [29] (consult also their book [30]), homogenization problems involving sieves have been analyzed predominantly in the realm of elliptic PDEs; see [10, 15, 16, 34, 38, 41, 43] and references therein for a non-exhaustive list of classic works. Analogous percolation processes, in the context of fluid mechanics of incompressible flows, were thoroughly examined in the benchmark papers of Allaire [2], Conca [11, 12], and Sánchez-Palencia [44] for the (linear) steady-state Stokes system with nonhomogeneous Dirichlet boundary conditions; further investigations in the linear setting were also conducted in [7, 8, 20, 31, 32]. It would appear that far less attention has been devoted to the nonlinear counterpart, and we highlight the work of Jäger and Mikelić [25] relating to the time-dependent Navier-Stokes equations in a bounded domain containing a *thick* grill. The present article intends to serve as a contribution in the analysis of the Navier-Stokes sieve problem, with the purpose of conferring a deeper understanding of the questions described before and motivating ulterior studies on this matter.

Let us now describe in detail the mathematical problem we are concerned with. Given $h > R > 0$, in the space $\mathbb{R}^3$, we consider an open straight cylinder Ω of radius R and length $2h$ whose axis of symmetry is directed along the z-axis:

$$\Omega \doteq \left\{(x, y, z) \in \mathbb{R}^3 \mid 0 \leq x^2 + y^2 < R^2 \, , \; -h < z < h \right\}$$

For any $\xi \in \mathbb{R}^2$ and $r > 0$, we denote by $D(\xi, r) \subset \mathbb{R}^2$ the open disk of radius r with center at ξ. In particular, define $\Sigma \doteq D((0, 0), R)$ as the middle cross section of Ω. Let $(K_n)_{n\in\mathbb{N}}$ be a sequence of open, bounded, and simply connected subsets of $\mathbb{R}^2$ with a C^2-boundary such that $(0, 0) \in K_n$, $\forall n \in \mathbb{N}$, and

$$\sup_{n\in\mathbb{N}} |K_n| < \infty \, .$$

Take $\varepsilon_* \in (0, 1)$ such that $\varepsilon_*|K_n| < \pi R^2$, $\forall n \in \mathbb{N}$. Following [36, 37], given $\alpha > 0$ and $\varepsilon \in (0, \varepsilon_*]$, define $r_\varepsilon \doteq \exp(-\varepsilon^{-\alpha})$, and suppose there exist an integer $N(\varepsilon) \geq 1$ and a collection of points $\xi_1^\varepsilon, \ldots, \xi_{N(\varepsilon)}^\varepsilon \in \mathbb{R}^2$ satisfying the following properties,

$$\begin{aligned}
&\xi_n^\varepsilon + r_\varepsilon \overline{K_n} \subset D(\xi_n^\varepsilon, \delta_0 \, r_\varepsilon) \subset D(\xi_n^\varepsilon, \delta_1 \, \varepsilon) \subset \Sigma \qquad \forall n \in \{1, \ldots, N(\varepsilon)\} \, , \\
&\partial D\left(\xi_n^\varepsilon, \delta_1 \, \varepsilon\right) \cap \partial D\left(\xi_m^\varepsilon, \delta_1 \, \varepsilon\right) = \emptyset \qquad \forall n, m \in \{1, \ldots, N(\varepsilon)\} \, , \; n \neq m \, , \qquad (1) \\
&\partial D\left(\xi_n^\varepsilon, \delta_1 \, \varepsilon\right) \cap \partial \Sigma = \emptyset \qquad \forall n \in \{1, \ldots, N(\varepsilon)\} \, ,
\end{aligned}$$

for some constants $\delta_0, \delta_1 > 0$ that are independent of $\varepsilon \in (0, \varepsilon_*]$. Setting $K_n^\varepsilon \doteq \xi_n^\varepsilon + r_\varepsilon \, K_n$ for every $n \in \{1, \ldots, N(\varepsilon)\}$, we will refer to the family $\{K_n^\varepsilon\}_{n=1}^{N(\varepsilon)}$

satisfying (1) as the *perforations* or *apertures*, while the set

$$\Gamma_\varepsilon \doteq \Sigma \setminus \overline{K_\varepsilon} \doteq \Sigma \setminus \bigcup_{n=1}^{N(\varepsilon)} \overline{K_n^\varepsilon} \tag{2}$$

represents the *sieve* or *grill* at the ε-level. We emphasize that, given $\varepsilon \in (0, \varepsilon_*]$, the family of apertures $\{K_n^\varepsilon\}_{n=1}^{N(\varepsilon)}$ is built in such a way that the *size* of each perforation is proportional to r_ε, while the mutual distance between any two consecutive holes is proportional to ε. Moreover, since we only consider those apertures that are *strictly* contained in Σ (in the sense of $(1)_3$), the following bound on the number $N(\varepsilon)$ holds:

$$N(\varepsilon) \leq \left(\frac{R}{\delta_1\,\varepsilon}\right)^2 \tag{3}$$

Notice, however, that the sets $\{K_n^\varepsilon\}_{n=1}^{N(\varepsilon)}$ may have different shapes and that they are not necessarily periodically distributed in Σ. We now define the *fluid domain* (open, connected, and bounded set) at the ε-level as

$$\Omega_\varepsilon \doteq \Omega \setminus \overline{\Gamma_\varepsilon} = \Omega_- \cup K_\varepsilon \cup \Omega_+\,, \tag{4}$$

where

$$\Omega_- \doteq \{(x,y,z) \in \Omega \mid -h < z < 0\}\,, \qquad \Omega_+ \doteq \{(x,y,z) \in \Omega \mid 0 < z < h\}\,. \tag{5}$$

The boundary of Ω_ε is decomposed as $\partial\Omega_\varepsilon = \Gamma_I \cup \Gamma_W^\varepsilon \cup \Gamma_O$, where

$$\begin{aligned}
\Gamma_I &= \Big\{(x,y,z) \in \mathbb{R}^3 \mid x^2 + y^2 < R^2,\ z = -h\Big\}\,,\\
\Gamma_W^\varepsilon &= \mathcal{L} \cup \Gamma_\varepsilon \doteq \Big\{(x,y,z) \in \mathbb{R}^3 \mid x^2 + y^2 = R^2,\ -h < z < h\Big\} \cup \Gamma_\varepsilon\,,\\
\Gamma_O &= \Big\{(x,y,z) \in \mathbb{R}^3 \mid x^2 + y^2 < R^2,\ z = h\Big\}\,.
\end{aligned} \tag{6}$$

The outward unit normal to $\partial\Omega_\varepsilon$ is denoted by ν (with some abuse of notation, as such vector also depends on ε). Henceforth, we will refer to Γ_I and Γ_O in (6) as the *inlet* and *outlet* of Ω, respectively, while Γ_W^ε includes all the *physical walls* of Ω_ε; see Fig. 2.

Given $\varepsilon \in (0, \varepsilon_*]$, we analyze the steady motion of a viscous incompressible fluid (having a constant kinematic viscosity equal to 1) along Ω_ε, which is characterized by its velocity vector field $u_\varepsilon : \Omega_\varepsilon \longrightarrow \mathbb{R}^3$ and its scalar pressure $p_\varepsilon : \Omega_\varepsilon \longrightarrow \mathbb{R}$, under the action of an external force $f : \Omega \longrightarrow \mathbb{R}^3$. Such stationary motion

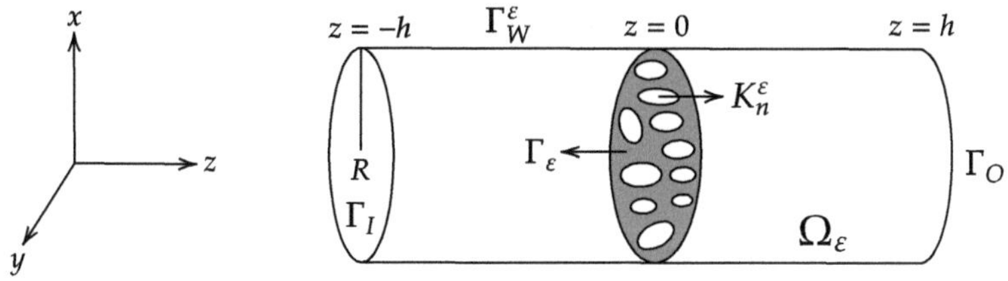

Fig. 2 Representation of the fluid domain Ω_ε and the sieve Γ_ε

will be modeled through the *prescribed pressure drop problem* in Ω_ε, that is, the following boundary-value problem (with mixed boundary conditions) associated with the steady-state Navier-Stokes equations in Ω_ε:

$$\begin{cases} -\Delta u_\varepsilon + (u_\varepsilon \cdot \nabla) u_\varepsilon + \nabla p_\varepsilon = f\,, \quad \nabla \cdot u_\varepsilon = 0 \ \text{ in } \ \Omega_\varepsilon\,, \\ u_\varepsilon = 0 \ \text{ on } \ \Gamma_W^\varepsilon\,, \\ u_\varepsilon \times \nu = 0\,, \quad p_\varepsilon + \dfrac{1}{2}|u_\varepsilon|^2 = p^- \ \text{ on } \ \Gamma_I\,, \\ u_\varepsilon \times \nu = 0\,, \quad p_\varepsilon + \dfrac{1}{2}|u_\varepsilon|^2 = p^+ \ \text{ on } \ \Gamma_O \end{cases} \tag{7}$$

While $(7)_2$ describes the usual no-slip boundary condition on the physical walls Γ_W^ε, the first equality in $(7)_3$-$(7)_4$ dictates that the fluid must enter and leave the domain Ω orthogonal to the inlet and outlet walls. The second identity in $(7)_3$-$(7)_4$ imposes that, respectively, on the inlet Γ_I and outlet Γ_O, the *Bernoulli pressure* defined as $\Phi_\varepsilon \doteq p_\varepsilon + |u_\varepsilon|^2/2$ must equal to some prescribed constants $p^\mp \in \mathbb{R}$ that represent the *pressure drop* $p^+ - p^-$ along the pipe containing the sieve. Given a velocity field $u_\varepsilon \in C^2(\Omega_\varepsilon) \cap C(\overline{\Omega_\varepsilon})$ solving (7), observe that the transversal flux rate

$$F_\varepsilon \doteq \int_{\Sigma(s)} u_\varepsilon \cdot \widehat{k} \qquad \forall s \in [-h, h] \tag{8}$$

is constant along the pipe but *depends* on the solution, where $\Sigma(0) \doteq \Sigma$ and

$$\Sigma(s) \doteq \left\{(x, y, z) \in \mathbb{R}^3 \mid 0 \le x^2 + y^2 < R^2,\ z = s\right\} \qquad \forall s \in [-h, h] \setminus \{0\}\,. \tag{9}$$

Indeed, given $s \in (0, h]$, we define the region

$$\Omega_\varepsilon(s) \doteq \{(x, y, z) \in \Omega_\varepsilon \mid -h < z < s\}\,.$$

Since u_ε vanishes on Γ_W^ε, after applying the divergence theorem (separately in Ω_- and $\Omega_\varepsilon(s) \cap \Omega_+$), we infer the following,

$$
\begin{aligned}
0 = \int_{\Omega_\varepsilon(s)} \nabla \cdot u_\varepsilon &= \int_{\Omega_-} \nabla \cdot u_\varepsilon + \int_{\Omega_\varepsilon(s)\cap\Omega_+} \nabla \cdot u_\varepsilon \\
&= -\int_{\Gamma_I} u_\varepsilon \cdot \widehat{k} + \int_{K_\varepsilon} u_\varepsilon \cdot \widehat{k} + \int_{\Sigma(s)} u_\varepsilon \cdot \widehat{k} - \int_{K_\varepsilon} u_\varepsilon \cdot \widehat{k} \\
&= \int_{\Sigma(s)} u_\varepsilon \cdot \widehat{k} - \int_{\Gamma_I} u_\varepsilon \cdot \widehat{k} \, ,
\end{aligned}
$$

so that

$$
\int_{\Sigma(s)} u_\varepsilon \cdot \widehat{k} = \int_{\Gamma_I} u_\varepsilon \cdot \widehat{k} \qquad \forall s \in (0, h] \, .
$$

In a similar way, one can prove (8) for every $s \in [-h, 0]$.

In view of the identity

$$
\nabla \left(\frac{1}{2} |u_\varepsilon|^2 \right) = (\nabla u_\varepsilon)^\top u_\varepsilon \quad \text{in} \quad \Omega_\varepsilon \, ,
$$

it is customary (see, for example, [24, 26]) to add the term $(\nabla u_\varepsilon)^\top u_\varepsilon$ to both sides of the equation of conservation of momentum $(7)_1$, thereby resulting in the problem

$$
\begin{cases}
-\Delta u_\varepsilon + (u_\varepsilon \cdot \nabla) u_\varepsilon - (\nabla u_\varepsilon)^\top u_\varepsilon + \nabla \Phi_\varepsilon = f \, , \quad \nabla \cdot u_\varepsilon = 0 \;\; \text{in} \;\; \Omega_\varepsilon \, , \\
u_\varepsilon = 0 \;\; \text{on} \;\; \Gamma_W^\varepsilon \, , \\
u_\varepsilon \times \nu = 0 \, , \quad \Phi_\varepsilon = p^- \;\; \text{on} \;\; \Gamma_I \, , \\
u_\varepsilon \times \nu = 0 \, , \quad \Phi_\varepsilon = p^+ \;\; \text{on} \;\; \Gamma_O \, .
\end{cases}
\tag{10}
$$

As already announced, the main goal of the present manuscript is to study the asymptotic behavior of the solutions of (10) as $\varepsilon \to 0^+$. Our main homogenization result reads as follows:

Theorem 1.1 *Let* $(\Omega_\varepsilon)_{\varepsilon \in I_*}$ *be the family of domains* (4) *and* $\alpha > 0$. *Given any* $p^\pm \in \mathbb{R}$ *and* $f \in L^2(\Omega)$, *let* $(u_\varepsilon, \Phi_\varepsilon) \in H^1(\Omega_\varepsilon) \times L^2(\Omega_\varepsilon)$ *be a weak solution of problem* (10). *Then, up to the extraction of a subsequence, the sequences of restrictions* $\{(u_\varepsilon^\pm, \Phi_\varepsilon^\pm)\}_{\varepsilon \in I_*} \subset H^1(\Omega_\pm) \times L^2(\Omega_\pm)$ *converge strongly to weak solutions*

$(u^{\pm}, \Phi^{\pm}) \in H^1(\Omega_{\pm}) \times L^2(\Omega_{\pm})$ *of the following boundary-value problems, in* $\Omega_{\pm}$, *as* $\varepsilon \to 0^+$:

$$\begin{cases} -\Delta u^- + (u^- \cdot \nabla)u^- - (\nabla u^-)^\top u^- + \nabla\Phi^- = f^- , & \nabla \cdot u^- = 0 \;\; in \;\; \Omega_- , \\ u^- \times \nu = 0 , \quad \Phi^- = p^- \;\; on \;\; \Gamma_I , \\ u^- = 0 \;\; on \;\; (\partial\Omega_- \cap \mathcal{L}) \cup \Sigma \end{cases} \tag{11}$$

and

$$\begin{cases} -\Delta u^+ + (u^+ \cdot \nabla)u^+ - (\nabla u^+)^\top u^+ + \nabla\Phi^+ = f^+ , & \nabla \cdot u^+ = 0 \;\; in \;\; \Omega_+ , \\ u^+ \times \nu = 0 , \quad \Phi^+ = p^+ \;\; on \;\; \Gamma_O , \\ u^+ = 0 \;\; on \;\; (\partial\Omega_+ \cap \mathcal{L}) \cup \Sigma \end{cases} \tag{12}$$

In order to prove Theorem 1.1, we firstly ensure the unrestricted weak solvability of problem (10) (concerning the size of the data of the problem, i.e., external force and prescribed pressure drop) for a fixed $\varepsilon \in (0, \varepsilon_*]$. This is achieved in Theorem 2.1 by means of the Leray-Schauder principle, following the outlines of [26, Theorem 3.2]. Alongside this result, in Sect. 2, we provide a series of lemmas that allow us, in turn, to derive the uniform ε-independent bounds for the solutions of (10) (Theorem 3.1) that are necessary to produce the homogenization limit in Theorem 1.1. Among those lemmas, we underline the outcome of Lemma 2.3 guaranteeing the unbounded growth, as $\varepsilon \to 0^+$, of the continuity constant of the right inverse of the divergence operator in Ω_ε. One of the main difficulties of our problem arises as a consequence of this, since such continuity constant is essential for the obtainment of pressure estimates; we are then led to separate our analysis into each of the sub-domains $\Omega_{\pm}$, determining the projection of the Bernoulli pressure Φ_ε onto the spaces of square-integrable functions having zero mean value in $\Omega_{\pm}$. Subsequently, applying the classical energy method in homogenization theory [42, Appendix], in Sect. 4 (specifically, Theorem 4.1), we show that, as $\varepsilon \to 0^+$, the sieve Γ_ε asymptotically becomes the *physical* wall Σ, meaning that the effective equations are the two, independent, stationary Navier-Stokes systems (11)–(12) with a no-slip boundary condition on Σ. Quite surprisingly, this result holds for any value of $\alpha > 0$, that is, regardless of the initial size of the perforations. This is in clear contrast with other existing results in the homogenization literature, where three asymptotic outcomes can be obtained depending upon the value of $\alpha > 0$; see [2, 11, 17, 20, 31, 41, 47]. In broad terms, a priori one would expect to generate the following effective laws:[2]

- In the subcritical case, when $0 < \alpha < 2$, the perforations are too large, and therefore the presence of the sieve does not affect the fluid motion. The homogenized equations then coincide with those of the original prescribed pressure drop problem in the whole pipe Ω.
- In the critical case, when $\alpha = 2$, the homogenized equations exhibit an *unexpected* nonhomogeneous Dirichlet boundary condition on Σ (leakage); see [11, Section 2.2].
- In the supercritical case, when $\alpha > 2$, the perforations are too small, so that the sieve *converges* to the wall Σ, and the homogenized equations concur precisely with (11)–(12).

Nevertheless, special attention should be drawn to [11, Théorème 2.3 bis], where the author proves a result analogous to Theorem 1.1 in the critical case $\alpha = 2$ but assuming that the transverse flux across Σ (which, under nonhomogeneous Dirichlet boundary conditions, is independent of $\varepsilon \in (0, \varepsilon_*]$) is null. In our setting, the ε-uniform bounds obtained in Theorem 3.1, combined with an ε-dependent trace inequality on Σ satisfied by every weak solution of (10), enable us to show that F_ε decays as $\sqrt{r_\varepsilon}$ as $\varepsilon \to 0^+$; see Lemma 2.2 and Corollary 3.1.

2 Solvability of the Boundary-Value Problem at the ε-Level

Let $\varepsilon \in I_*$ be a fixed parameter, with $I_* \doteq (0, \varepsilon_*]$. As in [45, Section 2], we introduce the functional spaces (of vector fields) that will be employed hereafter,

$$\mathcal{V}(\Omega_\varepsilon) \doteq \left\{ v \in H^1(\Omega_\varepsilon) \mid v \times \nu = 0 \text{ on } \Gamma_I \cup \Gamma_O ; \qquad v = 0 \text{ on } \Gamma_W^\varepsilon \right\}$$

and

$$\mathcal{V}_\sigma(\Omega_\varepsilon) \doteq \{ v \in \mathcal{V}(\Omega_\varepsilon) \mid \nabla \cdot v = 0 \text{ in } \Omega_\varepsilon \} ,$$

which are Hilbert spaces if endowed with the Dirichlet scalar product of the gradients, denoted by

$$[v, w]_{\mathcal{V}(\Omega_\varepsilon)} \doteq \int_{\Omega_\varepsilon} \nabla v \cdot \nabla w \qquad \forall v, w \in \mathcal{V}(\Omega_\varepsilon) . \tag{13}$$

[2] The determination of $\alpha = 2$ as the critical value follows from the estimates, with respect to $\varepsilon \in (0, \varepsilon_*]$, concerning the two-dimensional *relative capacity* of the perforations K_ε inside the disk Σ, as in [40, Lemma 2.1].

Remark 2.1 Notice that Ω_ε is *not* a locally Lipschitz domain: in a neighborhood of the perforated wall Γ_ε, Ω_ε is not locally placed on one side of its boundary. However, the domains $\Omega_\pm$ are Lipschitz (see (5)), so that the elements of $H^1(\Omega_\pm)$ have a well-defined trace on Σ. Accordingly, to fix notation and following [11, 12], given any function $v \in L^1(\Omega_\varepsilon)$ (scalar or vector), we denote by $v^\pm \in L^1(\Omega_\pm)$ its restriction to $\Omega_\pm$. It can then be easily proved (as in [21, Theorem 1.7.1]) that for any $v \in L^2(\Omega_\varepsilon)$ (scalar or vector), there holds

$$v \in H^1(\Omega_\varepsilon) \qquad \Longleftrightarrow \qquad v^\pm \in H^1(\Omega_\pm) \quad \text{and} \quad v^+|_{K_\varepsilon} = v^-|_{K_\varepsilon} \,.$$

In particular, for any $v \in L^2(\Omega_\varepsilon)$ (scalar or vector), it follows that

$$v \in H^1_0(\Omega_\varepsilon) \quad \Longleftrightarrow \quad v^\pm \in H^1(\Omega_\pm)\,, \quad v^+|_{K_\varepsilon} = v^-|_{K_\varepsilon}\,, \quad v|_{\partial\Omega} = v^\pm|_{\Gamma_\varepsilon} = 0\,.$$

Nevertheless, given $v \in H^1(\Omega_\varepsilon)$ (scalar or vector), the traces $v^\pm|_{\Gamma_\varepsilon} \in H^{1/2}(\Gamma_\varepsilon)$ do not necessarily coincide almost everywhere on Γ_ε.

Before proving the weak solvability of problem (10) (equivalently, of problem (7)), we will state a series of lemmas that shall play a key role in the sequel. The first (elementary) one concerns the uniform boundedness (with respect to $\varepsilon \in I_*$) of some Sobolev embedding constants:

Lemma 2.1 *Let Ω_ε be as in* (4). *For any $p \in [1, 6]$ and any function (scalar or vector) $\varphi \in H^1(\Omega_\varepsilon)$ that vanishes on $\mathcal{L}$, there holds the estimate*

$$\|\varphi\|_{L^p(\Omega_\varepsilon)} \leq C_* \|\nabla\varphi\|_{L^2(\Omega_\varepsilon)} \qquad \forall \varepsilon \in I_*\,, \tag{14}$$

for some constant $C_* > 0$ that depends on Ω, $p \in [1, 6]$, and $\{\delta_0, \delta_1\}$ but is independent of $\varepsilon \in I_*$.

Proof Given $p \in [1, 6]$, in what follows, $C > 0$ will always denote a generic constant that depends on Ω, $p \in [1, 6]$, and $\{\delta_0, \delta_1\}$ (independent of $\varepsilon \in I_*$) but that may change from line to line.

Given any function (scalar or vector) $\varphi \in H^1(\Omega_\varepsilon)$ vanishing on $\mathcal{L}$, since $\varphi^\pm \in H^1(\Omega_\pm)$ vanishes on $\mathcal{L} \cap \partial\Omega_\pm$, we can apply the Sobolev inequality, separately in $\Omega_\pm$, to deduce that

$$\begin{aligned}\|\varphi\|^p_{L^p(\Omega_\varepsilon)} &= \|\varphi^-\|^p_{L^p(\Omega_-)} + \|\varphi^+\|^p_{L^p(\Omega_+)} \\ &\leq C\left(\|\nabla\varphi^-\|^p_{L^2(\Omega_-)} + \|\nabla\varphi^+\|^p_{L^2(\Omega_+)}\right) \leq 2C\|\nabla\varphi\|^p_{L^2(\Omega_\varepsilon)}\,,\end{aligned}$$

which finishes the proof. □

Our second lemma in this section, in the spirit of [11, Lemme 3.2], involves a ε-dependent trace inequality for functions that vanish on the wall Γ_ε.

Lemma 2.2 *Let Ω_ε be as in* (4). *For any function (scalar or vector) $\varphi \in H^1(\Omega_\varepsilon)$ that vanishes on Γ_ε, there holds the estimate*

$$\|\varphi\|_{L^2(\Sigma)} \le C_* \sqrt{r_\varepsilon}\, \|\nabla\varphi\|_{L^2(\Omega_\varepsilon)} \qquad \forall \varepsilon \in I_* \,, \tag{15}$$

for some constant $C_* > 0$ that depends on Ω and $\{\delta_0, \delta_1\}$ but is independent of $\varepsilon \in I_*$.

Proof In what follows, $C > 0$ will always denote a generic constant that depends on Ω and $\{\delta_0, \delta_1\}$ (independent of $\varepsilon \in I_*$), but that may change from line to line.

Given any $n \in \{1, \dots, N(\varepsilon)\}$, let $\{(a_n^\varepsilon, b_n^\varepsilon)\}_{n=1}^{N(\varepsilon)} \subset \mathbb{R}^2$ be such that $\xi_n^\varepsilon = (a_n^\varepsilon, b_n^\varepsilon)$; see again (1). For $n \in \{1, \dots, N(\varepsilon)\}$, define then the (open) cylinders

$$\mathcal{P}_n^\varepsilon \doteq \left\{ (x, y, z) \in \mathbb{R}^3 \;\middle|\; \left(\frac{x - a_n^\varepsilon}{r_\varepsilon}\right)^2 + \left(\frac{y - b_n^\varepsilon}{r_\varepsilon}\right)^2 < \delta_0^2 \,,\; 0 < z < r_\varepsilon\, h \right\}$$

and also $\mathcal{P}_0 \doteq D((0,0), \delta_0) \times (0, h)$. Therefore, given any $\varphi \in H^1(\Omega_\varepsilon)$ (scalar or vector) vanishing on Γ_ε, the trace inequality

$$\|\varphi\|_{L^2(D((0,0),\delta_0))} \le C \|\nabla\varphi^+\|_{L^2(\mathcal{P}_0)} \tag{16}$$

certainly holds (see Remark 2.1). Applying (16) and the changes of variables

$$(x, y) \in D\left(\xi_n^\varepsilon, \delta_0\, r_\varepsilon\right) \longleftrightarrow \left(\frac{x - a_n^\varepsilon}{r_\varepsilon}, \frac{y - b_n^\varepsilon}{r_\varepsilon}\right) \in D((0,0), \delta_0) \,,$$

$$(x, y, z) \in \mathcal{P}_n^\varepsilon \longleftrightarrow \left(\frac{x - a_n^\varepsilon}{r_\varepsilon}, \frac{y - b_n^\varepsilon}{r_\varepsilon}, \frac{z}{r_\varepsilon}\right) \in \mathcal{P}_0 \,,$$

we obtain

$$\begin{aligned}
\|\varphi\|^2_{L^2(\Sigma)} &= \sum_{n=1}^{N(\varepsilon)} \left(\int_{K_n^\varepsilon} |\varphi(x, y, 0)|^2 \, dx\, dy \right) \\
&\le \sum_{n=1}^{N(\varepsilon)} \left(\int_{D(\xi_n^\varepsilon, \delta_0 r_\varepsilon)} |\varphi(x, y, 0)|^2 \, dx\, dy \right) \\
&= r_\varepsilon^2 \sum_{n=1}^{N(\varepsilon)} \left(\int_{D((0,0),\delta_0)} |\varphi(r_\varepsilon x + a_n^\varepsilon, r_\varepsilon y + b_n^\varepsilon, 0)|^2 \, dx\, dy \right)
\end{aligned}$$

$$\leq C\, r_\varepsilon^4 \sum_{n=1}^{N(\varepsilon)} \left(\int_{\mathcal{P}_0} |\nabla \varphi^+(r_\varepsilon\, x + a_n^\varepsilon, r_\varepsilon\, y + b_n^\varepsilon, r_\varepsilon\, z)|^2 \, dx\, dy\, dz \right)$$

$$= C\, r_\varepsilon \sum_{n=1}^{N(\varepsilon)} \left(\int_{\mathcal{P}_n^\varepsilon} |\nabla \varphi^+(x, y, z)|^2 \, dx\, dy\, dz \right) \leq C\, r_\varepsilon \|\nabla \varphi\|^2_{L^2(\Omega_\varepsilon)} ,$$

thus concluding the proof. □

Let $Q \subset \mathbb{R}^3$ be any bounded domain that can be decomposed as the union of a finite number of open sets, each one being star-shaped with respect to some ball strictly contained in it.[3] Consider the space of square-integrable functions in Q having zero mean value:

$$L_0^2(Q) \doteq \left\{ g \in L^2(Q) \;\middle|\; \int_Q g = 0 \right\} \tag{17}$$

We define the *Bogovskii constant* of Q as

$$C_B(Q) \doteq \sup_{g \in L_0^2(Q)\setminus\{0\}} \inf \left\{ \frac{\|\nabla v\|_{L^2(Q)}}{\|g\|_{L^2(Q)}} \;\middle|\; v \in H_0^1(Q)\,,\ \nabla \cdot v = g \ \text{ in } \ Q \right\} ; \tag{18}$$

see [18, Section 2] and also [19, Remark III.3.4]. Bogovskii [4] showed that, given any $q \in L_0^2(Q)$, there exists a vector field $X \in H_0^1(Q)$ such that $\nabla \cdot X = q$ in Q and

$$\|\nabla X\|_{L^2(Q)} \leq C_B(Q) \|q\|_{L^2(Q)} .$$

For a fixed $\varepsilon \in I_*$, the fluid domain Ω_ε in (4) can be certainly decomposed as the union of a finite number of open sets, each one being star-shaped with respect to some ball strictly contained in it. Our next result shows the unbounded growth of the Bogovskii constant of Ω_ε with respect to $\varepsilon \in I_*$.

Lemma 2.3 *Let Ω_ε be as in* (4). *There holds the lower bound*

$$C_B(\Omega_\varepsilon) \geq \frac{C_*}{\sqrt{r_\varepsilon}} \qquad \forall \varepsilon \in I_* , \tag{19}$$

for some constant $C_* > 0$ that depends on Ω and $\{\delta_0, \delta_1\}$ but is independent of $\varepsilon \in I_*$.

---

[3] This condition is satisfied, for example, by any bounded domain in $\mathbb{R}^3$ enjoying the *interior cone property*; consult [35, Section 2.3.4] and also [19, Remark III.3.4].

Proof In what follows, $C > 0$ will always denote a generic constant that depends on Ω and $\{\delta_0, \delta_1\}$ (independent of $\varepsilon \in I_*$), but that may change from line to line.

Define the function $g_* \in L^2_0(\Omega)$ (see (17)) by

$$g_*(x, y, z) = \begin{cases} -1 & \text{if } (x, y, z) \in \Omega_- \\ 0 & \text{if } (x, y, z) \in \Sigma \\ 1 & \text{if } (x, y, z) \in \Omega_+ , \end{cases}$$

and therefore, we also have $g_* \in L^2_0(\Omega_\varepsilon)$ with $\|g_*\|_{L^2(\Omega_\varepsilon)} = \sqrt{|\Omega|}$. It then suffices to prove that

$$\|\nabla J\|_{L^2(\Omega_\varepsilon)} \geq \frac{C}{\sqrt{r_\varepsilon}} \quad \text{for any vector field } J \in H^1_0(\Omega_\varepsilon) \text{ such that } \nabla \cdot J = g_* \text{ in } \Omega_\varepsilon .$$

Let $J \in H^1_0(\Omega_\varepsilon)$ be any vector field such that $\nabla \cdot J = g_*$ almost everywhere in Ω_ε. In view of the divergence theorem, we then have

$$\int_{K_\varepsilon} J \cdot \widehat{k} = \int_{\Sigma} J \cdot \widehat{k} = \int_{\Omega_-} \nabla \cdot J^- = \int_{\Omega_-} g_*^- = -\pi R^2 h .$$

Then, applying Hölder's inequality and (15), we deduce

$$\pi R^2 h = \left| \int_{\Sigma} J \cdot \widehat{k} \right| \leq \|J\|_{L^1(\Sigma)} \leq \sqrt{\pi}\, R\, \|J\|_{L^2(\Sigma)} \leq C\sqrt{r_\varepsilon}\, \|\nabla J\|_{L^2(\Omega_\varepsilon)} ,$$

which finishes the proof. □

Another essential preliminary result concerns the construction of a *Bogovskii-type* operator on the space $L^2(\Omega_\varepsilon)$, which exploits the corresponding operator on $L^2_0(\Omega_\varepsilon)$. Inspired by [26, Lemma 4.2] and [45, Lemma 2.2], we have the following:

Lemma 2.4 *Let Ω_ε be as in* (4) *and $q \in L^2(\Omega_\varepsilon)$. Then, there exists a vector field $Y_\varepsilon \in \mathcal{V}(\Omega_\varepsilon)$ such that*

$$\nabla \cdot Y_\varepsilon = q \ \text{ in } \Omega_\varepsilon , \qquad \|\nabla Y_\varepsilon\|_{L^2(\Omega_\varepsilon)} \leq C_*(1 + C_B(\Omega_\varepsilon))\|q\|_{L^2(\Omega_\varepsilon)} \quad \forall \varepsilon \in I_* , \tag{20}$$

for some constant $C_* > 0$ that depends on Ω and $\{\delta_0, \delta_1\}$ but is independent of $\varepsilon \in I_*$.

Proof In what follows, $C > 0$ will always denote a generic constant that depends on Ω and $\{\delta_0, \delta_1\}$ (independent of $\varepsilon \in I_*$), but that may change from line to line.

Consider a Hagen-Poiseuille flow having unit flow rate in Ω, that is,

$$U_0(x,y,z) \doteq \frac{2}{\pi R^4}(R^2 - x^2 - x^2)\widehat{k} \qquad \forall (x,y,z) \in \overline{\Omega}\,; \tag{21}$$

see [28, Chapter II] for further details. Clearly, $U_0 \in C^\infty(\overline{\Omega})$ is divergence-free; it vanishes on $\mathcal{L}$ and $U_0 \times \nu = 0$ on $\Gamma_I \cup \Gamma_O$. Moreover,

$$\int_{\Gamma_O} U_0 \cdot \widehat{k} = 1\,. \tag{22}$$

We then define the vector field

$$Q_\varepsilon(x,y,z) \doteq \left(\int_{\Omega_\varepsilon} q\right) \frac{z(z+h)}{2h^2}\, U_0(x,y,z) \qquad \forall (x,y,z) \in \Omega_\varepsilon\,, \tag{23}$$

which is clearly an element of $\mathcal{V}(\Omega_\varepsilon) \cap C^\infty(\overline{\Omega})$ such that $Q_\varepsilon = 0$ on $\Gamma_I \cup \Sigma$ and

$$\|\nabla Q_\varepsilon\|_{L^2(\Omega_\varepsilon)} \le C\|q\|_{L^2(\Omega_\varepsilon)}\,. \tag{24}$$

On the other hand, from the divergence theorem and (22), we obtain

$$\int_{\Omega_\varepsilon} \nabla \cdot Q_\varepsilon = \int_{\Omega_-} \nabla \cdot Q_\varepsilon^- + \int_{\Omega_+} \nabla \cdot Q_\varepsilon^+ = \int_{\Gamma_O} Q_\varepsilon^+ \cdot \widehat{k} = \int_{\Omega_\varepsilon} q\,,$$

so that $q - \nabla \cdot Q_\varepsilon \in L^2_0(\Omega_\varepsilon)$ (see (17)). Then, there exists a vector field $X_\varepsilon \in H^1_0(\Omega_\varepsilon)$ such that

$$\begin{cases} \nabla \cdot X_\varepsilon = q - \nabla \cdot Q_\varepsilon \quad \text{in} \quad \Omega_\varepsilon\,, \\ \|\nabla X_\varepsilon\|_{L^2(\Omega_\varepsilon)} \le C_B(\Omega_\varepsilon)\|q - \nabla \cdot Q_\varepsilon\|_{L^2(\Omega_\varepsilon)}\,. \end{cases} \tag{25}$$

We finally define $Y_\varepsilon \doteq X_\varepsilon + Q_\varepsilon$ which, in view of (24), is an element of $\mathcal{V}(\Omega_\varepsilon)$ satisfying (20). □

We give the following definition for the weak solutions of problem (10) (equally of problem (7)):

Definition 2.1 A vector field $u \in \mathcal{V}_\sigma(\Omega_\varepsilon)$ is called a *weak solution* of (10) if

$$\int_{\Omega_\varepsilon} \nabla u \cdot \nabla \varphi + \int_{\Omega_\varepsilon} \left[\nabla u - (\nabla u)^\top\right] u \cdot \varphi + p^+ \left(\int_{\Gamma_O} \varphi \cdot \widehat{k}\right) - p^- \left(\int_{\Gamma_I} \varphi \cdot \widehat{k}\right) = \int_{\Omega_\varepsilon} f \cdot \varphi\,,$$

for every vector field $\varphi \in \mathcal{V}_\sigma(\Omega_\varepsilon)$.

For the sake of clarity, let us explain how the weak formulation of Definition 2.1 arises. Suppose that $u = (u_1, u_2, u_3) \in C^2(\overline{\Omega_\varepsilon})$ and $\Phi \in C^1(\overline{\Omega_\varepsilon})$ solve (10) in the classical sense. Given any vector field $\varphi = (\varphi_1, \varphi_2, \varphi_3) \in H^1(\Omega_\varepsilon)$, we integrate by parts (separately in $\Omega_\pm$) in the following way:

$$\begin{aligned}\int_{\Omega_\varepsilon}(-\Delta u + \nabla\Phi)\cdot\varphi &= \int_{\Omega_-}(-\Delta u^- + \nabla\Phi^-)\cdot\varphi^- + \int_{\Omega_+}(-\Delta u^+ + \nabla\Phi^+)\cdot\varphi^+\\ &= \int_{\Omega_\varepsilon}\nabla u\cdot\nabla\varphi - \int_{\Omega_\varepsilon}\Phi(\nabla\cdot\varphi)\\ &\quad + \int_{\partial\Omega_-}\left(\Phi^-\nu - \frac{\partial u^-}{\partial\nu}\right)\cdot\varphi^- + \int_{\partial\Omega_+}\left(\Phi^+\nu - \frac{\partial u^+}{\partial\nu}\right)\cdot\varphi^+\end{aligned} \tag{26}$$

If, in addition, we assume that φ is divergence-free and vanishes on Γ_W^ε, from (26), we obtain

$$\begin{aligned}&\int_{\Omega_\varepsilon}(-\Delta u + \nabla\Phi)\cdot\varphi\\ &= \int_{\Omega_\varepsilon}\nabla u\cdot\nabla\varphi + \int_{\Gamma_I}\left(\Phi^-\nu - \frac{\partial u^-}{\partial\nu}\right)\cdot\varphi^- + \int_{K_\varepsilon}\left(\Phi^-\nu - \frac{\partial u^-}{\partial\nu}\right)\cdot\varphi^-\\ &\quad + \int_{\Gamma_O}\left(\Phi^+\nu - \frac{\partial u^+}{\partial\nu}\right)\cdot\varphi^+ + \int_{K_\varepsilon}\left(\Phi^+\nu - \frac{\partial u^+}{\partial\nu}\right)\cdot\varphi^+\\ &= \int_{\Omega_\varepsilon}\nabla u\cdot\nabla\varphi + \int_{\Gamma_I}\left(\Phi^-\nu - \frac{\partial u^-}{\partial\nu}\right)\cdot\varphi^- + \int_{\Gamma_O}\left(\Phi^+\nu - \frac{\partial u^+}{\partial\nu}\right)\cdot\varphi^+,\end{aligned} \tag{27}$$

where the integrals over K_ε cancel each other because of the opposite signs of the outward unit normal to $\partial\Omega_\pm$ on Σ (see Remark 2.1). Notice that $\nu = \mp\widehat{k}$ on Γ_I and Γ_O, respectively; thus, $u_1 = u_2 = 0$ on $\Gamma_I \cup \Gamma_O$, in view of $(10)_3$-$(10)_4$. The regularity and incompressibility condition of u then imply that

$$\frac{\partial u_3^-}{\partial z} = -\left(\frac{\partial u_1^-}{\partial x} + \frac{\partial u_2^-}{\partial y}\right) = 0 \quad \text{on} \quad \Gamma_I\,;$$

$$\frac{\partial u_3^+}{\partial z} = -\left(\frac{\partial u_1^+}{\partial x} + \frac{\partial u_2^+}{\partial y}\right) = 0 \quad \text{on} \quad \Gamma_O\,.$$

If we further impose that $\varphi \times \nu = 0$ on $\Gamma_I \cup \Gamma_O$ (so that $\varphi_1 = \varphi_2 = 0$ on $\Gamma_I \cup \Gamma_O$), we get

$$\frac{\partial u^-}{\partial \nu} \cdot \varphi^- = -\frac{\partial u_3^-}{\partial z} \varphi_3^- = 0 \quad \text{on} \quad \Gamma_I \, ; \quad \frac{\partial u^+}{\partial \nu} \cdot \varphi^+ = \frac{\partial u_3^+}{\partial z} \varphi_3^+ = 0 \quad \text{on} \quad \Gamma_O \, . \tag{28}$$

Inserting (28) into (27), together with the boundary conditions verified by the Bernoulli pressure in $(10)_3$-$(10)_4$, yields the identity

$$\int_{\Omega_\varepsilon} (-\Delta u + \nabla \Phi) \cdot \varphi = \int_{\Omega_\varepsilon} \nabla u \cdot \nabla \varphi + p^+ \left(\int_{\Gamma_O} \varphi \cdot \widehat{k} \right) - p^- \left(\int_{\Gamma_I} \varphi \cdot \widehat{k} \right) . \tag{29}$$

Finally, multiplying the equation of conservation of momentum $(10)_1$ by a vector field $\varphi \in \mathcal{V}_\sigma(\Omega_\varepsilon)$ and integrating by parts in $\Omega_\pm$ (enforcing (29)) provide the weak formulation of Definition 2.1. Conversely, we refer to [26, Section 2] for an explanation of the fact that the boundary conditions involving the Bernoulli pressure in (10) are implicitly contained in the variational formulation of Definition 2.1.

The main result of this section guarantees the unrestricted solvability (concerning the size of the data) of problem (10).

Theorem 2.1 *Let Ω_ε be as in* (4)*. For any $p^\pm \in \mathbb{R}$ and $f \in L^2(\Omega)$, there exists at least one weak solution $u_\varepsilon \in \mathcal{V}_\sigma(\Omega_\varepsilon)$ of the prescribed pressure drop problem* (10) *and a unique associated pressure $\Phi_\varepsilon \in L^2(\Omega_\varepsilon)$ such that the pair $(u_\varepsilon, \Phi_\varepsilon)$ satisfies the following identity:*

$$\begin{aligned} &\int_{\Omega_\varepsilon} \nabla u_\varepsilon \cdot \nabla \varphi + \int_{\Omega_\varepsilon} \left[\nabla u_\varepsilon - (\nabla u_\varepsilon)^\top \right] u_\varepsilon \cdot \varphi + p^+ \left(\int_{\Gamma_O} \varphi \cdot \widehat{k} \right) \\ &- p^- \left(\int_{\Gamma_I} \varphi \cdot \widehat{k} \right) - \int_{\Omega_\varepsilon} \Phi_\varepsilon (\nabla \cdot \varphi) = \int_{\Omega_\varepsilon} f \cdot \varphi \qquad \forall \varphi \in \mathcal{V}(\Omega_\varepsilon) \end{aligned} \tag{30}$$

Proof Since $\varepsilon \in I_*$ is a fixed parameter in this section, the corresponding subscript will be omitted sometimes for the sake of simplicity. Accordingly, throughout this proof, $C > 0$ will denote a generic constant that depends on Ω_ε, but that may change from line to line.

Given any $\varphi \in \mathcal{V}_\sigma(\Omega_\varepsilon)$, by arguing as in (8), one can easily prove that

$$\int_{\Gamma_I} \varphi \cdot \widehat{k} = \int_{\Gamma_O} \varphi \cdot \widehat{k} \, . \tag{31}$$

In view of Definition 2.1, we then search for a vector field $u \in \mathcal{V}_\sigma(\Omega_\varepsilon)$ such that

$$\int_{\Omega_\varepsilon} \nabla u \cdot \nabla \varphi + \int_{\Omega_\varepsilon} \mathcal{E}(u) u \cdot \varphi + (p^+ - p^-) \int_{\Gamma_O} \varphi \cdot \widehat{k} = \int_{\Omega_\varepsilon} f \cdot \varphi \, , \tag{32}$$

for every vector field $\varphi \in \mathcal{V}_\sigma(\Omega_\varepsilon)$. In (32), we denote by $\mathcal{E}(w) \doteq \nabla w - (\nabla w)^\top$ the skew-symmetric gradient of any vector field $w \in H^1(\Omega_\varepsilon)$. For a given and fixed $u \in \mathcal{V}_\sigma(\Omega_\varepsilon)$, the applications

$$\varphi \in \mathcal{V}_\sigma(\Omega_\varepsilon) \longmapsto \int_{\Omega_\varepsilon} \mathcal{E}(u)u \cdot \varphi \,,$$

$$\varphi \in \mathcal{V}_\sigma(\Omega_\varepsilon) \longmapsto \int_{\Omega_\varepsilon} f \cdot \varphi - (p^+ - p^-) \int_{\Gamma_O} \varphi \cdot \widehat{k}$$

clearly define linear continuous functions on $\mathcal{V}_\sigma(\Omega_\varepsilon)$. Then, in view of the Riesz representation theorem, the identity (32) may be written as

$$[u + \mathcal{P}(u) - \mathcal{F}, \varphi]_{\mathcal{V}_\sigma(\Omega_\varepsilon)} = 0 \qquad \forall \varphi \in \mathcal{V}_\sigma(\Omega_\varepsilon) \,,$$

(see (13)), for some (unique) elements $\mathcal{P}(u), \mathcal{F} \in \mathcal{V}_\sigma(\Omega_\varepsilon)$ such that

$$[\mathcal{P}(u), \varphi]_{\mathcal{V}(\Omega_\varepsilon)} = \int_{\Omega_\varepsilon} \mathcal{E}(u)u \cdot \varphi \qquad \forall \varphi \in \mathcal{V}_\sigma(\Omega_\varepsilon) \,,$$

$$[\mathcal{F}, \varphi]_{\mathcal{V}(\Omega_\varepsilon)} = \int_{\Omega_\varepsilon} f \cdot \varphi - (p^+ - p^-) \int_{\Gamma_O} \varphi \cdot \widehat{k} \qquad \forall \varphi \in \mathcal{V}_\sigma(\Omega_\varepsilon) \,.$$

We have so defined a linear operator $\mathcal{P} : \mathcal{V}_\sigma(\Omega_\varepsilon) \longrightarrow \mathcal{V}_\sigma(\Omega_\varepsilon)$, and we are led to find a solution $u \in \mathcal{V}_\sigma(\Omega_\varepsilon)$ of the following nonlinear operator equation:

$$u + \mathcal{P}(u) - \mathcal{F} = 0 \quad \text{in } \mathcal{V}_\sigma(\Omega_\varepsilon) \tag{33}$$

Exactly as in [27, Chapter 5, Theorem 1], one can show that the operator $\mathcal{P}$ is compact. Therefore, as a consequence of the Leray-Schauder principle [49, Chapter 6], in order to prove that (33) possesses at least one solution, it suffices to guarantee that any $v^\lambda \in \mathcal{V}_\sigma(\Omega_\varepsilon)$ such that

$$v^\lambda + \lambda(\mathcal{P}(v^\lambda) - \mathcal{F}) = 0 \quad \text{in } \mathcal{V}_\sigma(\Omega_\varepsilon) \tag{34}$$

is uniformly bounded with respect to $\lambda \in [0, 1]$. Given any $\lambda \in [0, 1]$ and a vector field $v^\lambda \in \mathcal{V}_\sigma(\Omega_\varepsilon) \setminus \{0\}$ such that (34) holds, we clearly have

$$\int_{\Omega_\varepsilon} \nabla v^\lambda \cdot \nabla \varphi + \lambda \int_{\Omega_\varepsilon} \mathcal{E}(v^\lambda) v^\lambda \cdot \varphi + \lambda(p^+ - p^-) \int_{\Gamma_O} \varphi \cdot \widehat{k} = \lambda \int_{\Omega_\varepsilon} f \cdot \varphi \,, \tag{35}$$

for every $\varphi \in \mathcal{V}_\sigma(\Omega_\varepsilon)$. By putting $\varphi = v^\lambda$ in (35), we obtain

$$\|\nabla v^\lambda\|^2_{L^2(\Omega_\varepsilon)} = \lambda \int_{\Omega_\varepsilon} f \cdot v^\lambda - \lambda(p^+ - p^-) \int_{\Gamma_O} v^\lambda \cdot \widehat{k} \,. \tag{36}$$

Applying the Hölder, Poincaré, and trace inequalities, in Ω_ε (see Lemma 2.1), on the right-hand side of (36) gives us the bound

$$\|\nabla v^\lambda\|_{L^2(\Omega_\varepsilon)} \leq C\left(\|f\|_{L^2(\Omega_\varepsilon)} + |p^+ - p^-|\right) \qquad \forall \lambda \in [0,1]\,,$$

which shows that problem (10) admits, at least, one weak solution $u_\varepsilon \in \mathcal{V}_\sigma(\Omega_\varepsilon)$. Denote by $\mathcal{V}^{-1}(\Omega_\varepsilon)$ the dual space of $\mathcal{V}(\Omega_\varepsilon)$ and by $\langle\cdot,\cdot\rangle_{\mathcal{V}(\Omega_\varepsilon)}$ the duality product between $\mathcal{V}^{-1}(\Omega_\varepsilon)$ and $\mathcal{V}(\Omega_\varepsilon)$. The gradient $-\operatorname{grad}_{\Omega_\varepsilon} : L^2(\Omega_\varepsilon) \longrightarrow \mathcal{V}^{-1}(\Omega_\varepsilon)$ of a scalar function $q \in L^2(\Omega_\varepsilon)$ is defined by

$$\langle -\operatorname{grad}_{\Omega_\varepsilon}(q), \varphi\rangle_{\mathcal{V}(\Omega_\varepsilon)} \doteq \int_{\Omega_\varepsilon} q(\nabla\cdot\varphi) \qquad \forall \varphi \in \mathcal{V}(\Omega_\varepsilon)\,,$$

thus being the adjoint of the divergence operator $\operatorname{div}_{\Omega_\varepsilon} : \mathcal{V}(\Omega_\varepsilon) \longrightarrow L^2(\Omega_\varepsilon)$. Therefore, the closed range theorem of Banach [48, Chapter VII] can be applied to deduce that

$$\begin{aligned}
&\operatorname{Range}(-\operatorname{grad}_{\Omega_\varepsilon}) = \operatorname{Ker}(\operatorname{div}_{\Omega_\varepsilon})^\perp \\
&= \mathcal{V}_\sigma(\Omega_\varepsilon)^\perp \doteq \{\mathcal{F} \in \mathcal{V}^{-1}(\Omega_\varepsilon) \mid \langle \mathcal{F},\varphi\rangle_{\mathcal{V}(\Omega_\varepsilon)} = 0 \quad \forall \varphi \in \mathcal{V}_\sigma(\Omega_\varepsilon)\}\,.
\end{aligned}$$

We then define $Q_\varepsilon \in \mathcal{V}^{-1}(\Omega_\varepsilon)$ by

$$\begin{aligned}
Q_\varepsilon(\varphi) \doteq& \int_{\Omega_\varepsilon} \nabla u_\varepsilon\cdot\nabla\varphi + \int_{\Omega_\varepsilon} \mathcal{E}(u_\varepsilon)u_\varepsilon\cdot\varphi \\
&+ p^+\left(\int_{\Gamma_O}\varphi\cdot\widehat{k}\right) - p^-\left(\int_{\Gamma_I}\varphi\cdot\widehat{k}\right) - \int_{\Omega_\varepsilon} f\cdot\varphi \qquad \forall\varphi \in \mathcal{V}(\Omega_\varepsilon)\,.
\end{aligned}$$

In view of (31)–(32), we deduce that $Q_\varepsilon \in \operatorname{Range}(-\operatorname{grad}_{\Omega_\varepsilon})$, thus guaranteeing the existence of a scalar pressure $\Phi_\varepsilon \in L^2(\Omega_\varepsilon)$ such that

$$Q_\varepsilon(\varphi) = \int_{\Omega_\varepsilon} \Phi_\varepsilon(\nabla\cdot\varphi) \qquad \forall\varphi \in \mathcal{V}(\Omega_\varepsilon)\,. \tag{37}$$

Suppose there exists another scalar function $q_\varepsilon \in L^2(\Omega_\varepsilon)$ verifying (37), thereby implying

$$\int_{\Omega_\varepsilon} (\Phi_\varepsilon - q_\varepsilon)(\nabla\cdot\varphi) = 0 \qquad \forall\varphi \in \mathcal{V}(\Omega_\varepsilon)\,. \tag{38}$$

Lemma 2.4 provides a vector field $Y_\varepsilon \in \mathcal{V}(\Omega_\varepsilon)$ such that $\nabla\cdot Y_\varepsilon = \Phi_\varepsilon - q_\varepsilon$ in Ω_ε. Taking $\varphi = Y_\varepsilon$ in (38) ensures that $\Phi_\varepsilon = q_\varepsilon$ almost everywhere in Ω_ε. This concludes the proof. □

Remark 2.2 In virtue of the divergence theorem, we observe that

$$\int_{\Omega_\varepsilon} \nabla \cdot \varphi = \int_{\Gamma_O} \varphi \cdot \widehat{k} - \int_{\Gamma_I} \varphi \cdot \widehat{k} \qquad \forall \varphi \in \mathcal{V}(\Omega_\varepsilon)\,.$$

Thus, the (unique) Bernoulli pressure associated with any weak solution of problem (10) can be modified, up to an additive constant, at the expense of altering the prescribed pressure drop. In particular, given a weak solution $u_\varepsilon \in \mathcal{V}_\sigma(\Omega_\varepsilon)$ of (10), its associated pressure $\Phi_\varepsilon \in L^2(\Omega_\varepsilon)$ can be recast in such a way that $\Phi_\varepsilon \in L^2_0(\Omega_\varepsilon)$, at the cost of introducing an ε-dependent term in the pressure drop, namely, the mean value of Φ_ε over Ω_ε. That is the reason why we do not project the Bernoulli pressure onto $L^2_0(\Omega_\varepsilon)$.

Remark 2.3 In view of Theorem 2.1, in the sequel, any pair $(u, \Phi) \in \mathcal{V}_\sigma(\Omega_\varepsilon) \times L^2(\Omega_\varepsilon)$ will be called a *weak solution* of problem (10) if it satisfies the integral identity (30) for every $\varphi \in \mathcal{V}(\Omega_\varepsilon)$.

3 Boundary-Value Problem at the ε-Level: Uniform Bounds

Let $\varepsilon \in I_*$ be a fixed parameter. Denoting $\mathcal{L}_\pm \doteq \mathcal{L} \cap \partial\Omega_\pm$, we further introduce the following two spaces of vector fields:

$$\mathcal{V}(\Omega_-) \doteq \Big\{ v \in H^1(\Omega_-) \mid v \times \nu = 0 \ \text{ on } \ \Gamma_I\,; \qquad v = 0 \ \text{ on } \ \mathcal{L}_- \cup \Sigma \Big\} \ \textit{and}$$

$$\mathcal{V}(\Omega_+) \doteq \Big\{ v \in H^1(\Omega_+) \mid v \times \nu = 0 \ \text{ on } \ \Gamma_O\,; \qquad v = 0 \ \text{ on } \ \mathcal{L}_+ \cup \Sigma \Big\} \tag{39}$$

The main result of this section provides uniform bounds (with respect to $\varepsilon \in I_*$) for the solutions of problem (10).

Theorem 3.1 *Let Ω_ε be as in* (4). *For any $p^\pm \in \mathbb{R}$ and $f \in L^2(\Omega)$, let $(u_\varepsilon, \Phi_\varepsilon) \in \mathcal{V}_\sigma(\Omega_\varepsilon) \times L^2(\Omega_\varepsilon)$ be a weak solution of problem* (10). *Then, the uniform bound*

$$\sup_{\varepsilon \in I_*} \left(\|\nabla u_\varepsilon\|_{L^2(\Omega_\varepsilon)} + \|\Phi_\varepsilon\|_{L^2(\Omega_\varepsilon)} \right) \leq C_* \tag{40}$$

holds for some constant $C_* > 0$ that depends on Ω, $p^\pm$, f, and $\{\delta_0, \delta_1\}$ but is independent of $\varepsilon \in I_*$.

Proof In what follows, $C > 0$ will always denote a generic constant that depends on Ω and $\{\delta_0, \delta_1\}$ (independent of $\varepsilon \in I_*$), but that may change from line to line.

We start by noticing that, in view of (8), we have

$$\int_{\Gamma_I} u_\varepsilon \cdot \widehat{k} = \int_{\Gamma_O} u_\varepsilon \cdot \widehat{k} = \int_{\Sigma} u_\varepsilon \cdot \widehat{k}\,.$$

Then, by taking $\varphi = u_\varepsilon$ as a test function in the weak formulation of Definition 2.1, we get

$$\|\nabla u_\varepsilon\|^2_{L^2(\Omega_\varepsilon)} = \int_{\Omega_\varepsilon} f \cdot u_\varepsilon - (p^+ - p^-) \int_{\Sigma} u_\varepsilon \cdot \widehat{k}\,. \tag{41}$$

Applying the Hölder and Poincaré inequalities (see Lemma 2.1), together with (15), on the right-hand side of (41), gives us

$$\begin{aligned}\|\nabla u_\varepsilon\|^2_{L^2(\Omega_\varepsilon)} &\leq \|f\|_{L^2(\Omega_\varepsilon)} \|u_\varepsilon\|_{L^2(\Omega_\varepsilon)} + \sqrt{\pi}\, R\, |p^+ - p^-|\, \|u_\varepsilon\|_{L^2(\Sigma)} \\ &\leq C \left(\|f\|_{L^2(\Omega)} + \sqrt{r_\varepsilon}\, |p^+ - p^-| \right) \|\nabla u_\varepsilon\|_{L^2(\Omega_\varepsilon)}\,,\end{aligned}$$

so that

$$\|\nabla u_\varepsilon\|_{L^2(\Omega_\varepsilon)} \leq C \left(\|f\|_{L^2(\Omega)} + |p^+ - p^-| \right) \qquad \forall \varepsilon \in I_*\,. \tag{42}$$

Now, given any vector field $\varphi \in \mathcal{V}(\Omega_\pm)$, by trivial extension to the complement $\Omega_\mp$, we have that $\varphi \in \mathcal{V}(\Omega_\varepsilon)$. Once taken as a test function in the weak formulation of (10) (see Remark 2.3), this gives

$$\begin{aligned}&\int_{\Omega_-} \nabla u_\varepsilon^- \cdot \nabla \varphi + \int_{\Omega_-} \mathcal{E}(u_\varepsilon^-) u_\varepsilon^- \cdot \varphi - p^- \left(\int_{\Gamma_I} \varphi \cdot \widehat{k} \right) - \int_{\Omega_-} \Phi_\varepsilon^-(\nabla \cdot \varphi) \\ &= \int_{\Omega_-} f^- \cdot \varphi \quad \forall \varphi \in \mathcal{V}(\Omega_-)\end{aligned} \tag{43}$$

and also

$$\begin{aligned}&\int_{\Omega_+} \nabla u_\varepsilon^+ \cdot \nabla \varphi + \int_{\Omega_+} \mathcal{E}(u_\varepsilon^+) u_\varepsilon^+ \cdot \varphi + p^+ \left(\int_{\Gamma_O} \varphi \cdot \widehat{k} \right) - \int_{\Omega_+} \Phi_\varepsilon^+(\nabla \cdot \varphi) \\ &= \int_{\Omega_+} f^+ \cdot \varphi \quad \forall \varphi \in \mathcal{V}(\Omega_+)\,.\end{aligned} \tag{44}$$

On the other hand, the divergence theorem trivially implies that

$$\int_{\Omega_-} \nabla \cdot \varphi = -\int_{\Gamma_I} \varphi \cdot \widehat{k} \quad \forall \varphi \in \mathcal{V}(\Omega_-)\,; \quad \int_{\Omega_+} \nabla \cdot \varphi = \int_{\Gamma_O} \varphi \cdot \widehat{k} \quad \forall \varphi \in \mathcal{V}(\Omega_+)\,. \tag{45}$$

Therefore, defining the functions $P_\varepsilon^\pm \in L_0^2(\Omega_\pm)$ by

$$P_\varepsilon^\pm \doteq \Phi_\varepsilon^\pm - \phi_\varepsilon^\pm \doteq \Phi_\varepsilon^\pm - \frac{1}{|\Omega_\pm|}\int_{\Omega_\pm} \Phi_\varepsilon^\pm \qquad \text{in } \Omega_\pm \,, \tag{46}$$

the identities (43)–(44) become, owing to (45),

$$\begin{aligned} &\int_{\Omega_-} \nabla u_\varepsilon^- \cdot \nabla\varphi + \int_{\Omega_-} \mathcal{E}(u_\varepsilon^-)u_\varepsilon^- \cdot \varphi - \int_{\Omega_-} P_\varepsilon^-(\nabla\cdot\varphi) \\ &= \int_{\Omega_-} f^- \cdot \varphi - (\phi_\varepsilon^- - p^-)\left(\int_{\Gamma_I} \varphi\cdot\widehat{k}\right) \qquad \forall\varphi \in \mathcal{V}(\Omega_-) \end{aligned} \tag{47}$$

and also

$$\begin{aligned} &\int_{\Omega_+} \nabla u_\varepsilon^+ \cdot \nabla\varphi + \int_{\Omega_+} \mathcal{E}(u_\varepsilon^+)u_\varepsilon^+ \cdot \varphi - \int_{\Omega_+} P_\varepsilon^+(\nabla\cdot\varphi) \\ &= \int_{\Omega_+} f^+ \cdot \varphi - (p^+ - \phi_\varepsilon^+)\left(\int_{\Gamma_O} \varphi\cdot\widehat{k}\right) \qquad \forall\varphi \in \mathcal{V}(\Omega_+) \,. \end{aligned} \tag{48}$$

In view of (42)–(46), it now suffices to show the uniform bound

$$\sup_{\varepsilon\in I_*}\left(\|P_\varepsilon^\pm\|_{L^2(\Omega_\pm)} + |\phi_\varepsilon^\pm|\right) \le C \,. \tag{49}$$

Since $P_\varepsilon^\pm \in L_0^2(\Omega_\pm)$ and both domains $\Omega_\pm$ have Lipschitz-continuous boundaries, there exist vector fields $J_\varepsilon^\pm \in H_0^1(\Omega_\pm)$ such that

$$\nabla\cdot J_\varepsilon^\pm = P_\varepsilon^\pm \quad \text{in} \quad \Omega_\pm \qquad \text{and} \qquad \|\nabla J_\varepsilon^\pm\|_{L^2(\Omega_\pm)} \le C\|P_\varepsilon^\pm\|_{L^2(\Omega_\pm)} \,; \tag{50}$$

see again [4]. As $H_0^1(\Omega_\pm) \subset \mathcal{V}(\Omega_\pm)$, we can take $\varphi = J_\varepsilon^\pm$ as test functions in (47)–(48), thereby obtaining

$$\int_{\Omega_\pm} \nabla u_\varepsilon^\pm \cdot \nabla J_\varepsilon^\pm + \int_{\Omega_\pm} \mathcal{E}(u_\varepsilon^\pm)u_\varepsilon^\pm \cdot J_\varepsilon^\pm - \|P_\varepsilon^\pm\|_{L^2(\Omega_\pm)}^2 = \int_{\Omega_\pm} f^\pm \cdot J_\varepsilon^\pm \,.$$

We then apply the inequalities of Hölder, Poincaré, and Sobolev (in $\Omega_\pm$), together with (50), to estimate

$$\begin{aligned} \|P_\varepsilon^\pm\|_{L^2(\Omega_\pm)}^2 &= \int_{\Omega_\pm} \nabla u_\varepsilon^\pm \cdot \nabla J_\varepsilon^\pm + \int_{\Omega_\pm} \mathcal{E}(u_\varepsilon^\pm)u_\varepsilon^\pm \cdot J_\varepsilon^\pm - \int_{\Omega_\pm} f^\pm \cdot J_\varepsilon^\pm \\ &\le \|\nabla u_\varepsilon^\pm\|_{L^2(\Omega_\pm)}\|\nabla J_\varepsilon^\pm\|_{L^2(\Omega_\pm)} \end{aligned}$$

$$
\begin{aligned}
&+ \|\mathcal{E}(u_\varepsilon^\pm)\|_{L^2(\Omega_\pm)} \|u_\varepsilon^\pm\|_{L^4(\Omega_\pm)} \|J_\varepsilon^\pm\|_{L^4(\Omega_\pm)} \\
&+ \|f\|_{L^2(\Omega)} \|J_\varepsilon^\pm\|_{L^2(\Omega_\pm)} \\
\leq\ & \|\nabla u_\varepsilon^\pm\|_{L^2(\Omega_\pm)} \|\nabla J_\varepsilon^\pm\|_{L^2(\Omega_\pm)} + C\|\nabla u_\varepsilon^\pm\|^2_{L^2(\Omega_\pm)} \|\nabla J_\varepsilon^\pm\|_{L^2(\Omega_\pm)} \\
&+ C\|f\|_{L^2(\Omega)} \|\nabla J_\varepsilon^\pm\|_{L^2(\Omega_\pm)} \\
\leq\ & C\left(1 + \|\nabla u_\varepsilon^\pm\|^2_{L^2(\Omega_\pm)} + \|f\|_{L^2(\Omega)}\right) \|P_\varepsilon^\pm\|_{L^2(\Omega_\pm)} ,
\end{aligned}
$$

and thus, after further imposing (42), we get

$$
\|P_\varepsilon^\pm\|_{L^2(\Omega_\pm)} \leq C\left(1 + \|f\|^2_{L^2(\Omega)} + |p^+ - p^-|^2\right) \qquad \forall \varepsilon \in I_* . \tag{51}
$$

Define the vector field $\varphi_* \in C^\infty(\overline{\Omega})$ by

$$
\varphi_*(x, y, z) \doteq \frac{2}{\pi R^4}\left(\frac{z}{h}\right)^2 (R^2 - x^2 - y^2)\widehat{k} \qquad \forall (x, y, z) \in \overline{\Omega} , \tag{52}
$$

so that clearly $\varphi_*^\pm \in \mathcal{V}(\Omega_\pm)$ with

$$
\int_{\Gamma_I} \varphi_*^- \cdot \widehat{k} = \int_{\Gamma_O} \varphi_*^+ \cdot \widehat{k} = 1 .
$$

Taking $\varphi = \varphi_*^\pm$ as test functions in (47)–(48), we obtain, respectively,

$$
\begin{aligned}
\phi_\varepsilon^- = p^- &+ \int_{\Omega_-} f^- \cdot \varphi_*^- - \int_{\Omega_-} \nabla u_\varepsilon^- \cdot \nabla \varphi_*^- - \int_{\Omega_-} \mathcal{E}(u_\varepsilon^-) u_\varepsilon^- \cdot \varphi_*^- \\
&+ \int_{\Omega_-} P_\varepsilon^- (\nabla \cdot \varphi_*^-)
\end{aligned} \tag{53}
$$

and also

$$
\begin{aligned}
\phi_\varepsilon^+ = p^+ &- \int_{\Omega_+} f^+ \cdot \varphi_*^+ + \int_{\Omega_+} \nabla u_\varepsilon^+ \cdot \nabla \varphi_*^+ + \int_{\Omega_+} \mathcal{E}(u_\varepsilon^+) u_\varepsilon^+ \cdot \varphi_*^+ \\
&- \int_{\Omega_+} P_\varepsilon^+ (\nabla \cdot \varphi_*^+) .
\end{aligned} \tag{54}
$$

Then, applying Hölder's inequality and (14) to both identities (53) and (54) gives us the bounds

$$|\phi_\varepsilon^\pm| \leq C\left(|p^\pm| + \|f\|_{L^2(\Omega)} + \|\nabla u_\varepsilon^\pm\|^2_{L^2(\Omega_\pm)} + \|P_\varepsilon^\pm\|_{L^2(\Omega_\pm)}\right),$$

and again from (42)–(51), we conclude

$$|\phi_\varepsilon^\pm| \leq C\left(1 + |p^\pm|^2 + \|f\|^2_{L^2(\Omega)}\right) \qquad \forall \varepsilon \in I_* . \tag{55}$$

Putting together (51)–(55) provides the uniform bound (49), thus finishing the proof. □

Remark 3.1 Let $\varepsilon \in I_*$ and $(u_\varepsilon, \Phi_\varepsilon) \in \mathcal{V}_\sigma(\Omega_\varepsilon) \times L^2(\Omega_\varepsilon)$ be a weak solution of problem (10). From Lemma 2.4, we deduce the existence of a vector field $Y_\varepsilon \in \mathcal{V}(\Omega_\varepsilon)$ such that

$$\nabla \cdot Y_\varepsilon = \Phi_\varepsilon \quad \text{in} \quad \Omega_\varepsilon, \qquad \|\nabla Y_\varepsilon\|_{L^2(\Omega_\varepsilon)} \leq C_*(1+C_B(\Omega_\varepsilon))\|\Phi_\varepsilon\|_{L^2(\Omega_\varepsilon)}, \tag{56}$$

where $C_B(\Omega_\varepsilon) > 0$ is the Bogovskii constant of Ω_ε (as in Lemma 2.3) and $C_* > 0$ is a constant depending on Ω and $\{\delta_0, \delta_1\}$ (independent of $\varepsilon \in I_*$). Then, arguing as in the proof of Theorem (3.1) (see the estimates prior to (51)), we can derive the bound

$$\|\Phi_\varepsilon\|_{L^2(\Omega_\varepsilon)} \leq C_*(1 + C_B(\Omega_\varepsilon))\left(1 + \|\nabla u_\varepsilon\|^2_{L^2(\Omega_\varepsilon)} + \|f\|_{L^2(\Omega)}\right) \qquad \forall \varepsilon \in I_* .$$

Nevertheless, in view of (19)–(42), this last inequality prevents us from reaching any conclusion concerning the growth of the $L^2(\Omega_\varepsilon)$-norm of the Bernoulli pressure Φ_ε with respect to $\varepsilon \in I_*$. That is the reason why, in the proof of Theorem (3.1), we separated the analysis into each sub-domain $\Omega_\pm$.

Combined with Lemma 2.2, the uniform bounds obtained in Theorem (3.1) allow us to state the following decay rate for the transverse flux rate and trace over Σ of any weak solution of problem (10). More precisely,

Corollary 3.1 *Let Ω_ε be as in* (4). *For any $p^\pm \in \mathbb{R}$ and $f \in L^2(\Omega)$, let $u_\varepsilon \in \mathcal{V}_\sigma(\Omega_\varepsilon)$ be a weak solution of problem* (10), *whose transverse flux rate along Ω_ε is defined as in* (8). *Then, the estimate*

$$\|u_\varepsilon\|_{L^2(\Sigma)} + |F_\varepsilon| \leq C_*\sqrt{r_\varepsilon} \qquad \forall \varepsilon \in I_* \tag{57}$$

holds for some constant $C_* > 0$ that depends on Ω, $p^\pm$, f, and $\{\delta_0, \delta_1\}$ but is independent of $\varepsilon \in I_*$.

Proof In what follows, $C > 0$ will always denote a generic constant that depends on Ω and $\{\delta_0, \delta_1\}$ (independent of $\varepsilon \in I_*$), but that may change from line to line.

It suffices to observe that since $u_\varepsilon \in \mathcal{V}_\sigma(\Omega_\varepsilon)$ vanishes on Γ_ε, as a consequence of Hölder's inequality, the trace inequality (15), and the uniform bound (40), we deduce

$$|F_\varepsilon| = \left| \int_\Sigma u_\varepsilon \cdot \widehat{k} \right| \leq \|u_\varepsilon\|_{L^1(\Sigma)} \leq \sqrt{\pi}\, R\, \|u_\varepsilon\|_{L^2(\Sigma)} \leq C\sqrt{r_\varepsilon} \qquad \forall \varepsilon \in I_* ,$$

which concludes the proof. □

4 Asymptotic Behavior as $\varepsilon \to 0^+$: Homogenized Equations

By employing the uniform bounds obtained in Theorem (3.1), through a compactness argument (which can be seen as a simplified version of the renowned *energy method* of Tartar [42, Appendix]), we derive in this section the effective (or *homogenized*) equations satisfied by the solutions of (10) as $\varepsilon \to 0^+$.

Theorem 4.1 *Let $(\Omega_\varepsilon)_{\varepsilon \in I_*}$ be the family of domains* (4)*. Given any $p^\pm \in \mathbb{R}$ and $f \in L^2(\Omega)$, let $(u_\varepsilon, \Phi_\varepsilon) \in \mathcal{V}_\sigma(\Omega_\varepsilon) \times L^2(\Omega_\varepsilon)$ be a weak solution of problem* (10)*. Then, up to the extraction of a subsequence, the sequences of restrictions $\{(u_\varepsilon^\pm, \Phi_\varepsilon^\pm)\}_{\varepsilon \in I_*} \subset H^1(\Omega_\pm) \times L^2(\Omega_\pm)$ converge strongly to weak solutions $(u^\pm, \Phi^\pm) \in H^1(\Omega_\pm) \times L^2(\Omega_\pm)$ of the following boundary-value problems, in $\Omega_\pm$, as $\varepsilon \to 0^+$:*

$$\begin{cases} -\Delta u^- + (u^- \cdot \nabla)u^- - (\nabla u^-)^\top u^- + \nabla \Phi^- = f^- , \quad \nabla \cdot u^- = 0 \ \ in \ \ \Omega_- , \\ u^- \times \nu = 0 , \quad \Phi^- = p^- \ \ on \ \ \Gamma_I , \\ u^- = 0 \ \ on \ \ \mathcal{L}_- \cup \Sigma \end{cases} \tag{58}$$

and

$$\begin{cases} -\Delta u^+ + (u^+ \cdot \nabla)u^+ - (\nabla u^+)^\top u^+ + \nabla \Phi^+ = f^+ , \quad \nabla \cdot u^+ = 0 \ \ in \ \ \Omega_+ , \\ u^+ \times \nu = 0 , \quad \Phi^+ = p^+ \ \ on \ \ \Gamma_O , \\ u^+ = 0 \ \ on \ \ \mathcal{L}_+ \cup \Sigma . \end{cases} \tag{59}$$

Furthermore, $(u^\pm, \Phi^\pm) \in H^2(\Omega_\pm) \times H^1(\Omega_\pm)$, and they satisfy in strong form the systems (58) *and* (59)*, respectively.*

Proof In what follows, $C > 0$ will always denote a generic constant that depends on Ω and $\{\delta_0, \delta_1\}$ (independent of $\varepsilon \in I_*$), but that may change from line to line.

Let us introduce the following (closed) spaces of solenoidal vector fields:

$$\mathcal{Y}_\sigma(\Omega_-) \doteq \left\{ v \in H^1(\Omega_-) \mid \nabla \cdot v = 0 \ \text{ in } \ \Omega_- ; \ \ v \times \nu = 0 \ \text{ on } \ \Gamma_I ; \ \ v = 0 \ \text{ on } \ \mathcal{L}_- \right\},$$

$$\mathcal{Y}_\sigma(\Omega_+) \doteq \left\{ v \in H^1(\Omega_+) \mid \nabla \cdot v = 0 \ \text{ in } \ \Omega_+ ; \ \ v \times \nu = 0 \ \text{ on } \ \Gamma_O ; \ \ v = 0 \ \text{ on } \ \mathcal{L}_+ \right\}$$

The bounds in (40) entail that the sequences $\{(u_\varepsilon^\pm, \Phi_\varepsilon^\pm)\}_{\varepsilon \in I_*} \subset \mathcal{Y}_\sigma(\Omega_\pm) \times L^2(\Omega_\pm)$ are uniformly bounded, so there exist $(u^\pm, \Phi^\pm) \in \mathcal{Y}_\sigma(\Omega_\pm) \times L^2(\Omega_\pm)$ such that the following convergences hold as $\varepsilon \to 0^+$,

$$\begin{gathered} u_\varepsilon^\pm \rightharpoonup u^\pm \ \text{ weakly in } \mathcal{Y}_\sigma(\Omega_\pm) ; \quad u_\varepsilon^\pm \to u^\pm \ \text{ strongly in } L^4(\Omega_\pm) ; \\ \Phi_\varepsilon^\pm \rightharpoonup \Phi^\pm \ \text{ weakly in } L^2(\Omega_\pm) , \end{gathered} \tag{60}$$

along subsequences that are not being relabeled (see [35, Theorem 6.2]). Moreover, in (57), we deduce that, along the subsequences (60),

$$\lim_{\varepsilon \to 0^+} \|u_\varepsilon^\pm\|_{L^2(\Sigma)} = 0 , \tag{61}$$

so that, in fact, $u^\pm \in \mathcal{Y}_\sigma(\Omega_\pm) \cap \mathcal{V}(\Omega_\pm)$. Now, given any test function $\varphi \in \mathcal{V}(\Omega_\pm)$, the identities (43)–(44) are valid for every $\varepsilon \in I_*$, along the subsequences (60). With the help of the convergences in (60), we can easily prove that

$$\begin{aligned} &\lim_{\varepsilon \to 0^+} \int_{\Omega_\pm} \nabla u_\varepsilon^\pm \cdot \nabla \varphi = \int_{\Omega_\pm} \nabla u^\pm \cdot \nabla \varphi , \\ &\lim_{\varepsilon \to 0^+} \int_{\Omega_\pm} \mathcal{E}(u_\varepsilon^\pm) u_\varepsilon^\pm \cdot \varphi = \int_{\Omega_\pm} \mathcal{E}(u^\pm) u^\pm \cdot \varphi , \\ &\lim_{\varepsilon \to 0^+} \int_{\Omega_\pm} \Phi_\varepsilon^\pm (\nabla \cdot \varphi) = \int_{\Omega_\pm} \Phi^\pm (\nabla \cdot \varphi) , \end{aligned} \tag{62}$$

thereby implying

$$\begin{aligned} &\int_{\Omega_-} \nabla u^- \cdot \nabla \varphi + \int_{\Omega_-} \mathcal{E}(u^-) u^- \cdot \varphi - p^- \left(\int_{\Gamma_I} \varphi \cdot \widehat{k} \right) - \int_{\Omega_-} \Phi^- (\nabla \cdot \varphi) \\ &= \int_{\Omega_-} f^- \cdot \varphi \quad \forall \varphi \in \mathcal{V}(\Omega_-) \end{aligned} \tag{63}$$

and also

$$\int_{\Omega_+} \nabla u^+ \cdot \nabla\varphi + \int_{\Omega_+} \mathcal{E}(u^+)u^+ \cdot \varphi + p^+ \left(\int_{\Gamma_O} \varphi \cdot \widehat{k}\right) - \int_{\Omega_+} \Phi^+(\nabla \cdot \varphi)$$
$$= \int_{\Omega_+} f^+ \cdot \varphi \quad \forall \varphi \in \mathcal{V}(\Omega_+)\,. \tag{64}$$

Notice that, by reasoning as in (8), we immediately deduce that

$$\int_{\Sigma(s)} u^- \cdot \widehat{k} = 0 \qquad \forall s \in [-h, 0]\,; \qquad \int_{\Sigma(s)} u^+ \cdot \widehat{k} = 0 \qquad \forall s \in [0, h]\,. \tag{65}$$

Now, consider the domains

$$\Omega^*_- \doteq \left\{(x, y, z) \in \Omega \;\middle|\; -h < z < -\frac{h}{2}\right\} \quad \text{and}$$
$$\Omega^*_+ \doteq \left\{(x, y, z) \in \Omega \;\middle|\; \frac{h}{2} < z < h\right\}.$$

Fix any $\varepsilon \in I_*$ along the subsequences (60). The same extension argument of [46, Theorem 3.2] (which employs the usual interior regularity results for the steady-state Navier-Stokes equations; see [19, Theorem IX.5.1] or [26, Section 2]) allows us to deduce that $(u^\pm_\varepsilon, \Phi^\pm_\varepsilon) \in H^2(\Omega^*_\pm) \times H^1(\Omega^*_\pm)$ and so

$$\begin{cases} -\Delta u^\pm_\varepsilon + \nabla \Phi^\pm_\varepsilon = f^\pm - \mathcal{E}(u^\pm_\varepsilon)(u^\pm_\varepsilon) \quad \text{almost everywhere in } \Omega^*_\pm\,; \\ \Phi_\varepsilon = p^- \ \text{ a. e. on } \Gamma_I\,; \qquad \Phi_\varepsilon = p^+ \ \text{ a.e. on } \Gamma_O\,. \end{cases} \tag{66}$$

The Sobolev embedding $H^1(\Omega_\pm) \subset L^6(\Omega_\pm)$ implies $f^\pm - \mathcal{E}(u^\pm_\varepsilon)(u^\pm_\varepsilon) \in L^{3/2}(\Omega_\pm)$. Moreover, from the inequalities of Hölder and Sobolev (applied in $\Omega_\pm$), we derive the estimate

$$\|f^\pm - \mathcal{E}(u^\pm_\varepsilon)(u^\pm_\varepsilon)\|_{L^{3/2}(\Omega_\pm)} \le C\left(\|f\|_{L^{3/2}(\Omega)} + \|\nabla u^\pm_\varepsilon\|^2_{L^2(\Omega_\pm)}\right). \tag{67}$$

The pair $(u^\pm_\varepsilon, \Phi^\pm_\varepsilon) \in W^{2,3/2}(\Omega^*_\pm) \times W^{1,3/2}(\Omega^*_\pm)$ is also a strong solution to the Stokes system $(66)_1$ in $\Omega^*_\pm$, with a right-hand side given by $f - \mathcal{E}(u^\pm_\varepsilon)(u^\pm_\varepsilon)$. If we apply the same extension argument of the proof of [46, Theorem 3.2], we can then invoke local regularity results for the Stokes equations (see [19, Theorem IV.4.1] or the *Cattabriga estimates* of [9, Teorema, page 311]) and (67) to obtain the bound

$$\begin{aligned}
&\|u_\varepsilon^\pm\|_{W^{2,3/2}(\Omega_\pm^*)} + \|\Phi_\varepsilon^\pm\|_{W^{1,3/2}(\Omega_\pm^*)} \\
&\leq C\left(\|f^\pm - \mathcal{E}(u_\varepsilon^\pm)(u_\varepsilon^\pm)\|_{L^{3/2}(\Omega_\pm)} + \|\nabla u_\varepsilon^\pm\|_{L^{3/2}(\Omega_\pm)} + \|\Phi_\varepsilon^\pm\|_{L^{3/2}(\Omega_\pm)}\right) \qquad (68)\\
&\leq C\left(1 + \|f\|_{L^{3/2}(\Omega)} + \|\nabla u_\varepsilon^\pm\|^2_{L^2(\Omega_\pm)} + \|\Phi_\varepsilon^\pm\|^2_{L^2(\Omega_\pm)}\right) \qquad \forall \varepsilon \in I_* .
\end{aligned}$$

Once inserted into the right-hand side of (68), the uniform bounds in (40) entail that the sequences $(\Phi_\varepsilon^\pm)_{\varepsilon\in I_*} \subset W^{1,3/2}(\Omega_\pm^*)$ are uniformly bounded. Therefore, there exist $\widehat{\Phi}_\pm \in W^{1,3/2}(\Omega_\pm^*)$ such that the following convergences hold as $\varepsilon \to 0^+$,

$$\Phi_\varepsilon^\pm \rightharpoonup \widehat{\Phi}_\pm \quad \text{weakly in } W^{1,3/2}(\Omega_\pm^*) \quad \text{and} \quad \Phi_\varepsilon^\pm \to \widehat{\Phi}_\pm \quad \text{strongly in } L^1(\partial\Omega_\pm^*)\,, \tag{69}$$

along subsequences that are not being relabeled; see again [35, Theorem 6.2]. We readily notice that since also $\Phi_\varepsilon^\pm \rightharpoonup \Phi^\pm$ weakly in $L^2(\Omega_\pm^*)$ as $\varepsilon \to 0^+$, by uniqueness of the weak limit, there must hold $\widehat{\Phi}_\pm = \Phi^\pm$ in $\Omega_\pm^*$, respectively. Moreover, in view of $(66)_2$ and the strong convergence in (69), we deduce that

$$\Phi^- = p^- \quad \text{a. e. on } \Gamma_I\,; \qquad \Phi^+ = p^+ \quad \text{a. e. on } \Gamma_O\,. \tag{70}$$

The identities in (63)–(64)–(70) allow us to conclude that $(u^\pm, \Phi^\pm) \in H^1(\Omega_\pm) \times L^2(\Omega_\pm)$ are weak solutions of systems (58)–(59), respectively. The improved regularity $(u^\pm, \Phi^\pm) \in H^2(\Omega_\pm) \times H^1(\Omega_\pm)$ follows afterward by repeating, again, the extension argument of [46, Theorem 3.2].

In order to show the strong convergence in $H^1(\Omega_\pm) \times L^2(\Omega_\pm)$, we start by defining the vector field $u_* \in H^1(\Omega)$ as

$$u_* \doteq \begin{cases} u^- \text{ in } \Omega_- \\ u^+ \text{ in } \Omega_+\,, \end{cases}$$

(recall that $u^- = u^+ = 0$ on Σ) so that u_* is divergence-free separately in Ω_- and Ω_+. Fix any $\varepsilon \in I_*$ along the subsequences (60). Given any scalar function $\psi \in C_0^\infty(\Omega)$, since $u^- = u^+ = 0$ on Σ, an integration by parts gives us

$$\begin{aligned}
\int_\Omega u_* \cdot \nabla\psi = \int_{\Omega_-} u^- \cdot \nabla\psi^- + \int_{\Omega_+} u^+ \cdot \nabla\psi^+ = \int_{\partial\Omega_-} \psi^-(u^- \cdot \nu) \\
+ \int_{\partial\Omega_+} \psi^+(u^+ \cdot \nu) = 0\,,
\end{aligned}$$

and, therefore, u_* is divergence-free in the whole Ω, implying that $u_* \in \mathcal{V}_\sigma(\Omega_\varepsilon)$. Taking $\varphi = u_\varepsilon - u_*$ as a test function in Definition 2.1 (enforcing (65)) gives us

$$\int_{\Omega_\varepsilon} \nabla u_\varepsilon \cdot \nabla(u_\varepsilon - u_*) - \int_{\Omega_\varepsilon} \mathcal{E}(u_\varepsilon) u_\varepsilon \cdot u_* + (p^+ - p^-) F_\varepsilon = \int_{\Omega_\varepsilon} f \cdot (u_\varepsilon - u_*)\,,$$

implying that

$$\begin{aligned}
&\|\nabla(u_\varepsilon - u_*)\|^2_{L^2(\Omega_\varepsilon)} \\
&= (p^- - p^+) F_\varepsilon + \int_{\Omega_\varepsilon} \nabla u_* \cdot \nabla(u_* - u_\varepsilon) + \int_{\Omega_\varepsilon} \mathcal{E}(u_\varepsilon) u_\varepsilon \cdot u_* + \int_{\Omega_\varepsilon} f \cdot (u_\varepsilon - u_*) \\
&= (p^- - p^+) F_\varepsilon + \int_{\Omega_-} \nabla u^- \cdot \nabla(u^- - u_\varepsilon^-) + \int_{\Omega_-} \mathcal{E}(u_\varepsilon^-) u_\varepsilon^- \cdot u^- \\
&\quad + \int_{\Omega_-} f^- \cdot (u_\varepsilon^- - u^-) \\
&\quad + \int_{\Omega_+} \nabla u^+ \cdot \nabla(u^+ - u_\varepsilon^+) + \int_{\Omega_+} \mathcal{E}(u_\varepsilon^+) u_\varepsilon^+ \cdot u^+ + \int_{\Omega_+} f^+ \cdot (u_\varepsilon^+ - u^+)\,.
\end{aligned}$$

Taking the limit as $\varepsilon \to 0^+$ in this last identity, observing (57)–(60), we conclude that

$$\lim_{\varepsilon \to 0^+} \|\nabla(u_\varepsilon^- - u^-)\|_{L^2(\Omega_-)} = \lim_{\varepsilon \to 0^+} \|\nabla(u_\varepsilon^+ - u^+)\|_{L^2(\Omega_+)} = 0\,,$$

that is,

$$u_\varepsilon^\pm \to u^\pm \;\text{ strongly in } H^1(\Omega_\pm) \;\text{ as }\; \varepsilon \to 0^+\,. \tag{71}$$

Concerning the Bernoulli pressure, recall the sequences $(P_\varepsilon^\pm)_{\varepsilon \in I_*} \subset L_0^2(\Omega_\pm)$ and $(\phi_\varepsilon^\pm)_{\varepsilon \in I_*} \subset \mathbb{R}$ introduced in (46). Likewise, define the functions $P^\pm \in L_0^2(\Omega_\pm)$ by

$$P^\pm \doteq \Phi^\pm - \phi^\pm \doteq \Phi^\pm - \frac{1}{|\Omega_\pm|} \int_{\Omega_\pm} \Phi^\pm \qquad \text{in } \Omega_\pm\,. \tag{72}$$

In view of the $L^2(\Omega_\pm)$ weak convergence in (60) and the identities

$$\begin{aligned}
&\|\Phi_\varepsilon^\pm\|^2_{L^2(\Omega_\pm)} = \|P_\varepsilon^\pm\|^2_{L^2(\Omega_\pm)} + |\Omega_\pm| |\phi_\varepsilon^\pm|^2 \qquad \forall \varepsilon \in I_*\,; \\
&\|\Phi^\pm\|^2_{L^2(\Omega_\pm)} = \|P^\pm\|^2_{L^2(\Omega_\pm)} + |\Omega_\pm| |\phi^\pm|^2\,,
\end{aligned}$$

it then suffices to show that

$$\lim_{\varepsilon\to 0^+} \|P_\varepsilon^\pm\|_{L^2(\Omega_\pm)} = \|P^\pm\|_{L^2(\Omega_\pm)} \qquad \text{and} \qquad \lim_{\varepsilon\to 0^+} \phi_\varepsilon^\pm = \phi^\pm\,. \tag{73}$$

Let $(J_\varepsilon^\pm)_{\varepsilon\in I_*} \subset H_0^1(\Omega_\pm)$ be the sequences of vector fields satisfying (50), for every $\varepsilon \in I_*$ along the subsequences (60). Then, inequality (51) entails that the sequences $(J_\varepsilon^\pm)_{\varepsilon\in I_*} \subset H_0^1(\Omega_\pm)$ must be uniformly bounded, and so we deduce the existence of vector fields $J^\pm \in H_0^1(\Omega_\pm)$ such that

$$J_\varepsilon^\pm \rightharpoonup J^\pm \quad \text{weakly in } H^1(\Omega_\pm)\,, \; J_\varepsilon^\pm \to J^\pm \quad \text{strongly in } L^4(\Omega_\pm) \quad \text{as } \varepsilon \to 0^+\,, \tag{74}$$

along a (not relabeled) subsequence. Given any scalar functions $\psi^\pm \in C_0^\infty(\Omega_\pm)$, since $\nabla\cdot J_\varepsilon^\pm = P_\varepsilon^\pm$ in $\Omega_\pm$ for any $\varepsilon \in I_*$, an integration by parts gives us

$$\int_{\Omega_\pm} J_\varepsilon^\pm \cdot \nabla\psi^\pm = -\int_{\Omega_\pm} P_\varepsilon^\pm\,\psi^\pm \qquad \forall \varepsilon \in I_*\,,$$

along the subsequences (60)–(74). Taking the limit in this last equality as $\varepsilon \to 0^+$ implies

$$\int_{\Omega_\pm} J^\pm \cdot \nabla\psi^\pm = -\int_{\Omega_\pm} P^\pm\,\psi^\pm \qquad \forall \psi^\pm \in C_0^\infty(\Omega_\pm;\mathbb{R})\,,$$

that is, $\nabla\cdot J^\pm = P^\pm$ in $\Omega_\pm$. As in the proof of Theorem 3.1, we derive the identity

$$\|P_\varepsilon^\pm\|^2_{L^2(\Omega_\pm)} = \int_{\Omega_\pm} \nabla u_\varepsilon^\pm \cdot \nabla J_\varepsilon^\pm + \int_{\Omega_\pm} \mathcal{E}(u_\varepsilon^\pm)u_\varepsilon^\pm \cdot J_\varepsilon^\pm - \int_{\Omega_\pm} f^\pm \cdot J_\varepsilon^\pm \quad \forall \varepsilon \in I_*\,, \tag{75}$$

along the subsequences (60)–(74). Letting $\varepsilon \to 0^+$ in (75), enforcing (71)–(74), we infer

$$\lim_{\varepsilon\to 0^+} \|P_\varepsilon^\pm\|^2_{L^2(\Omega_\pm)} = \int_{\Omega_\pm} \nabla u^\pm \cdot \nabla J^\pm + \int_{\Omega_\pm} \mathcal{E}(u^\pm)u^\pm \cdot J^\pm - \int_{\Omega_\pm} f^\pm \cdot J^\pm\,. \tag{76}$$

On the other hand, observe that the identities (63)–(64) become, owing again to (45),

$$\begin{aligned} &\int_{\Omega_-} \nabla u^- \cdot \nabla\varphi + \int_{\Omega_-} \mathcal{E}(u^-)u^- \cdot \varphi - \int_{\Omega_-} P^-(\nabla\cdot\varphi) \\ &= \int_{\Omega_-} f^- \cdot \varphi - (\phi^- - p^-)\left(\int_{\Gamma_I} \varphi\cdot\widehat{k}\right) \qquad \forall \varphi \in \mathcal{V}(\Omega_-) \end{aligned} \tag{77}$$

and also

$$\begin{aligned}&\int_{\Omega_+} \nabla u^+ \cdot \nabla \varphi + \int_{\Omega_+} \mathcal{E}(u^+)u^+ \cdot \varphi - \int_{\Omega_+} P^+(\nabla \cdot \varphi)\\ &= \int_{\Omega_+} f^+ \cdot \varphi - (p^+ - \phi^+)\left(\int_{\Gamma_O} \varphi \cdot \widehat{k}\right) \qquad \forall \varphi \in \mathcal{V}(\Omega_+)\,.\end{aligned} \tag{78}$$

Then by taking $\varphi = J^{\pm}$ as test functions in (77)–(78) (recall that $H_0^1(\Omega_\pm) \subset \mathcal{V}(\Omega_\pm)$), respectively, we immediately obtain

$$\|P^{\pm}\|^2_{L^2(\Omega_\pm)} = \int_{\Omega_\pm} \nabla u^{\pm} \cdot \nabla J^{\pm} + \int_{\Omega_\pm} \mathcal{E}(u^{\pm})u^{\pm} \cdot J^{\pm} - \int_{\Omega_\pm} f^{\pm} \cdot J^{\pm}\,. \tag{79}$$

Identities (76)–(79) secure the first limit in (73), that is,

$$P^{\pm}_{\varepsilon} \to P^{\pm} \ \text{ strongly in } L^2(\Omega_\pm) \ \text{ as } \varepsilon \to 0^+\,. \tag{80}$$

Now, let us go back to the vector field $\varphi_* \in C^\infty(\overline{\Omega})$ introduced in (52) and formulas (53)–(54), which are valid for every $\varepsilon \in I_*$ along the subsequences (60). Then, letting $\varepsilon \to 0^+$ in (53)–(54), enforcing (71)–(80), yields

$$\begin{aligned}\lim_{\varepsilon\to 0^+} \phi^-_{\varepsilon} &= p^- + \int_{\Omega_-} f^- \cdot \varphi^-_* - \int_{\Omega_-} \nabla u^- \cdot \nabla \varphi^-_* - \int_{\Omega_-} \mathcal{E}(u^-)u^- \cdot \varphi^-_*\\ &\quad + \int_{\Omega_-} P^-(\nabla \cdot \varphi^-_*)\,,\end{aligned}$$

$$\begin{aligned}\lim_{\varepsilon\to 0^+} \phi^+_{\varepsilon} &= p^+ - \int_{\Omega_+} f^+ \cdot \varphi^+_* + \int_{\Omega_+} \nabla u^+ \cdot \nabla \varphi^+_* + \int_{\Omega_+} \mathcal{E}(u^+)u^+ \cdot \varphi^+_*\\ &\quad - \int_{\Omega_+} P^+(\nabla \cdot \varphi^+_*)\,.\end{aligned}$$

Choosing $\varphi = \varphi^{\pm}_*$ as test functions in (77)–(78) (recall that $\varphi^{\pm}_* \in \mathcal{V}(\Omega_\pm)$), respectively, we derive

$$\begin{aligned}\phi^- &= p^- + \int_{\Omega_-} f^- \cdot \varphi^-_* - \int_{\Omega_-} \nabla u^- \cdot \nabla \varphi^-_* - \int_{\Omega_-} \mathcal{E}(u^-)u^- \cdot \varphi^-_*\\ &\quad + \int_{\Omega_-} P^-(\nabla \cdot \varphi^-_*)\,,\end{aligned}$$

$$\phi^+ = p^+ - \int_{\Omega_+} f^+ \cdot \varphi_*^+ + \int_{\Omega_+} \nabla u^+ \cdot \nabla \varphi_*^+ + \int_{\Omega_+} \mathcal{E}(u^+) u^+ \cdot \varphi_*^+$$
$$- \int_{\Omega_+} P^+ (\nabla \cdot \varphi_*^+) \, ,$$

thereby ensuring the second limit in (73). This concludes the proof. □

In the absence of external force, the result of Theorem 4.1 yields a rather surprising statement:

Corollary 4.1 *Let $(\Omega_\varepsilon)_{\varepsilon \in I_*}$ be the family of domains (4). Given any pressure drop $p^\pm \in \mathbb{R}$, let $(u_\varepsilon, \Phi_\varepsilon) \in \mathcal{V}_\sigma(\Omega_\varepsilon) \times L^2(\Omega_\varepsilon)$ be a weak solution of problem (10) (with $f = 0$). Then, up to the extraction of a subsequence, the sequences of restrictions $\{(u_\varepsilon^\pm, \Phi_\varepsilon^\pm)\}_{\varepsilon \in I_*} \subset H^1(\Omega_\pm) \times L^2(\Omega_\pm)$ converge strongly to the constant vectors $(0, p^\pm)$ as $\varepsilon \to 0^+$.*

Proof From Theorem 4.1, we know that the sequences $\{(u_\varepsilon^\pm, \Phi_\varepsilon^\pm)\}_{\varepsilon \in I_*} \subset \mathcal{Y}_\sigma(\Omega_\pm) \times L^2(\Omega_\pm)$ converge strongly to weak solutions $(u^\pm, \Phi^\pm) \in \mathcal{Y}_\sigma(\Omega_\pm) \times L^2(\Omega_\pm)$ of the following boundary-value problems, in $\Omega_\pm$, as $\varepsilon \to 0^+$:

$$\begin{cases} -\Delta u^- + (u^- \cdot \nabla) u^- - (\nabla u^-)^\top u^- + \nabla \Phi^- = 0 \, , \quad \nabla \cdot u^- = 0 \text{ in } \Omega_- \, , \\ u^- \times \nu = 0 \, , \quad \Phi^- = p^- \text{ on } \Gamma_I \, , \\ u^- = 0 \text{ on } \mathcal{L}_- \cup \Sigma \end{cases} \tag{81}$$

and

$$\begin{cases} -\Delta u^+ + (u^+ \cdot \nabla) u^+ - (\nabla u^+)^\top u^+ + \nabla \Phi^+ = 0 \, , \quad \nabla \cdot u^+ = 0 \text{ in } \Omega_+ \, , \\ u^+ \times \nu = 0 \, , \quad \Phi^+ = p^+ \text{ on } \Gamma_O \, , \\ u^+ = 0 \text{ on } \mathcal{L}_+ \cup \Sigma \end{cases} \tag{82}$$

Furthermore, $(u^\pm, \Phi^\pm) \in H^2(\Omega_\pm) \times H^1(\Omega_\pm)$. The weak formulations of problems (81)–(82) read, respectively,

$$\begin{aligned} &\int_{\Omega_-} \nabla u^- \cdot \nabla \varphi + \int_{\Omega_-} \mathcal{E}(u^-) u^- \cdot \varphi \\ &- p^- \left(\int_{\Gamma_I} \varphi \cdot \widehat{k} \right) - \int_{\Omega_-} \Phi^- (\nabla \cdot \varphi) = 0 \qquad \forall \varphi \in \mathcal{V}(\Omega_-) \end{aligned} \tag{83}$$

and also

$$\int_{\Omega_+} \nabla u^+ \cdot \nabla\varphi + \int_{\Omega_+} \mathcal{E}(u^+)u^+ \cdot \varphi$$
$$+ p^+ \left(\int_{\Gamma_O} \varphi \cdot \widehat{k}\right) - \int_{\Omega_+} \Phi^+(\nabla \cdot \varphi) = 0 \qquad \forall \varphi \in \mathcal{V}(\Omega_+) . \tag{84}$$

Recalling, from the proof of Theorem 4.1, that $u^\pm \in \mathcal{V}(\Omega_\pm)$ have zero flux across $\Omega_\pm$, it suffices to select $\varphi = u^\pm$ as test functions in (83)–(84), respectively, to conclude the proof. □

Remark 4.1 The result of Corollary 4.1 dictates that even under the action of an arbitrarily large prescribed pressure drop, the fluid motion becomes quiescent in the homogenization limit.

Remark 4.2 All the results presented in this manuscript hold whenever the container Ω is a smooth and distorted pipe of finite length, provided that its inlet and outlet are parallel to the XY-plane. To be precise, for some real numbers $0 < a < b < h$, we can take

$$\Omega = \left\{(x, y, z) \in \mathbb{R}^3 \mid (x, y) \in \Theta_1 \, , \ -h < z < -a\right\} \cup \Omega_c$$
$$\cup \left\{(x, y, z) \in \mathbb{R}^3 \mid (x, y) \in \Theta_2 \, , \ b < z < h\right\} ,$$

for some smooth bounded domains Θ_1, $\Theta_2 \subset \mathbb{R}^2$, $\Omega_c \subset \mathbb{R}^3$ being an open, bounded, smooth, and simply connected set; this geometry, depicted in Fig. 3, constitutes a truncation (parallel to the XY-plane) of the domain considered in the celebrated *Leray problem*, as described in [3, Definition 1.1] or [19, Chapter XIII]. In this case, assuming that the cross section $\Sigma \subset \Omega$ determining the sieve (which is also parallel to the XY-plane) is located at $z = 0$, the truncated Hagen-Poiseuille flow (52) can

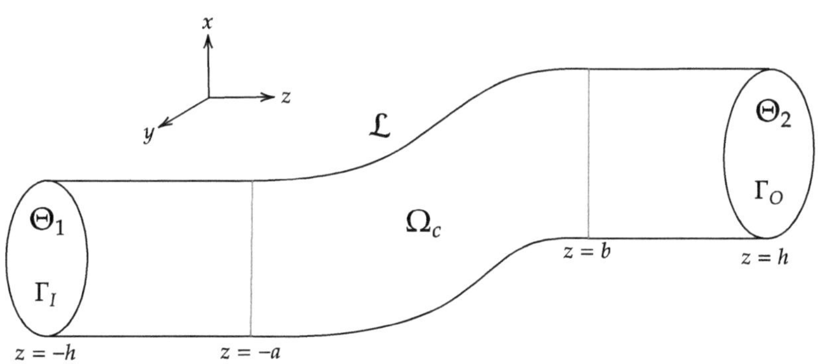

Fig. 3 Representation of an admissible, smooth, and distorted pipe of finite length

be defined as

$$\varphi_*(x, y, z) = \left(\frac{z}{h}\right)^2 V_*(x, y, z) \qquad \forall (x, y, z) \in \Omega \,,$$

where the vector field $V_* \in H^1(\Omega)$ is the unique weak solution to the following Stokes problem in Ω:

$$\begin{cases} -\Delta V_* + \nabla \Pi_* = 0 \,, \quad & \nabla \cdot V_* = 0 \ \text{ in } \ \Omega \,, \\ V_* = V_1 \ \text{ on } \ \Gamma_I \,, & V_* = V_2 \ \text{ on } \ \Gamma_O \,, \\ V_* = 0 \ \text{ on } \ \mathcal{L} \end{cases} \tag{85}$$

In $(85)_2$, for $j \in \{1, 2\}$, the boundary velocity $V_j \in H^1_0(\Theta_j)$ is given by

$$V_j(x, y, z) \doteq \frac{1}{\ell_j} v_j(x, y) \widehat{k} \qquad \forall (x, y, z) \in \Theta_j \,,$$

where $v_j \in H^1_0(\Theta_j; \mathbb{R})$ is the unique weak solution of the following torsion problem in Θ_j:

$$-\Delta v_j = 1 \ \text{ in } \ \Theta_j \,, \qquad v_j = 0 \ \text{ on } \ \partial\Theta_j$$

and

$$\ell_j \doteq \int_{\Theta_j} v_j = \int_{\Theta_j} |\nabla v_j|^2 > 0$$

Competing Interests The author declares that there is no conflict of interest.

Ethics Approval Data sharing not applicable to this article as no datasets were generated or analyzed during the current study.

References

1. H. Al Baba, C. Amrouche, N. Seloula. Instationary Stokes problem with pressure boundary condition in L^p-spaces. *Journal of Evolution Equations*, **17** (2017), 641–667.
2. G. Allaire. Homogenization of the Navier-Stokes equations in open sets perforated with tiny holes II: non-critical sizes of the holes for a volume distribution and a surface distribution of holes. *Archive for Rational Mechanics and Analysis*, **113** (1991), 261–298.
3. C. J. Amick. Steady solutions of the Navier-Stokes equations in unbounded channels and pipes. *Annali della Scuola Normale Superiore di Pisa - Classe di Scienze*, **4(3)** (1977), 473–513.
4. M. Bogovskii. Solution of the first boundary-value problem for the equation of continuity of an incompressible medium. *Doklady Akademii Nauk USSR*, **248(5)** (1979), 1037–1040.

5. W. Borchers, K. Pileckas. Existence, uniqueness and asymptotics of steady jets. *Archive for Rational Mechanics and Analysis*, **120** (1992), 1–49.
6. W. Borchers, G. P. Galdi, K. Pileckas. On the uniqueness of Leray-Hopf solutions for the flow through an aperture. *Archive for Rational Mechanics and Analysis*, **122** (1993), 19–33.
7. A. Bourgeat, E. Marušić-Paloka, A. Mikelić. Effective fluid flow in a porous medium containing a thin fissure. *Asymptotic Analysis*, **11(3)** (1995), 241–262.
8. A. Brillard, D. Gómez, M. Lobo, E. Pérez, T. Shaposhnikova. Boundary homogenization in perforated domains for adsorption problems with an advection term. *Applicable Analysis*, **95(7)** (2016), 1517–1533.
9. L. Cattabriga. Su un problema al contorno relativo al sistema di equazioni di Stokes. *Rendiconti del Seminario Matematico della Università di Padova*, **31** (1961), 308–340.
10. D. Cioranescu, A. Damlamian, G. Griso, D. Onofrei. The periodic unfolding method for perforated domains and Neumann sieve models. *Journal de Mathématiques Pures et Appliquées*, **89(3)** (2008), 248–277.
11. C. Conca. Étude d'un fluide traversant une paroi perforée I: comportement limite pres de la paroi. *Journal de Mathématiques Pures et Appliquêes*, **66(1)** (1987), 1–43.
12. C. Conca. Étude d'un fluide traversant une paroi perforée II: comportement limite loin de la paroi. *Journal de Mathématiques Pures et Appliquées*, **66(1)** (1987), 45–70.
13. C. Conca, F. Murat, O. Pironneau. The Stokes and Navier-Stokes equations with boundary conditions involving the pressure. *Japanese Journal of Mathematics*, **20(2)** (1994), 279–318.
14. C. Conca, C. Parés, O. Pironneau, M. Thiriet. Navier-Stokes equations with imposed pressure and velocity fluxes. *International Journal for Numerical Methods in Fluids*, **20(4)** (1995), 267–287.
15. A. Damlamian. Le problème de la passoire de Neumann. *Rendiconti del Seminario Matematico. Università e Politecnico di Torino*, **43(3)** (1985), 427–450.
16. T. Del Vecchio. The thick Neumann's sieve.*Annali di Matematica Pura ed Applicata*, **147(1)** (1987), 363–402.
17. M. A. Fernández, J.-F. Gerbeau, V. Martin. Numerical simulation of blood flows through a porous interface. *ESAIM: Mathematical Modelling and Numerical Analysis*, **42(6)** (2008), 961–990.
18. I. Fragalà, F. Gazzola, G. Sperone. Solenoidal extensions in domains with obstacles: explicit bounds and applications to Navier–Stokes equations. *Calculus of Variations and Partial Differential Equations*, **59(6)** (2020), Paper nº196, 1–24.
19. G. P. Galdi. *An Introduction to the Mathematical Theory of the Navier-Stokes Equations: Steady- State Problems*. Springer Science & Business Media, 2011.
20. D. Gómez, M. Lobo, E. Pérez, E. Sánchez-Palencia. Homogenization in perforated domains: a Stokes grill and an adsorption process. *Applicable Analysis*, **97(16)** (2018), 2893–2919.
21. P. Grisvard. *Elliptic Problems in Nonsmooth Domains*. Society for Industrial and Applied Mathe- matics, Philadelphia, 2011.
22. J. G. Heywood. On uniqueness questions in the theory of viscous flow. *Acta Mathematica*, **136** (1976), 61–102.
23. J. G. Heywood. Auxiliary flux and pressure conditions for Navier-Stokes problems. In *Approximation Methods for Navier-Stokes Problems: Proceedings of the Symposium Held by the International Union of Theoretical and Applied Mechanics (IUTAM) at the University of Paderborn, Germany, September 9–15, 1979*, pages 223–234. Springer Heidelberg, 2006.
24. J. G. Heywood, R. Rannacher, S. Turek. Artificial boundaries and flux and pressure conditions for the incompressible Navier–Stokes equations. *International Journal for Numerical Methods in Fluids*, **22(5)** (1996), 325–352.
25. W. Jäger, A. Mikelić. On the effective equations of a viscous incompressible fluid flow through a filter of finite thickness. *Communications on Pure and Applied Mathematics*, **51(9–10)** (1998), 1073–1121.
26. M. V. Korobkov, K. Pileckas, R. Russo. Solvability in a finite pipe of steady-state Navier–Stokes equations with boundary conditions involving Bernoulli pressure. *Calculus of Variations and Partial Differential Equations*, **59(1)** (2020), Paper nº32, 1–22.

27. O. A. Ladyzhenskaya. *The Mathematical Theory of Viscous Incompressible Flow*, volume 76. Gordon and Breach New York, 1969.
28. L. Landau and E. Lifshitz. *Theoretical Physics: Fluid Mechanics*, volume 6. Pergamon Press, 1987.
29. V. A. Marchenko, E. Y. Khruslov. Boundary-value problems with fine-grained boundary. *Matematicheskii Sbornik*, **107(3)** (1964), 458–472.
30. V. A. Marchenko, E. Y. Khruslov. *Boundary Value Problems in Domains with Fine-Grained Boundaries*. Naukova Dumka, Academy of Sciences of the Ukrainian SSR, Kiev, 1974.
31. F. Maris, B. Vernescu. Effective leak conditions across a membrane. *Complex Variables and Elliptic Equations*, **57(2–4)** (2012), 437–453.
32. S. Marušić. Low concentration limit for a fluid flow through a filter. *Mathematical Models and Methods in Applied Sciences*, **8(04)** (1998), 623–643.
33. S. Marušić. On the Navier–Stokes system with pressure boundary condition. *Annali dell'Università di Ferrara*, **53** (2007), 319–331.
34. F. Murat. The Neumann sieve. In *Proceedings of the Symposium on Nonlinear Variational Problems held at the Isola d'Elba, Italy, 1983*, volume 127 of *Research Notes in Mathematics*, pages 24–32, 2006.
35. J. Nečas. *Direct Methods in the Theory of Elliptic Equations*. Springer Science & Business Media, 2011.
36. Š. Nečasová, F. Oschmann. Homogenization of the two-dimensional evolutionary compressible Navier–Stokes equations. *Calculus of Variations and Partial Differential Equations*, **62(6)** (2023), Paper nº184, 13 pages.
37. Š. Nečasová, J. Pan. Homogenization problems for the compressible Navier–Stokes system in 2D perforated domains. *Mathematical Methods in the Applied Sciences*, **45(12)** (2022), 7859–7873.
38. G. Nguetseng. Problèmes d'écrans perforés pour l'èquation de Laplace. *ESAIM: Mathematical Modelling and Numerical Analysis*, **19(1)** (1985), 33–63.
39. B. Nowakowski, G. Ströhmer. In-flow and out-flow problem for the Stokes system. *Journal of Mathematical Fluid Mechanics*, **22** (2020), 1–15.
40. C. Patriarca, G. Sperone. Homogenization of Leray's flux problem for the steady-state Navier-Stokes equations in a multiply-connected planar domain. *Preprint arXiv:2312.00945*, (2023).
41. C. Picard. Analyse limite d'équations variationnelles dans un domaine contenant une grille. *ESAIM: Mathematical Modelling and Numerical Analysis*, **21(2)** (1987), 293–326.
42. E. Sánchez-Palencia. *Non-Homogeneous Media and Vibration Theory*, volume 127 of *Lecture Note in Physics*. Springer-Verlag, 1980.
43. E. Sánchez-Palencia. Boundary-value problems in domains containing perforated walls. In *Nonlinear Partial Differential Equations and their Applications. Collège de France Seminar*, volume 3, pages 309–325, 1982.
44. E. Sánchez-Palencia. Un problème d'écoulement lent d'une fluide visqueux incompressible au travers d'une paroi finement perforée. In *Les Méthodes de l'Homogénéisation: Theorie et Applications en Physique*, volume 57 of *Collection de la Direction des Études et Recherches d'Électricité de France*, pages 371–400. 1985.
45. G. Sperone. Homogenization of the steady-state Navier-Stokes equations with prescribed flux rate or pressure drop in a perforated pipe. *Journal of Differential Equations*, **375** (2023), 653–681.
46. G. Sperone. Steady-state Navier-Stokes flow in an obstructed pipe under mixed boundary conditions and with a prescribed transversal flux rate. *Calculus of Variations and Partial Differential Equations*, **62(9)** (2023), Paper nº236, 57 pages.
47. J. Sánchez-Hubert, E. Sánchez-Palencia. Acoustic fluid flow through holes and permeability of perforated walls. *Journal of Mathematical Analysis and Applications*, **87(2)** (1982), 427–453.
48. K. Yosida. *Functional Analysis*. Springer Science & Business Media, 1995.
49. E. Zeidler. *Nonlinear Functional Analysis and its Applications I: Fixed-Point Theorems*. Springer Science & Business Media, 2013.

On the Derivation of the Compressible Primitive Equations

Tong Tang and Šárka Nečasová

Abstract The chapter deals with the problem of the derivation of the compressible primitive equation (viscous and inviscid) rigorously from compressible anisotropic Navier-Stokes equations and Euler equations, respectively.

1 Introduction

The geophysical fluid dynamics is a fundamental subject to understand the atmosphere and ocean, whereas, from the mathematical point of view and numerical perspective, it is very complicated to use the full hydrodynamical and thermodynamical equations to analyze and simulate atmospheric flows and oceanic flows. Therefore, scientists have introduced the **primitive equation (PE)** model in geophysical fluid dynamics. It was Richardson that derived originally PE model in the 1920s for weather prediction. But lacking stability of calculations, this model was not so successful. Then, Bryan [9] improved the PE model by applying the hydrostatic approximation in 1969.

It is conventional to use the Boussinesq and hydrostatic approximation to derive the PE model from incompressible Navier-Stokes equations or Euler equations. Therefore, how to derive it rigorously is an interesting question in the mathematical community. During the last decades, there has been a large amount of literature concerning the rigorous mathematical justification of deriving the PE model. More precisely, for an incompressible (PE) system, Azérad and Guillén [2] proved that the incompressible Navier-Stokes equations converge to PE in the sense of weak solutions. Further, Li and Titi [31] showed the convergence of the weak solutions of incompressible Navier-Stokes equations to the strong solutions of PE. Based

T. Tang (✉)
Yang Zhou University, Jiangsu, China
e-mail: tt0507010156@126.com

Š. Nečasová
Institute of Mathematics, Czech Academy of Sciences, Prague, Czech Republic
e-mail: matus@math.cas.cz

T. Bodnár et al. (eds.), *Stochastics in Fluids*, Lecture Notes in Mathematical Fluid Mechanics, https://doi.org/10.1007/978-3-032-02299-8_7

on [2, 31], Donatelli and Juhasz [13] gave a justification that the PE model with the pollution effect is the hydrostatic limit of the Navier-Stokes equations with an advection-diffusion equation. Grenier [24] used the energy estimates, and Brenier [5] used the relative entropy inequality to prove that the smooth solutions of the incompressible Euler system converge to smooth solutions of inviscid PE.

One of the typical features of the PE model is that there is no information for the vertical velocity in the momentum equation, and the vertical velocity is determined by the horizontal velocity through the continuity equation that here reduces to incompressible constraint. Therefore, the mathematical and numerical study of the PE model was unsolved until the 1990s when Lions, Teman, and Wang [32, 33] established fundamental results in this field.

After that, Guillén-González, Masmoudi, and Rodríguez-Bellido [23] proved the local existence of strong solutions and uniqueness with some interesting anisotropic estimates. The **global well-posedness of PE** in the three-dimensional case was shown by Cao and Titi, [10]. Using the semigroup method, Hieber and Kashiwabara [26] extended the result to the case of $L^p - L^q$ maximal regularity, relaxing the smoothness on the initial data. Regarding the inviscid PE system, Brenier [4] proved the existence of smooth solutions in two dimensions under the convex horizontal velocity assumptions. Later, Masmoudi and Wong [38] extended Brenier's result and proved uniqueness.

Readers can refer to [6, 7, 11, 19, 25, 30, 31, 34, 40, 43] and the references therein for more physical background and other interesting mathematical results.

In order to understand the mechanism of long-term weather prediction and climate changes, some mathematicians begin to study the mathematical equations and models of PEs in compressible case as the atmosphere is a specific compressible fluid. Under the assumption that the PE is a compressible case, Gatapov and Kazhikhov [21] proved the global existence of weak solutions with the constant viscosity coefficients in the 2D case for *compressible primitive equations* (CPE) system. However, they did not show the rigorous derivation. Recently, Ersoy et al. [15] derived a 3D CPE model with degenerate viscosity, which is deduced from the 3D Navier-Stokes equations by nondimensional approximation. Then, Tang and Gao [42] showed the stability of weak solution in the 2D case. The stability means that a subsequence of weak solutions will converge to another weak solution if it satisfies some uniform bounds. For the corresponding result of well-posedness, readers can refer to [27, 36, 44].

The goal of this chapter is to derive CPE rigorously from compressible Navier-Stokes equations and Euler equations, respectively. Section 2 is devoted to deriving the viscous CPE from Navier-Stokes equations. We show the inviscid CPE from Euler equations in Sect. 3. Finally, we give some comments in Sect. 4.

2 Viscous CPE

In this section, we consider a model describing the compressible anisotropic Navier-Stokes equations

$$\begin{cases} \partial_t \rho + \operatorname{div}(\rho \mathbf{u}) = 0, \\ (\rho \mathbf{u})_t + \operatorname{div}(\rho \mathbf{u} \otimes \mathbf{u}) + \nabla p(\rho) = \mu_x \Delta_x \mathbf{u} + \mu_z \partial_{zz} \mathbf{u}, \end{cases} \tag{1}$$

in the thin domain $(0, T) \times \Omega_\epsilon$, where $\Omega_\epsilon = \{(x, z) | x \in \mathbb{T}^2, -\epsilon < z < \epsilon\}$.

By x, we denote the horizontal direction, and by z, we denote the vertical direction, while μ_x and μ_z are given constant horizontal viscous coefficient and vertical viscous coefficient. The velocity $\mathbf{u} = (\mathbf{v}, w)$, where $\mathbf{v}(t, x, z) \in \mathbb{R}^2$ and $w(t, x, z) \in \mathbb{R}$ represent the horizonal velocity and vertical velocity, respectively. Throughout this chapter, we use $\operatorname{div}\mathbf{u} = div_x \mathbf{v} + \partial z w$ and $\nabla = (\nabla_x, \partial_z)$ to denote the three-dimensional spatial divergence and gradient, respectively, and Δ_x stands for horizontal Laplacian.

As the atmosphere and ocean are the thin layers, where the fluid layer depth is small compared to the radius of sphere, Pedlosky [40] pointed out that "the pressure difference between any two points on the same vertical line depends only on the weight of the fluid between these points...." Here, we neglect the gravity and suppose the pressure $p(\rho)$ satisfies the barotropic pressure law where the pressure and the density are related by the formula $p(\rho) = \rho^\gamma$ $(\gamma > 1)$. *Therefore, we assume the density* ρ *is independent of* z, that is, $\rho = \rho(t, x)$. This plausible assumption agrees well with experiment and is frequently taken as a hypothesis in geophysical fluid dynamics.

Similar to the assumptions by Azérad and Guillén [2], Li and Titi [31], we suppose $\mu_x = 1$ and $\mu_z = \epsilon^2$. As stressed by Azérad and Guillén [2], it is necessary to consider the above anisotropic viscosity scaling, which is fundamental for the derivation of (PE). Under this assumption, the system is rewritten as the following:

$$\begin{cases} \partial_t \rho + \operatorname{div}_x(\rho \mathbf{v}) + \partial_z(\rho w) = 0, \\ \rho \partial_t \mathbf{v} + \rho(\mathbf{u} \cdot \nabla)\mathbf{v} - \Delta_x \mathbf{v} - \epsilon^2 \partial_{zz} \mathbf{v} + \nabla_x p(\rho) = 0, \\ \rho \partial_t w + \rho \mathbf{u} \cdot \nabla w - \Delta_x \mathbf{w} - \epsilon^2 \partial_{zz} w + \partial_z p(\rho) = 0 \end{cases} \tag{2}$$

Inspired by Azérad and Guillén [2], Li and Titi [31], we introduce the following new unknowns,

$$\mathbf{u}_\epsilon = (\mathbf{v}_\epsilon, w_\epsilon), \ \mathbf{v}_\epsilon(x, z, t) = \mathbf{v}(x, \epsilon z, t), \ w_\epsilon = \frac{1}{\epsilon} w(x, \epsilon z, t), \ \rho_\epsilon = \rho(x, t),$$

for any $(x, z) \in \Omega := \mathbb{T}^2 \times (-1, 1)$. Then, the system (2) becomes the following compressible scaled Navier-Stokes equations (CNS):

$$\begin{cases} \partial_t \rho_\epsilon + \operatorname{div}_x(\rho_\epsilon \mathbf{v}_\epsilon) + \partial_z(\rho_\epsilon w_\epsilon) = 0, \\ \rho_\epsilon \partial_t \mathbf{v}_\epsilon + \rho_\epsilon(\mathbf{u}_\epsilon \cdot \nabla)\mathbf{v}_\epsilon - \Delta_x \mathbf{v}_\epsilon - \partial_{zz}\mathbf{v}_\epsilon + \nabla_x p(\rho_\epsilon) = 0, \\ \epsilon^2(\rho_\epsilon \partial_t w_\epsilon + \rho_\epsilon \mathbf{u}_\epsilon \cdot \nabla w_\epsilon - \Delta_x \mathbf{w}_\epsilon - \partial_{zz} w_\epsilon) + \partial_z p(\rho_\epsilon) = 0 \end{cases} \tag{3}$$

We supplement the CNS with the following boundary and initial conditions.

The goal of this work is to investigate the limit process $\epsilon \to 0$ in the system of (3) that converges in a certain sense to the following (CPE):

$$\begin{cases} \partial_t \rho + \operatorname{div}_x(\rho \mathbf{v}) + \partial_z(\rho w) = 0, \\ \partial_t(\rho \mathbf{v}) + \operatorname{div}_x(\rho \mathbf{v} \otimes \mathbf{v}) + \partial_z(\rho \mathbf{v} w) + \nabla_x p(\rho) = \Delta_x \mathbf{v} + \partial_{zz}\mathbf{v}, \\ \partial_z p(\rho) = 0 \end{cases} \tag{4}$$

As stressed by Azérad and Guillén [2], Li and Titi [31], the hydrostatic approximation is one of the important features of the PE model. A rigorous justification of the limit passage from anisotropic Navier-Stokes equations to its hydrostatic approximation via the small aspect limit seems to be of obvious practical importance. Recently, Bella, Feireisl, and Novotný [3] and Maltese and Novotný [37] proved the limit passage from 3D compressible Navier-Stokes equations to 1D and 2D compressible Navier-Stokes equations in the thin domain. See also the result by Ducomet et al. [14]. Heuristically, inspired by their works, we develop and adapt the corresponding idea of relative entropy inequality for compressible Navier-Stokes equations. However, we should point out that *there are significant differences in the mathematical structure between the Navier-Stokes equations and the CPE model.* Due to the hydrostatic approximation, there is no information for the vertical velocity in the momentum equation of the CPE model. Moreover, opposite with respect to the incompressible case, the mathematical analysis to CPEs is much more complicated, as the oscillation of the density and the density change along with pressure in the (CPE) model, which is the apparent difference with the incompressible case, and brings new difficult and challenge. Therefore, the classical method used in the Navier-Stokes system and PE system cannot be applied straightforwardly to CPE. Luckily, based on our previous work [20] of weak-strong uniqueness to CPE, we prove the aspect ratio limit of compressible anisotropic Navier-Stokes equations. This is the first work to use the relative entropy inequality for proving the hydrostatic approximation in the compressible case.

2.1 Preliminaries

In this section, we first introduce some basic inequalities needed in the later proof. The first inequality is the so-called the generalized Poincaré inequality.

Lemma 2.1 *Let $2 \le p \le 6$ and $\rho \ge 0$ such that $0 < M \le \int_\Omega \rho dx$, $\int_\Omega \rho^\gamma dx \le E_0$ for some ($\gamma > 1$), then*

$$\|f\|_{L^p(\Omega)} \le C\|\nabla f\|_{L^2(\Omega)} + \|\rho^{\frac{1}{2}} f\|_{L^2(\Omega)},$$

where C depends on M and E_0.

For the details of proof, see, for example, [16].

Let us also mention the Gagliardo-Nirenberg inequality (see [39]):

Lemma 2.2 *For a function $u : \Omega \to \mathbb{R}$ defined on a bounded Lipschitz domain $\Omega \subset \mathbb{R}^n$, $\forall 1 \le q, r \le \infty$, and a natural number m, suppose that a real number θ and a natural number j are such that*

$$\frac{1}{p} = \frac{j}{n} + (\frac{1}{r} - \frac{m}{n})\theta + \frac{1-\theta}{q}$$

and

$$\frac{j}{m} \le \theta \le 1;$$

then there exists constant C independent of u such that

$$\|D^j u\|_{L^p(\Omega)} \le C\|D^m u\|^{\theta}_{L^r(\Omega)}\|u\|^{1-\theta}_{L^q(\Omega)} + +C\|u\|_{L^s},$$

where $s > 0$ is arbitrary naturally.

Before stating our main result, we give the definition of a weak solution for CNS and a strong solution for CPE.

Remark 2.1 Recently, Bresch and Jabin [8] introduced a different compactness method from Lions or Feireisl, which can be applied to anisotropical stress tensor.[1] Bresch and Jabin have proved the global existence of weak solutions for non-monotone pressure and for anisotropical stress tensor. Therefore, following their theory, we can obtain the existence of a weak solution for monotone pressure and anisotropic viscous stress tensor.

Let us recall their definitions here.

[1] Let us emphasize the result of Bresch and Jabin is valid only for small coefficients of viscosities.

2.2 Dissipative Weak Solutions of CNS

Definition 2.1 We say that $[\rho_\epsilon, \mathbf{u}_\epsilon]$ with $\mathbf{u}_\epsilon = (\mathbf{v}_\epsilon, w_\epsilon)$ is a finite energy weak solution to the system of (3) if $\rho_\epsilon = \rho_\epsilon(x,t)$ and

$$\mathbf{u}_\epsilon \in L^2(0,T;H^1(\Omega)),\ \ \rho|\mathbf{u}_\epsilon|^2 \in L^\infty(0,T;L^1(\Omega)),$$
$$\rho_\epsilon \in L^\infty(0,T;L^\gamma(\Omega)) \cap C([0,T],L^1(\Omega)),\ \ \rho_\epsilon \geq 0. \tag{5}$$

- The continuity equation

$$[\int_\Omega \rho_\epsilon \psi dxdz]_{t=0}^{t=\tau} = \int_0^\tau \int_\Omega \rho_\epsilon \partial_t \psi + \rho_\epsilon \mathbf{v}_\epsilon \cdot \nabla_x \psi + \rho_\epsilon w_\epsilon \partial_z \psi dxdzdt \tag{6}$$

holds for all $\psi \in C_c^\infty([0,T) \times \overline{\Omega})$.
- The momentum equations

$$\begin{aligned}
&[\int_\Omega \rho_\epsilon \mathbf{v}_\epsilon \boldsymbol{\varphi}_\mathbf{H}\, dxdz]_{t=0}^{t=\tau} - \int_0^\tau \int_\Omega \rho_\epsilon \mathbf{v}_\epsilon \partial_t \boldsymbol{\varphi}_\mathbf{H} dxdzdt \\
&- \int_0^\tau \int_\Omega \rho_\epsilon \mathbf{u}_\epsilon \mathbf{v}_\epsilon \cdot \nabla \boldsymbol{\varphi}_\mathbf{H} dxdzdt \\
&+ \int_0^\tau \int_\Omega \nabla \mathbf{v}_\epsilon : \nabla \boldsymbol{\varphi}_\mathbf{H} dxdzdt \\
&- \int_0^\tau \int_\Omega p(\rho_\epsilon) \mathrm{div}_x \boldsymbol{\varphi}_\mathbf{H} dxdzdt = 0
\end{aligned} \tag{7}$$

and

$$\begin{aligned}
&\epsilon^2 [\int_\Omega \rho_\epsilon w_\epsilon \varphi_3 dxdz]_{t=0}^{t=\tau} - \epsilon^2 \int_0^\tau \int_\Omega \rho_\epsilon w_\epsilon \partial_t \varphi_3 dxdzdt \\
&- \epsilon^2 \int_0^\tau \int_\Omega \rho_\epsilon \mathbf{u}_\epsilon w_\epsilon \cdot \nabla \varphi_3 dxdzdt \\
&+ \epsilon^2 \int_0^\tau \int_\Omega \nabla w_\epsilon \cdot \nabla \varphi_3 dxdzdt - \int_0^\tau \int_\Omega p(\rho_\epsilon) \partial_z \varphi_3 dxdzdt = 0,
\end{aligned} \tag{8}$$

hold for any spatially periodic function $\boldsymbol{\varphi}_{\boldsymbol{H}}, \varphi_3 \in C_c^\infty([0,T) \times \overline{\Omega})$. Combining (7)–(8), we obtain

$$\begin{aligned}
&[\int_\Omega \rho_\epsilon \mathbf{v}_\epsilon \boldsymbol{\varphi}_{\boldsymbol{H}} dxdz + \epsilon^2 \int_\Omega \rho_\epsilon w_\epsilon \varphi_3 dxdz]_{t=0}^{t=\tau} \\
&\quad - \int_0^\tau \int_\Omega \rho_\epsilon \mathbf{v}_\epsilon \partial_t \boldsymbol{\varphi}_{\boldsymbol{H}} dxdzdt - \epsilon^2 \int_0^\tau \int_\Omega \rho_\epsilon w_\epsilon \partial_t \varphi_3 dxdzdt
\end{aligned}$$

$$-\int_0^\tau\int_\Omega \rho_\epsilon \mathbf{v}_\epsilon \otimes \mathbf{v}_\epsilon : \nabla_x \boldsymbol{\varphi}_{\boldsymbol{H}} dxdzdt - \int_0^\tau\int_\Omega \rho_\epsilon \mathbf{v}_\epsilon w_\epsilon \cdot \partial_z \varphi_H dxdzdt$$
$$-\epsilon^2 \int_0^\tau\int_\Omega \rho_\epsilon \mathbf{v}_\epsilon w_\epsilon \cdot \nabla_x \varphi_3 dxdzdt - \epsilon^2 \int_0^\tau\int_\Omega \rho_\epsilon w_\epsilon^2 \partial_z \varphi_3 dxdzdt$$
$$+\int_0^\tau\int_\Omega \nabla \mathbf{v}_\epsilon : \nabla \boldsymbol{\varphi}_{\boldsymbol{H}} dxdzdt + \epsilon^2 \int_0^\tau\int_\Omega \nabla w_\epsilon \cdot \nabla \varphi_3 dxdzdt$$
$$-\int_0^\tau\int_\Omega p(\rho_\epsilon)\mathrm{div}\boldsymbol{\varphi} dxdzdt = 0, \tag{9}$$

where spatially periodic function $\varphi = (\boldsymbol{\varphi}_{\mathbf{H}}, \varphi_3) \in C_c^\infty([0,T) \times \overline{\Omega})$ and $\mathrm{div}\boldsymbol{\varphi} = \mathrm{div}_x \boldsymbol{\varphi}_{\mathbf{H}} + \partial_z \varphi_3$.

- The energy inequality

$$[\int_\Omega \frac{1}{2}\rho_\epsilon |\mathbf{v}_\epsilon|^2 + \frac{\epsilon^2}{2}\rho_\epsilon |w_\epsilon|^2 + P(\rho_\epsilon))dxdz]|_{t=0}^{t=\tau}$$
$$+\int_0^\tau\int_\Omega (|\nabla \mathbf{v}_\epsilon|^2 + \epsilon^2 |\nabla w_\epsilon|^2) dxdzdt \le 0 \tag{10}$$

holds for a.a $\tau \in (0,T)$, where $P(\rho) = \rho \int_1^\rho \frac{p(z)}{z^2} dz$.

Remark 2.2 Recently, Abbatiello, Feireisl, and Novotny [1] proposed a new concept-dissipative turbulent solutions, which is so far the largest class of generalized solutions. We will prove the existence of dissipative turbulent solutions for CNS in future work, and it is easy to prove that every dissipative turbulent solution is a bounded energy weak solution.

2.3 *Strong Solution of CPE*

We say that $(r, \mathbf{U})$, $\mathbf{U} = (\mathbf{V}, W)$ is a strong solution to the CPE system (4) in $(0,T)\times \Omega$ if

$$r^{\frac{1}{2}} \in L^\infty(0,T; H^2(\Omega)),\ \partial_t r^{\frac{1}{2}} \in L^\infty(0,T; H^1(\Omega)),\ r > 0 \text{ for all } (t,x),$$
$$\mathbf{V} \in L^\infty(0,T; H^3(\Omega)) \cap L^2(0,T; H^4(\Omega)),\ \partial_t \mathbf{V} \in L^2(0,T; H^2(\Omega)),$$

with initial data $r_0^{\frac{1}{2}} \in H^2(\Omega)$, $r_0 > 0$, and $\mathbf{V}_0 \in H^3(\Omega)$. Liu and Titi [35] have proved the local existence of strong solutions to the CPE system (4).

Remark 2.3 Let us mention that in the definition of strong solution we consider the regularity $\mathbf{V} \in L^\infty(0,T; H^3)$, which is higher than the regularity considered by

Liu and Titi (see [35]), which means $\mathbf{V} \in L^\infty(0, T; H^2)$. The explanation how to get a higher regularity of solution can be found in [35].

Remark 2.4 As the density is independent of z, we can obtain the following information of vertical velocity for the weak solution of CNS,

$$\rho w(x, z, t) = -\mathrm{div}_x(\rho\widetilde{\mathbf{v}}) + z\mathrm{div}_x(\rho\overline{\mathbf{v}}), \text{ in the sense of } H^{-1}(\Omega), \tag{11}$$

where

$$\widetilde{\mathbf{v}}(x, z, t) = \int_0^z v(x, s, t)ds, \quad \overline{v}(x, t) = \int_0^1 v(x, z, t)dz.$$

Similarly, we can obtain the same equation for the strong solution of CPE in the classical sense. There is no information about w, so we need to derive its information. We should emphasize that (11) is the key step to obtain the existence of weak solution for CPE in [36, 44], which is inspired by incompressible case.

2.4 Relative Entropy Inequality

Motivated by Feireisl et al. [17], Feireisl and Novotný [18], for any finite energy weak solution $(\rho, \mathbf{u})$, where $\mathbf{u} = (\mathbf{v}, w)$, to the CNS system, we introduce the relative energy functional

$$\begin{aligned}
&\mathcal{E}(\rho, \mathbf{u}|r, \mathbf{U})\\
&= \int_\Omega [\frac{1}{2}\rho|\mathbf{v} - \mathbf{V}|^2 + \frac{\epsilon^2}{2}\rho|w - W|^2 + P(\rho) - P'(r)(\rho - r) - P(r)]dxdz\\
&= \int_\Omega (\frac{1}{2}\rho|\mathbf{v}|^2 + \frac{\epsilon^2}{2}\rho|w|^2 + P(\rho))dxdz - \int_\Omega (\rho\mathbf{v}\cdot\mathbf{V} + \epsilon^2\rho w W)dxdz\\
&+ \int_\Omega [\rho\frac{|\mathbf{V}|^2}{2} + \frac{\epsilon^2}{2}\rho|W|^2 - \rho P'(r)]dxdz + \int_\Omega p(r)dxdz\\
&= \sum_{i=1}^4 I_i,
\end{aligned} \tag{12}$$

where $r > 0$, $\mathbf{U} = (\mathbf{V}, W)$ are smooth "test" functions and spatially periodic function $\mathbf{V}, W \in C_c^\infty([0, T) \times \overline{\Omega})$. Here, we have used $rP'(r) - P(r) = p(r)$.

Lemma 2.3 *Let $(\rho, \mathbf{v}, w)$ be a dissipative weak solution introduced in Definition 2.1. Then, $(\rho, \mathbf{v}, w)$ satisfy the versatile relative entropy inequality*

$$\begin{aligned}
&\mathcal{E}(\rho, \mathbf{u}|r, \mathbf{U})|_{t=0}^{t=\tau} + \int_0^\tau \int_\Omega \big(\nabla \mathbf{v} : (\nabla \mathbf{v} - \nabla \mathbf{V}) + \epsilon^2 |\nabla w|^2\big) dxdzdt \\
&\leq \int_0^\tau \int_\Omega \rho(\partial_t \mathbf{V} + \mathbf{v} \cdot \nabla_x \mathbf{V} + w\partial_z \mathbf{V})(\mathbf{V} - \mathbf{v}) dxdt \\
&\quad + \epsilon^2 \int_0^\tau \int_\Omega \rho(\partial_t W + \mathbf{v} \cdot \nabla_x W + w\partial_z W)(W - w) dxdzdt \\
&\quad + \epsilon^2 \int_0^\tau \int_\Omega \nabla w \cdot \nabla W dxdzdt \\
&\quad - \int_0^\tau \int_\Omega P''(r)((\rho - r)\partial_t r + \rho \mathbf{v} \cdot \nabla_x r) dxdzdt - \int_0^\tau \int_\Omega p(r) \mathrm{div}_x \mathbf{V} dxdzdt.
\end{aligned} \tag{13}$$

Proof From the weak formulation and energy inequality (6)–(9), we deduce

$$I_1|_{t=0}^{t=\tau} + \int_0^\tau \int_\Omega (|\nabla \mathbf{v}|^2 + \epsilon^2 |\nabla w|^2) dxdzdt \leq 0, \tag{14}$$

$$\begin{aligned}
I_2|_{t=0}^{t=\tau} &= -\int_0^\tau \int_\Omega \rho \mathbf{v} \partial_t \mathbf{V} + \rho \mathbf{v} \otimes \mathbf{v} : \nabla_x \mathbf{v} + \rho \mathbf{v} w \cdot \partial_z \mathbf{V} dxdzdt \\
&\quad - \int_0^\tau \int_\Omega \epsilon^2 \rho w \partial_t W + \epsilon^2 \rho w (\mathbf{v} \cdot \nabla_x) W + \epsilon^2 \rho w^2 \partial_z W \\
&\quad + p(\rho) \mathrm{div}_x \mathbf{V} dxdzdt \\
&\quad + \int_0^\tau \int_\Omega \nabla \mathbf{v} : \nabla \mathbf{V} + \epsilon^2 \nabla w \cdot \nabla W dxdzdt,
\end{aligned} \tag{15}$$

$$\begin{aligned}
I_3|_{t=0}^{t=\tau} &= \int_0^\tau \int_\Omega \rho \partial_t \frac{|\mathbf{V}|^2}{2} + \rho \mathbf{v} \cdot \nabla_x \frac{|\mathbf{V}|^2}{2} + \rho w \partial_z \frac{|\mathbf{V}|^2}{2} dxdzdt \\
&\quad + \epsilon^2 \int_0^\tau \int_\Omega \rho \partial_t \frac{|W|^2}{2} + \rho \mathbf{v} \cdot \nabla_x \frac{|W|^2}{2} + \rho w \partial_z \frac{|W|^2}{2} dxdzdt \\
&\quad - \int_0^\tau \int_\Omega \rho \partial_t P'(r) + \rho \mathbf{v} \cdot \nabla_x P'(r) + \rho w \partial_z P'(r) dxdzdt \\
&= \int_0^\tau \int_\Omega \rho \mathbf{V} \partial_t \mathbf{V} + \rho \mathbf{v} (\mathbf{V} \cdot \nabla_x) \mathbf{V} + \rho w \mathbf{V} \partial_z \mathbf{V} dxdzdt \\
&\quad + \epsilon^2 \int_0^\tau \int_\Omega (\rho W \partial_t W + \rho W \mathbf{v} \cdot \nabla_x W + \rho w W \partial_z W) dxdzdt
\end{aligned}$$

$$- \int_0^\tau \int_\Omega \rho P''(r)\partial_t r + P''(r)\rho \mathbf{v} \cdot \nabla_x r dx dz dt, \tag{16}$$

$$I_4|_{t=0}^{t=\tau} = \int_0^\tau \int_\Omega \partial_t p(r) dx dz dt. \tag{17}$$

Summing (14)–(17) together, we obtain Lemma 2.3. □

Remark 2.5 In the context of continuum mechanics, the concept of relative entropy has many successful applications, providing weak-strong uniqueness results, as well as justification of singular limits, for both incompressible and compressible fluid models. However, it is important to point out that there are distinguished differences for relative entropy inequality between incompressible and compressible cases. The idea of relative entropy inequality was introduced by Dafermos [12] for hyperbolic equations and by Germain [22] to fluid dynamics. After that, Feireisl and his coauthors proved that the relative entropy inequality is valid for all smooth test functions with appropriate boundary conditions (see [17]). Precisely, the test functions $randU$ in the relative entropy inequality are arbitrarily smooth, r is strictly positive, and U satisfies the corresponding boundary conditions. Moreover, it is easy to check that the relative entropy inequality is satisfied as an equality as soon as the solution ρ, u is smooth enough.

Moreover, there is shown the so-called weak-strong uniqueness principle, which means that if the initial data of weak solutions and strong solutions coincide, weak solutions coincide with the unique strong solution on the time interval where the strong solution lives. This principle gives us the possibility to use relative entropy inequality and weak-strong principle to show, for example, singular limits from weak solutions to strong solution with corresponding scaling if the corresponding reminder can be estimated in appropriate way. For more details, see [17, 18].

Based on the relative entropy inequality, we can obtain the following lemma from [17]

Lemma 2.4 *Let* $0 < a < b < \infty$. *Then, there exists* $c = c(a, b) > 0$ *such that for all* $\rho \in [0, \infty)$ *and* $r \in [a, b]$ *there holds*

$$P(\rho) - P'(r)(\rho - r) - P(r) \geq \begin{cases} C|\rho - r|^2, & \text{when } \frac{r}{2} < \rho < r, \\ C(1 + \rho^\gamma), & \text{otherwise}, \end{cases}$$

where $C = C(a, b)$.

Moreover, from [17], we learn that

$$\mathcal{E}(\rho, \mathbf{u}|r, \mathbf{U})(t) \in L^\infty(0, T), \ \int_\Omega \chi_{\rho \geq r} \rho^\gamma dx dz \leq C\mathcal{E}(\rho, \mathbf{u}|r, \mathbf{U})(t),$$

$$\int_\Omega \chi_{\rho \leq \frac{r}{2}} 1 dx dz \leq C\mathcal{E}(\rho, \mathbf{u}|r, \mathbf{U})(t), \ \int_\Omega \chi_{\frac{r}{2} < \rho < r} (\rho - r)^2 dx dz \leq C\mathcal{E}(\rho, \mathbf{u}|r, \mathbf{U})(t). \tag{18}$$

For a rigorous proof of Lemma 2.4 and (18), the reader is referred to [17].

2.5 Main Result and Proof

Now, we are ready to state our main result.

Theorem 2.1 *Let $\gamma > 6$, $T_{max} > 0$ be the lifetime of strong solution to CPE system (4) corresponding to initial data $[r_0, \mathbf{V}_0]$. Let $(\rho_\epsilon, \mathbf{u}_\epsilon)$, $\mathbf{u}_\epsilon = (\mathbf{v}_\epsilon, w_\epsilon)$ be a sequence of dissipative weak solutions to the CNS system (3) from the initial data $(\rho_{0,\epsilon}, \mathbf{u}_{0,\epsilon})$. Supposing that $\rho_{0,\epsilon} > 0$, $r_0 > 0$, and*

$$\mathcal{E}(\rho_{0,\epsilon}, \mathbf{u}_{0,\epsilon} | r_0, \mathbf{U}_0) \to 0,$$

where $\mathbf{U}_0 = (\mathbf{V}_0, W_0)$, then we have

$$ess \sup_{t\in(0,T_{max})} \mathcal{E}(\rho_\epsilon, \mathbf{u}_\epsilon | r, \mathbf{U}) \to 0,$$

where $\mathbf{U} = (\mathbf{V}, W)$ and the couple $(r, \mathbf{U})$ satisfy the CPE system (4) on the time interval $[0, T_{max})$.

In the following part, we will prove the Theorem 2.1. First, we will explain the idea of the proof: The proof of Theorem 2.1 relies on the versatile relative energy inequality by considering the strong solution $(r, \mathbf{U})$, where $\mathbf{U} = (\mathbf{V}, W)$, as test function in the relative energy inequality (13). Firstly, let us recall the versatile relative entropy inequality

$$\begin{aligned}
&\mathcal{E}(\rho_\epsilon, \mathbf{u}_\epsilon | r, \mathbf{U})|_{t=0}^{t=\tau} + \int_0^\tau \int_\Omega \left(\nabla \mathbf{v}_\epsilon \cdot (\nabla \mathbf{v}_\epsilon - \nabla \mathbf{V}) + \epsilon^2 |\nabla w_\epsilon|^2\right) dxdzdt \\
&\leq \int_0^\tau \int_\Omega \rho_\epsilon (\partial_t \mathbf{V} + \mathbf{v}_\epsilon \nabla_x \mathbf{V} + w_\epsilon \partial_z \mathbf{V})(\mathbf{V} - \mathbf{v}_\epsilon) dxdt \\
&+ \epsilon^2 \int_0^\tau \int_\Omega \rho_\epsilon (\partial_t W + \mathbf{v}_\epsilon \nabla_x W + w_\epsilon \partial_z W)(W - w_\epsilon) dxdzdt \\
&\quad + \epsilon^2 \int_0^\tau \int_\Omega \nabla w_\epsilon \cdot \nabla W dxdzdt \\
&\quad - \int_0^\tau \int_\Omega P''(r)((\rho_\epsilon - r)\partial_t r + \rho_\epsilon \mathbf{v}_\epsilon \nabla_x r) dxdzdt \\
&\quad - \int_0^\tau \int_\Omega p(r) \mathrm{div}_x \mathbf{V} dxdzdt.
\end{aligned} \tag{19}$$

The goal now is to find a lower bound of (19) in the following form

$$\mathcal{E}(\rho_\epsilon, \mathbf{u}_\epsilon | r, \mathbf{U})(t) + C \int_0^t \|\nabla \mathbf{v}_\epsilon - \nabla \mathbf{V}\|_{L^2}^2 dt + \epsilon^2 \int_0^t \|\nabla \mathbf{w}_\epsilon\|_{L^2}^2 dt$$

and an upper bound of (19) in the form

$$C(\delta)\int_0^t h(t)\mathcal{E}(\rho_\epsilon, \mathbf{u}_\epsilon|r, \mathbf{U})dt + \delta\int_0^t \|\nabla \mathbf{v}_\epsilon - \nabla \mathbf{V}\|^2_{W^{1,2}}dt + o(\epsilon^2),$$

with any $\delta > 0$, where C is independent of δ and ϵ, $h \in L^1(0,T)$ and $o(\epsilon^2) \to 0$ when $\epsilon \to 0$.

Using the above bounds, we can deduce

$$\mathcal{E}(\rho_\epsilon, \mathbf{u}_\epsilon|r, \mathbf{U})(\tau) \le C\int_0^\tau h(t)\mathcal{E}(\rho_\epsilon, \mathbf{u}_\epsilon|r, \mathbf{U})(t)dt + o(\epsilon^2),$$

and applying the Gronwall inequality implies the assertion. In the rest of this section, we will follow this way:

Step 1 We write

$$\int_\Omega \rho_\epsilon \mathbf{v}_\epsilon(\mathbf{V} - \mathbf{v}_\epsilon)\cdot \nabla_x \mathbf{V} dxdz =$$
$$\int_\Omega \rho_\epsilon(\mathbf{v}_\epsilon - \mathbf{V})(\mathbf{V} - \mathbf{v}_\epsilon)\cdot \nabla_x \mathbf{V} dxdz + \int_\Omega \rho_\epsilon \mathbf{V}(\mathbf{V} - \mathbf{v}_\epsilon)\cdot \nabla_x \mathbf{V} dxdz.$$

As $[r, \mathbf{V}, W]$ is a strong solution, it is obvious to obtain that

$$\int_\Omega \rho_\epsilon(\mathbf{v}_\epsilon - \mathbf{V})(\mathbf{V} - \mathbf{v}_\epsilon)\cdot \nabla_x \mathbf{V} dxdz \le C\mathcal{E}(\rho_\epsilon, \mathbf{u}_\epsilon|r, \mathbf{U}). \tag{20}$$

Moreover, the momentum equation reads as

$$(r\mathbf{V})_t + \mathrm{div}_x(r\mathbf{V}\otimes\mathbf{V}) + \partial_z(r\mathbf{V}W) + \nabla_x p(r) = \Delta\mathbf{V} = \Delta_x\mathbf{V} + \partial_{zz}\mathbf{V},$$

which implies that

$$\mathbf{V}_t + \mathbf{V}\cdot\nabla_x\mathbf{V} + W\partial_z\mathbf{V} = -\frac{1}{r}\nabla_x p(r) + \frac{1}{r}\Delta_x\mathbf{V} + \frac{1}{r}\partial_{zz}\mathbf{V}.$$

So, we rewrite the preceding two items on the right side of (19) as

$$\int_\Omega \rho_\epsilon[\partial_t\mathbf{V} + \mathbf{V}\nabla_x\mathbf{V} + W\partial_z\mathbf{V} + (\mathbf{v}_\epsilon - \mathbf{V})\nabla_x\mathbf{V} + (w_\epsilon - W)\partial_z\mathbf{V}](\mathbf{V} - \mathbf{v}_\epsilon)dxdz$$
$$= \int_\Omega \frac{\rho_\epsilon}{r}(\mathbf{V} - \mathbf{v}_\epsilon)(\Delta_x\mathbf{V} + \partial_{zz}\mathbf{V} - \nabla_x p(r))dxdz$$
$$+ \int_\Omega \rho_\epsilon(w_\epsilon - W)(\mathbf{V} - \mathbf{v}_\epsilon)\cdot\partial_z\mathbf{V}dxdz - \int_\Omega \rho_\epsilon(\mathbf{V} - \mathbf{v}_\epsilon)^2\nabla_x\mathbf{V}$$

and

$$\begin{aligned}
&\epsilon^2 \int_0^\tau \int_\Omega \rho_\epsilon(\partial_t W + \mathbf{v}_\epsilon \nabla_x W + w_\epsilon \partial_z W)(W - w_\epsilon) dxdzdt \\
&\quad \le \int_0^\tau \mathcal{E}(\rho_\epsilon, \mathbf{u}_\epsilon | r, \mathbf{U}) + \epsilon^4 \int_0^\tau \int_\Omega \rho_\epsilon(\partial_t W + \mathbf{v}_\epsilon \nabla_x W + w_\epsilon \partial_z W)^2 dxdzdt \\
&\quad = \int_0^\tau \mathcal{E}(\rho_\epsilon, \mathbf{u}_\epsilon | r, \mathbf{U}) + \epsilon^4 \int_0^\tau \int_\Omega \rho_\epsilon(\partial_t W + \mathbf{V} \nabla_x W + W \partial_z W)^2 dxdzdt \\
&\quad\quad + \epsilon^4 \int_0^\tau \int_\Omega \rho_\epsilon((\mathbf{v}_\epsilon - \mathbf{V}) \nabla_x W + (w_\epsilon - W)\partial_z W)^2 dxdzdt.
\end{aligned} \tag{21}$$

Noticing Lemma 2.3, we have

$$\begin{aligned}
&\int_0^\tau \int_\Omega \rho_\epsilon(\partial_t W + \mathbf{V} \nabla_x W + W \partial_z W)^2 dxdzdt \\
&\quad = \int_0^\tau \int_\Omega \chi_{\rho_\epsilon < \frac{r}{2}} \rho_\epsilon(\partial_t W + \mathbf{V} \nabla_x W + W \partial_z W)^2 dxdz \\
&\quad\quad + \int_0^\tau \int_\Omega \chi_{\frac{r}{2} \le \rho_\epsilon \le r} \rho_\epsilon(\partial_t W + \mathbf{V} \nabla_x W + W \partial_z W)^2 dxdz \\
&\quad\quad + \int_0^\tau \int_\Omega \chi_{\rho_\epsilon > r} \rho_\epsilon(\partial_t W + \mathbf{V} \nabla_x W + W \partial_z W)^2 dxdzdt \\
&\quad \le \int_0^\tau \int_\Omega \chi_{\rho_\epsilon < \frac{r}{2}} r(\partial_t W + \mathbf{V} \nabla_x W + W \partial_z W)^2 dxdz \\
&\quad\quad + \int_0^\tau \int_\Omega \chi_{\rho_\epsilon > r} \rho_\epsilon(\partial_t W + \mathbf{V} \nabla_x W + W \partial_z W)^2 dxdzdt \\
&\quad\quad + C \int_0^\tau \int_\Omega \chi_{\frac{r}{2} \le \rho_\epsilon \le r} (\rho_\epsilon - r)(\partial_t W + \mathbf{V} \nabla_x W + W \partial_z W)^2 dxdz + C \\
&\quad \le C \int_0^\tau \mathcal{E}(\rho_\epsilon, \mathbf{u}_\epsilon | r, \mathbf{U}) + C \int_0^\tau \int_\Omega \chi_{\frac{r}{2} \le \rho_\epsilon \le r} (\rho_\epsilon - r)^2 dxdz \\
&\quad\quad + C \int_0^\tau \int_\Omega \chi_{\rho_\epsilon > r} \rho_\epsilon^\gamma dxdz + C \\
&\quad \le C \int_0^\tau \mathcal{E}(\rho_\epsilon, \mathbf{u}_\epsilon | r, \mathbf{U}) + C.
\end{aligned} \tag{22}$$

It is easy to use the definition of relative entropy inequality and Cauchy inequality to obtain

$$\epsilon^4 \int_0^\tau \int_\Omega \rho_\epsilon((\mathbf{v}_\epsilon - \mathbf{V})\nabla_x W + (w_\epsilon - W)\partial_z W)^2 dxdzdt \\ \leq \epsilon^2 \mathcal{E}(\rho_\epsilon, \mathbf{u}_\epsilon | r, \mathbf{U}) + o(\epsilon^2). \tag{23}$$

Putting (22)–(23) into (21) yields

$$\epsilon^2 \int_0^\tau \int_\Omega \rho_\epsilon(\partial_t W + \mathbf{v}_\epsilon \nabla_x W + w_\epsilon \partial_z W)(W - w_\epsilon) dxdzdt \\ \leq C \int_0^\tau \mathcal{E}(\rho_\epsilon, \mathbf{u}_\epsilon | r, \mathbf{U}) + o(\epsilon^2).$$

Moreover, a simple application of the Cauchy inequality leads to the following:

$$\epsilon^2 \int_0^\tau \int_\Omega \nabla w_\epsilon \cdot \nabla W dxdzdt \leq \frac{\epsilon^2}{2} \int_0^\tau \int_\Omega |\nabla w_\epsilon|^2 dxdzdt + o(\epsilon^2)$$

Thus, we obtain that

$$\begin{aligned}
&\mathcal{E}(\rho_\epsilon, \mathbf{u}_\epsilon | r, \mathbf{U})|_{t=0}^{t=\tau} + \int_0^\tau \int_\Omega \left(\nabla \mathbf{v}_\epsilon \cdot (\nabla \mathbf{v}_\epsilon - \nabla \mathbf{V}) + \frac{\epsilon^2}{2} |\nabla w_\epsilon|^2\right) dxdzdt \\
&\leq C \int_0^\tau \mathcal{E}(\rho_\epsilon, \mathbf{u}_\epsilon | r, \mathbf{U}) dt - \int_0^\tau \int_\Omega P''(r)((\rho_\epsilon - r)\partial_t r + \rho_\epsilon \mathbf{v}_\epsilon \nabla_x r) dxdzdt \\
&\quad + \int_0^\tau \int_\Omega \frac{\rho_\epsilon}{r}(\mathbf{V} - \mathbf{v}_\epsilon)(\Delta_x \mathbf{V} + \partial_{zz}\mathbf{V}) dxdz \\
&\quad - \int_0^\tau \int_\Omega \frac{\rho_\epsilon}{r}(\mathbf{V} - \mathbf{v}_\epsilon)\nabla_x p(r) dxdz \\
&\quad + \int_0^\tau \int_\Omega \rho_\epsilon (w_\epsilon - W)(\mathbf{V} - \mathbf{v}_\epsilon) \cdot \partial_z \mathbf{V} dxdzdt \\
&\quad - \int_0^\tau \int_\Omega p(\rho_\epsilon) \mathrm{div}_x \mathbf{V} dxdzdt + o(\epsilon^2).
\end{aligned}$$

Step 2 The major challenge of the analysis is to estimate the complicated nonlinear term $\int_\Omega \rho_\epsilon (w_\epsilon - W)(\mathbf{V} - \mathbf{v}_\epsilon) \cdot \partial_z \mathbf{V} dxdz$. We rewrite it as

$$\int_\Omega \rho_\epsilon (w_\epsilon - W)(\mathbf{V} - \mathbf{v}_\epsilon) \cdot \partial_z \mathbf{V} dxdz \\ = \int_\Omega \rho_\epsilon w_\epsilon (\mathbf{V} - \mathbf{v}_\epsilon) \cdot \partial_z \mathbf{V} dxdz - \int_\Omega \rho_\epsilon W (\mathbf{V} - \mathbf{v}_\epsilon) \cdot \partial_z \mathbf{V} dxdz. \tag{24}$$

A similar heuristic argument from [17, 29] shows that the second term on the right side of (24) will be split into three parts

$$
\begin{aligned}
&\int_\Omega \rho_\epsilon W(\mathbf{V}-\mathbf{v}_\epsilon)\cdot\partial_z\mathbf{V}dxdz \\
&= \int_\Omega \chi_{\rho_\epsilon\le\frac{r}{2}}\rho_\epsilon W(\mathbf{V}-\mathbf{v}_\epsilon)\cdot\partial_z\mathbf{V}dxdz \\
&+ \int_\Omega \chi_{\frac{r}{2}<\rho_\epsilon<r}\rho_\epsilon W(\mathbf{V}-\mathbf{v}_\epsilon)\cdot\partial_z\mathbf{V}dxdz \\
&+ \int_\Omega \chi_{\rho_\epsilon\ge r}\rho_\epsilon W(\mathbf{V}-\mathbf{v}_\epsilon)\cdot\partial_z\mathbf{V}dxdz \\
&\le \|\chi_{\rho_\epsilon\le\frac{r}{2}}1\|_{L^2(\Omega)}\|r\|_{L^\infty}\|W\partial_z\mathbf{V}\|_{L^3}\|\mathbf{V}-\mathbf{v}_\epsilon\|_{L^6(\Omega)} \\
&+ \int_\Omega \chi_{\rho_\epsilon\ge r}\rho_\epsilon^{\frac{\gamma}{2}}|W\partial_z\mathbf{V}\cdot(\mathbf{V}-\mathbf{v}_\epsilon)|dxdz \\
&+ C\|\chi_{\frac{r}{2}<\rho_\epsilon<r}(\rho_\epsilon-r)\|_{L^2(\Omega)}\|W\partial_z\mathbf{V}\|_{L^3}\|\mathbf{V}-\mathbf{v}_\epsilon\|_{L^6(\Omega)} \\
&\le C\int_\Omega \chi_{\rho_\epsilon\le\frac{r}{2}}1dxdz + C\int_\Omega \chi_{\frac{r}{2}<\rho_\epsilon<r}(\rho_\epsilon-r)^2dxdz \\
&+ C\int_\Omega \chi_{\rho_\epsilon\ge r}\rho_\epsilon^\gamma dxdz + \delta\|\mathbf{V}-\mathbf{v}_\epsilon\|_{L^6(\Omega)}^2 \\
&\le C\mathcal{E}(\rho_\epsilon,\mathbf{u}_\epsilon|r,\mathbf{U}) + \delta\|\nabla_x\mathbf{V}-\nabla_x\mathbf{v}_\epsilon\|_{L^2(\Omega)}^2 + \delta\|\partial_z\mathbf{V}-\partial_z\mathbf{v}_\epsilon\|_{L^2(\Omega)}^2, \qquad (25)
\end{aligned}
$$

where in the last inequality we have used Lemma 2.1.

Remark 2.6 Here,

$$
\int_\Omega \chi_{\rho_\epsilon\ge r}\rho_\epsilon W(\mathbf{V}-\mathbf{v}_\epsilon)\cdot\partial_z\mathbf{V}dxdz = \int_{\rho_\epsilon\ge r}\rho_\epsilon W(\mathbf{V}-\mathbf{v}_\epsilon)\cdot\partial_z\mathbf{V}dxdz.
$$

The decomposition of ρ_ϵ comes from the essential set and residual set. For details, see [16, 17].

We now turn to analyze the first term on the right hand of (24), which is the crucial and difficult part in our proof. Taking (11) into it, we have

$$
\begin{aligned}
&\int_\Omega \rho_\epsilon w_\epsilon(\mathbf{V}-\mathbf{v}_\epsilon)\cdot\partial_z\mathbf{V}dxdz \\
&\qquad= \int_\Omega [-\mathrm{div}_x(\rho_\epsilon\widetilde{\mathbf{v}}_\epsilon) + z\mathrm{div}_x(\rho_\epsilon\overline{\mathbf{v}}_\epsilon)]\partial_z\mathbf{V}\cdot(\mathbf{V}-\mathbf{v}_\epsilon)dxdz
\end{aligned}
$$

$$= \int_\Omega (\rho_\epsilon \widetilde{\mathbf{v}}_\epsilon - z\rho_\epsilon \overline{\mathbf{v}}_\epsilon)\partial_z \nabla_x \mathbf{V} \cdot (\mathbf{V} - \mathbf{v}_\epsilon) dxdz$$

$$+ \int_\Omega (\rho_\epsilon \widetilde{\mathbf{v}}_\epsilon - z\rho_\epsilon \overline{\mathbf{v}}_\epsilon)\partial_z \mathbf{V} \cdot (\nabla_x \mathbf{V} - \nabla_x \mathbf{v}_\epsilon) dxdz. \tag{26}$$

In the following, we will estimate the terms on the right-hand side of (26). We only need to consider the most complicated terms; the remaining terms can be completed by the similar way. Firstly, we deal with $\int_\Omega \rho_\epsilon \widetilde{\mathbf{v}}_\epsilon \partial_z \nabla_x \mathbf{V} \cdot (\mathbf{V} - \mathbf{v}_\epsilon) dxdz$ as follows,

$$\int_\Omega \rho_\epsilon \widetilde{\mathbf{v}}_\epsilon \partial_z \nabla_x \mathbf{V} \cdot (\mathbf{V} - \mathbf{v}_\epsilon) dxdz$$

$$= \int_\Omega \rho_\epsilon (\widetilde{\mathbf{v}}_\epsilon - \widetilde{\mathbf{V}})\partial_z \nabla_x \mathbf{V} \cdot (\mathbf{V} - \mathbf{v}_\epsilon) dxdz + \int_\Omega \rho_\epsilon \widetilde{\mathbf{V}} \partial_z \nabla_x \mathbf{V} \cdot (\mathbf{V} - \mathbf{v}_\epsilon) dxdz$$

$$= J_1 + J_2,$$

where $\widetilde{\mathbf{V}} = \int_0^z \mathbf{V}(x, s, t) ds$.

Similar to the above analysis, we decompose the term J_2 into three parts:

$$\begin{aligned}
J_2 &= \int_\Omega \rho_\epsilon \widetilde{\mathbf{V}} \partial_z \nabla_x \mathbf{V} \cdot (\mathbf{V} - \mathbf{v}_\epsilon) dxdz \\
&= \int_\Omega \chi_{\rho_\epsilon \le \frac{r}{2}} \rho_\epsilon \widetilde{\mathbf{V}} \partial_z \nabla_x \mathbf{V} \cdot (\mathbf{V} - \mathbf{v}_\epsilon) dxdz \\
&\quad + \int_\Omega \chi_{\frac{r}{2} < \rho_\epsilon < r} \rho_\epsilon \widetilde{\mathbf{V}} \partial_z \nabla_x \mathbf{V} \cdot (\mathbf{V} - \mathbf{v}_\epsilon) dxdz \\
&\quad + \int_\Omega \chi_{\rho_\epsilon \ge r} \rho_\epsilon \widetilde{\mathbf{V}} \partial_z \nabla_x \mathbf{V} \cdot (\mathbf{V} - \mathbf{v}_\epsilon) dxdz \\
&\le \|\chi_{\rho_\epsilon \le \frac{r}{2}} 1\|_{L^2(\Omega)} \|r\|_{L^\infty} \|\widetilde{\mathbf{V}} \partial_z \nabla_x \mathbf{V}\|_{L^3} \|\mathbf{V} - \mathbf{v}_\epsilon\|_{L^6(\Omega)} \\
&\quad + \|\chi_{\rho_\epsilon \ge r} \rho_\epsilon^{\frac{\gamma}{2}}\|_{L^2(\Omega)} \|\widetilde{\mathbf{V}} \partial_z \nabla_x \mathbf{V}\|_{L^3(\Omega)} \|\mathbf{V} - \mathbf{v}_\epsilon\|_{L^6(\Omega)} \\
&\quad + C\|\chi_{\frac{r}{2} < \rho_\epsilon < r} (\rho_\epsilon - r)\|_{L^2(\Omega)} \|\widetilde{\mathbf{V}} \partial_z \nabla_x \mathbf{V}\|_{L^3(\Omega)} \|\mathbf{V} - \mathbf{v}_\epsilon\|_{L^6(\Omega)} \\
&\le C\mathcal{E}(\rho_\epsilon, \mathbf{u}_\epsilon | r, \mathbf{U})(t) + \delta \|\nabla_x \mathbf{V} - \nabla_x \mathbf{v}_\epsilon\|_{L^2(\Omega)}^2 + \delta \|\partial_z \mathbf{V} - \partial_z \mathbf{v}_\epsilon\|_{L^2(\Omega)}^2
\end{aligned}$$

On the other hand, by virtue of Cauchy inequality, it follows that

$$\begin{aligned}
J_1 &= \int_\Omega \left(\rho_\epsilon (\widetilde{\mathbf{v}}_\epsilon - \widetilde{\mathbf{V}})\partial_z \nabla_x \mathbf{V}\right) \cdot (\mathbf{V} - \mathbf{v}_\epsilon) dxdz \\
&\le \|\partial_z \nabla_x \mathbf{V}\|_{L^\infty}^2 \int_\Omega \rho_\epsilon |\widetilde{\mathbf{v}}_\epsilon - \widetilde{\mathbf{V}}|^2 dxdz + \int_\Omega \rho_\epsilon |\mathbf{V} - \mathbf{v}_\epsilon|^2 dxdz
\end{aligned}$$

$$
\begin{aligned}
&\le C\int_\Omega \rho_\epsilon \Big|\int_0^z (\mathbf{v}_\epsilon(s)-\mathbf{V}(s))ds\Big|^2 dxdz + \mathcal{E}(\rho_\epsilon,\mathbf{u}_\epsilon|r,\mathbf{U})\\
&\le C\int_\Omega \rho_\epsilon\Big(\int_0^1 |\mathbf{V}-\mathbf{v}_\epsilon|^2 ds\Big)dxdz + \mathcal{E}(\rho_\epsilon,\mathbf{u}_\epsilon|r,\mathbf{U})\\
&\le C\int_0^1\int_\Omega \rho_\epsilon|\mathbf{V}-\mathbf{v}_\epsilon|^2 dxdzds + \mathcal{E}(\rho_\epsilon,\mathbf{u}_\epsilon|r,\mathbf{U})\\
&\le C\int_\Omega \rho_\epsilon|\mathbf{V}-\mathbf{v}_\epsilon|^2 dxdz + \mathcal{E}(\rho_\epsilon,\mathbf{u}_\epsilon|r,\mathbf{U})\\
&\le C\mathcal{E}(\rho_\epsilon,\mathbf{u}_\epsilon|r,\mathbf{U}).
\end{aligned}
\tag{27}
$$

Secondly, we will investigate another complicated nonlinear term $\int_\Omega \rho_\epsilon \widetilde{\mathbf{v}}_\epsilon \partial_z \mathbf{V}\cdot(\nabla_x\mathbf{V}-\nabla_x\mathbf{v}_\epsilon)dxdz$. It is straightforward to show that

$$
\begin{aligned}
&\int_\Omega \rho_\epsilon \widetilde{\mathbf{v}}_\epsilon \partial_z \mathbf{V}\cdot(\nabla_x\mathbf{V}-\nabla_x\mathbf{v}_\epsilon)dxdz\\
&\qquad = \int_\Omega \chi_{\rho_\epsilon<r}\rho_\epsilon \widetilde{\mathbf{v}}_\epsilon \partial_z \mathbf{V}\cdot(\nabla_x\mathbf{V}-\nabla_x\mathbf{v}_\epsilon)dxdz\\
&\qquad + \int_\Omega \chi_{\rho_\epsilon\ge r}\rho_\epsilon \widetilde{\mathbf{v}}_\epsilon \partial_z \mathbf{V}\cdot(\nabla_x\mathbf{V}-\nabla_x\mathbf{v}_\epsilon)dxdz,
\end{aligned}
\tag{28}
$$

where the first term on the right side of (28) is split into two parts as

$$
\begin{aligned}
&\int_\Omega \chi_{\rho_\epsilon<r}\rho_\epsilon \widetilde{\mathbf{v}}_\epsilon \partial_z \mathbf{V}\cdot(\nabla_x\mathbf{V}-\nabla_x\mathbf{v}_\epsilon)dxdz\\
&\quad r = \int_\Omega \chi_{\rho_\epsilon<r}\rho_\epsilon (\widetilde{\mathbf{v}}_\epsilon-\widetilde{\mathbf{V}}) \partial_z \mathbf{V}\cdot(\nabla_x\mathbf{V}-\nabla_x\mathbf{v}_\epsilon)dxdz\\
&\qquad + \int_\Omega \chi_{\rho_\epsilon<r}\rho_\epsilon \widetilde{\mathbf{V}} \partial_z \mathbf{V}\cdot(\nabla_x\mathbf{V}-\nabla_x\mathbf{v}_\epsilon)dxdz\\
&\quad = \int_\Omega \chi_{\rho_\epsilon<r}\rho_\epsilon (\widetilde{\mathbf{v}}_\epsilon-\widetilde{\mathbf{V}}) \partial_z \mathbf{V}\cdot(\nabla_x\mathbf{V}-\nabla_x\mathbf{v}_\epsilon)dxdz\\
&\qquad + \int_\Omega \chi_{\frac{r}{2}<\rho_\epsilon<r}\rho_\epsilon \widetilde{\mathbf{V}} \partial_z \mathbf{V}\cdot(\nabla_x\mathbf{V}-\nabla_x\mathbf{v}_\epsilon)dxdz\\
&\qquad + \int_\Omega \chi_{\rho_\epsilon\le\frac{r}{2}}\rho_\epsilon \widetilde{\mathbf{V}} \partial_z \mathbf{V}\cdot(\nabla_x\mathbf{V}-\nabla_x\mathbf{v}_\epsilon)dxdz\\
&\quad \le \|\chi_{\rho_\epsilon<r}\rho_\epsilon^{\frac{1}{2}}\|_{L^\infty(\Omega)}\|\sqrt{\rho_\epsilon}(\widetilde{\mathbf{v}}_\epsilon-\widetilde{\mathbf{V}})\|_{L^2(\Omega)}\|\partial_z\mathbf{V}\|_{L^\infty(\Omega)}\|\nabla_x\mathbf{V}-\nabla_x\mathbf{v}_\epsilon\|_{L^2(\Omega)}\\
&\qquad + \|\chi_{\frac{r}{2}<\rho_\epsilon<r}\rho_\epsilon\|_{L^2(\Omega)}\|\widetilde{\mathbf{V}}\partial_z\mathbf{V}\|_{L^\infty(\Omega)}\|\nabla_x\mathbf{V}-\nabla_x\mathbf{v}_\epsilon\|_{L^2(\Omega)}
\end{aligned}
$$

$$+\|\chi_{\rho_\epsilon\le\frac{r}{2}}1\|_{L^2(\Omega)}\|r\|_{L^\infty(\Omega)}\|\widetilde{\mathbf{V}}\partial_z\mathbf{V}\|_{L^\infty(\Omega)}\|\nabla_x\mathbf{V}-\nabla_x\mathbf{v}_\epsilon\|_{L^2(\Omega)}$$
$$\le C\mathcal{E}(\rho_\epsilon,\mathbf{u}_\epsilon|r,\mathbf{U})(t)+\delta\|\nabla_x\mathbf{V}-\nabla_x\mathbf{v}_\epsilon\|^2_{L^2(\Omega)}.$$

The decomposition of the remainder of (28) is identical to the above as

$$\begin{aligned}
&\int_\Omega \chi_{\rho_\epsilon\ge r}\rho_\epsilon\widetilde{\mathbf{v}}_\epsilon\partial_z\mathbf{V}\cdot(\nabla_x\mathbf{V}-\nabla_x\mathbf{v}_\epsilon)dxdz\\
&\quad=\int_\Omega \chi_{\rho_\epsilon\ge r}\rho_\epsilon(\widetilde{\mathbf{v}}_\epsilon-\widetilde{\mathbf{V}})\partial_z\mathbf{V}\cdot(\nabla_x\mathbf{V}-\nabla_x\mathbf{v}_\epsilon)dxdz\\
&\quad+\int_\Omega \chi_{\rho_\epsilon\ge r}\rho_\epsilon\widetilde{\mathbf{V}}\partial_z\mathbf{V}\cdot(\nabla_x\mathbf{V}-\nabla_x\mathbf{v}_\epsilon)dxdz\\
&\quad=K_1+K_2,
\end{aligned}\tag{29}$$

where

$$\begin{aligned}
K_2&\le\int_\Omega \chi_{\rho_\epsilon\ge r}\rho_\epsilon^{\frac{\gamma}{2}}\widetilde{\mathbf{V}}\partial_z\mathbf{V}\cdot(\nabla_x\mathbf{V}-\nabla_x\mathbf{v}_\epsilon)dxdz\\
&\le\|\chi_{\rho_\epsilon\ge r}\rho_\epsilon^{\frac{\gamma}{2}}\|_{L^2(\Omega)}\|\widetilde{\mathbf{V}}\partial_z\mathbf{V}\|_{L^\infty(\Omega)}\|\nabla_x\mathbf{V}-\nabla_x\mathbf{v}_\epsilon\|_{L^2(\Omega)}\\
&\le C\|\chi_{\rho_\epsilon\ge r}\rho_\epsilon^{\frac{\gamma}{2}}\|^2_{L^2(\Omega)}+\delta\|\nabla_x\mathbf{V}-\nabla_x\mathbf{v}_\epsilon\|^2_{L^2(\Omega)}\\
&\le C\mathcal{E}(\rho_\epsilon,\mathbf{u}_\epsilon|r,\mathbf{U})(t)+\delta\|\nabla_x\mathbf{V}-\nabla_x\mathbf{v}_\epsilon\|^2_{L^2(\Omega)}.
\end{aligned}\tag{30}$$

It remains to estimate K_1. Due to Hölder inequality, it follows that

$$\begin{aligned}
K_1&\le\|\chi_{\rho_\epsilon\ge r}\rho_\epsilon\|_{L^6(\Omega)}\|\chi_{\rho_\epsilon\ge r}(\widetilde{\mathbf{v}}_\epsilon-\widetilde{\mathbf{V}})\|_{L^3(\Omega)}\|\partial_z\mathbf{V}\|_{L^\infty(\Omega)}\|\nabla_x\mathbf{V}-\nabla_x\mathbf{v}_\epsilon\|_{L^2(\Omega)}\\
&\le C\|\chi_{\rho_\epsilon\ge r}\rho_\epsilon\|_{L^6(\Omega)}\|\chi_{\rho_\epsilon\ge r}(\widetilde{\mathbf{v}}_\epsilon-\widetilde{\mathbf{V}})\|^{\frac{1}{2}}_{L^2(\Omega)}\|\chi_{\rho_\epsilon\ge r}(\widetilde{\mathbf{v}}_\epsilon-\widetilde{\mathbf{V}})\|^{\frac{1}{2}}_{L^6(\Omega)}\|\nabla_x\mathbf{V}\\
&\quad-\nabla_x\mathbf{v}_\epsilon\|_{L^2(\Omega)}\\
&\le C\|\chi_{\rho_\epsilon\ge r}\rho_\epsilon\|^2_{L^6(\Omega)}\|\chi_{\rho_\epsilon\ge r}(\widetilde{\mathbf{v}}_\epsilon-\widetilde{\mathbf{V}})\|_{L^2(\Omega)}\|\chi_{\rho_\epsilon\ge r}(\widetilde{\mathbf{v}}_\epsilon-\widetilde{\mathbf{V}})\|_{L^6(\Omega)}+\delta\|\nabla_x\mathbf{v}_\epsilon\\
&\quad-\nabla_x\mathbf{V}\|^2_{L^2(\Omega)}\\
&\le C\|\chi_{\rho_\epsilon\ge r}\rho_\epsilon\|^4_{L^6(\Omega)}\|\chi_{\rho_\epsilon\ge r}(\widetilde{\mathbf{v}}_\epsilon-\widetilde{\mathbf{V}})\|^2_{L^2(\Omega)}+\delta\|\widetilde{\mathbf{v}}_\epsilon-\widetilde{\mathbf{V}}\|^2_{L^6(\Omega)}+\delta\|\nabla_x\mathbf{v}_\epsilon\\
&\quad-\nabla_x\mathbf{V}\|^2_{L^2(\Omega)}\\
&\le C\|\chi_{\rho_\epsilon\ge r}\rho_\epsilon\|^4_{L^6(\Omega)}\|\chi_{\rho_\epsilon\ge r}(\widetilde{\mathbf{v}}_\epsilon-\widetilde{\mathbf{V}})\|^2_{L^2(\Omega)}+\delta\|\nabla_x\mathbf{v}_\epsilon-\nabla_x\mathbf{V}\|^2_{L^2(\Omega)}\\
&\quad+C\mathcal{E}(\rho_\epsilon,\mathbf{u}_\epsilon|r,\mathbf{U})+\delta\|\nabla_x\widetilde{\mathbf{v}}_\epsilon-\nabla_x\widetilde{\mathbf{V}}\|^2_{L^2(\Omega)}+\delta\|\partial_z\widetilde{\mathbf{v}}_\epsilon-\partial_z\widetilde{\mathbf{V}}\|^2_{L^2(\Omega)},
\end{aligned}$$

where we have used the Lemma 2.1 and the interpolation inequality

$$\|f\|_{L^3} \leq \|f\|_{L^2}^{\frac{1}{2}} \|f\|_{L^6}^{\frac{1}{2}}.$$

Recalling (18) and (27), we have

$$\|\chi_{\rho_\epsilon \geq r}\rho_\epsilon\|_{L^6(\Omega)}^4 = \Big(\int_{\rho_\epsilon \geq r} \rho_\epsilon^6 dxdz\Big)^{\frac{2}{3}} \leq C\Big(\int_\Omega \rho_\epsilon^\gamma dxdz\Big)^{\frac{4}{\gamma}} \leq \mathcal{E}(\rho_\epsilon, \mathbf{u}_\epsilon | r, \mathbf{U})^{\frac{4}{\gamma}}(t)$$

and

$$\begin{aligned}
\|\chi_{\rho_\epsilon \geq r}(\widetilde{\mathbf{v}}_\epsilon - \widetilde{\mathbf{V}})\|_{L^2(\Omega)}^2 &= \int_{\rho_\epsilon \geq r} |\widetilde{\mathbf{v}}_\epsilon - \widetilde{\mathbf{V}}|^2 dxdz = \int_{\rho_\epsilon \geq r} \frac{1}{\rho_\epsilon}\rho_\epsilon |\widetilde{\mathbf{v}}_\epsilon - \widetilde{\mathbf{V}}|^2 dxdz \\
&\leq \frac{1}{\inf_{t,x} r(t,x)} \mathcal{E}(\rho_\epsilon, \mathbf{u}_\epsilon | r, \mathbf{U}) \leq C\mathcal{E}(\rho_\epsilon, \mathbf{u}_\epsilon | r, \mathbf{U})(t).
\end{aligned}$$

An argument similar to the one used in (27) yields

$$\begin{aligned}
&\|\nabla_x \widetilde{\mathbf{v}}_\epsilon - \nabla_x \widetilde{\mathbf{V}}\|_{L^2(\Omega)}^2 \leq \|\nabla_x \mathbf{v}_\epsilon - \nabla_x \mathbf{V}\|_{L^2(\Omega)}^2, \quad \|\partial_z \widetilde{\mathbf{v}}_\epsilon - \partial_z \widetilde{\mathbf{V}}\|_{L^2(\Omega)}^2 \leq \|\partial_z \mathbf{v}_\epsilon \\
&- \partial_z \mathbf{V}\|_{L^2(\Omega)}^2.
\end{aligned}$$

Combining the above estimates, we arrive at the conclusion that

$$\begin{aligned}
\int_0^\tau K_1 dt \leq C \int_0^\tau h(t)\mathcal{E}(\rho_\epsilon, \mathbf{u}_\epsilon | r, \mathbf{U})(t)dt + \delta \int_0^\tau &\|\nabla_x \mathbf{v}_\epsilon - \nabla_x \mathbf{V}\|_{L^2(\Omega)}^2 \\
&+ \|\partial_z \mathbf{v}_\epsilon - \partial_z \mathbf{V}\|_{L^2(\Omega)}^2 dt,
\end{aligned}$$

where $h(t) \in L^1(0,T)$.

The estimate of the remainder in (26) can be completed by the analogous method. Therefore, we can summarize what we have proved as the following:

$$\begin{aligned}
&\mathcal{E}(\rho_\epsilon, \mathbf{u}_\epsilon | r, \mathbf{U})|_{t=0}^{t=\tau} + \int_0^\tau \int_\Omega \big(\nabla \mathbf{v}_\epsilon \cdot (\nabla \mathbf{v}_\epsilon - \nabla \mathbf{V}) + \epsilon^2 |\nabla w_\epsilon|^2\big) dxdzdt \\
&\leq C \int_0^\tau h(t)\mathcal{E}(\rho_\epsilon, \mathbf{u}_\epsilon | r, \mathbf{U})dt + \delta \int_0^\tau \|\nabla_x \mathbf{v}_\epsilon - \nabla_x \mathbf{V}\|_{L^2(\Omega)}^2 \\
&+ \|\partial_z \mathbf{v}_\epsilon - \partial_z \mathbf{V}\|_{L^2(\Omega)}^2 dt \\
&+ \int_0^\tau \int_\Omega \frac{\rho_\epsilon}{r}(\mathbf{V} - \mathbf{v}_\epsilon)(\Delta_x \mathbf{V} + \partial_{zz}\mathbf{V}) dxdzdt \\
&- \int_0^\tau \int_\Omega \frac{\rho_\epsilon}{r}(\mathbf{V} - \mathbf{v}_\epsilon)\nabla_x p(r) dxdzdt
\end{aligned}$$

$$
\begin{aligned}
&- \int_0^\tau \int_\Omega P''(r)((\rho_\epsilon - r)\partial_t r + \rho_\epsilon \mathbf{v}_\epsilon \nabla_x r) dxdzdt \\
&- \int_0^\tau \int_\Omega p(\rho_\epsilon) \mathrm{div}_x \mathbf{V} dxdzdt + o(\epsilon^2)
\end{aligned}
$$

Then, we deduce that

$$
\begin{aligned}
&\mathcal{E}(\rho_\epsilon, \mathbf{u}_\epsilon | r, \mathbf{U})|_{t=0}^{t=\tau} + \int_0^\tau \int_\Omega \big((\nabla_x \mathbf{v}_\epsilon - \nabla_x \mathbf{V}) : (\nabla_x \mathbf{v}_\epsilon - \nabla_x \mathbf{V}) \\
&\quad + |\partial_z \mathbf{v}_\epsilon - \partial_z \mathbf{V}_\epsilon|^2 + \epsilon^2 |\nabla w_\epsilon|^2\big) dxdzdt \\
&\quad \leq C \int_0^\tau h(t) \mathcal{E}(\rho_\epsilon, \mathbf{u}_\epsilon | r, \mathbf{U}) dt + \delta \int_0^\tau \|\nabla_x \mathbf{v}_\epsilon - \nabla_x \mathbf{V}\|_{L^2(\Omega)}^2 \\
&\quad + \|\partial_z \mathbf{v}_\epsilon - \partial_z \mathbf{V}\|_{L^2(\Omega)}^2 dt \\
&\quad + \int_0^\tau \int_\Omega (\frac{\rho_\epsilon}{r} - 1)(\mathbf{V} - \mathbf{v}_\epsilon)(\Delta_x \mathbf{V} + \partial_{zz} \mathbf{V}) dxdzdt \\
&\quad - \int_0^\tau \int_\Omega \frac{\rho_\epsilon}{r} (\mathbf{V} - \mathbf{v}_\epsilon) \nabla_x p(r) dxdzdt \\
&\quad - \int_0^\tau \int_\Omega P''(r)((\rho_\epsilon - r)\partial_t r + \rho_\epsilon \mathbf{v}_\epsilon \nabla_x r) dxdzdt \\
&\quad - \int_0^\tau \int_\Omega p(\rho_\epsilon) \mathrm{div}_x \mathbf{V} dxdzdt + o(\epsilon^2).
\end{aligned} \tag{31}
$$

Step 3 We are now in a position to estimate the remaining terms in the versatile relative energy inequality (31). It is clear to check that

$$
\begin{aligned}
&- \int_0^\tau \int_\Omega \frac{\rho_\epsilon}{r} (\mathbf{V} - \mathbf{v}_\epsilon) \nabla_x p(r) + p(\rho_\epsilon) \mathrm{div}_x \mathbf{V} + P''(r)((\rho_\epsilon - r)\partial_t r \\
&\quad + \rho_\epsilon \mathbf{v}_\epsilon \nabla_x r) dxdzdt \\
&= - \int_0^\tau \int_\Omega (\rho_\epsilon - r) P''(r) \partial_t r + P''(r) \rho_\epsilon \mathbf{v}_\epsilon \cdot \nabla_x r + \rho_\epsilon P''(r)(\mathbf{V} - \mathbf{v}_\epsilon) \cdot \nabla_x r \\
&\quad + p(\rho_\epsilon) \mathrm{div}_x \mathbf{V} dxdzdt \\
&= - \int_0^\tau \int_\Omega (\rho_\epsilon - r) P''(r) \partial_t r + P''(r) \rho_\epsilon \mathbf{V} \cdot \nabla_x r + p(\rho_\epsilon) \mathrm{div}_x \mathbf{V} dxdzdt \\
&= - \int_0^\tau \int_\Omega \rho_\epsilon P''(r)(\partial_t r + \mathbf{V} \cdot \nabla_x r) - r P''(r) \partial_t r + p(\rho_\epsilon) \mathrm{div}_x \mathbf{V} dxdzdt \\
&= - \int_0^\tau \int_\Omega \rho_\epsilon P''(r)(-r \mathrm{div}_x \mathbf{V} - r \partial_z W) - r P''(r) \partial_t r + p(\rho_\epsilon) \mathrm{div}_x \mathbf{V} dxdzdt
\end{aligned}
$$

$$
\begin{aligned}
&= -\int_0^\tau \int_\Omega \operatorname{div}_x \mathbf{V}\big(p(\rho_\epsilon) - p'(r)(\rho_\epsilon - r) - p(r)\big) dx dz dt \\
&\quad + \int_0^\tau \int_\Omega p'(r)(\rho_\epsilon - r)\partial_z W dx dz dt,
\end{aligned} \tag{32}
$$

where we have used the fact that $\partial_t r + \operatorname{div}_x \mathbf{V} r + \mathbf{V} \cdot \nabla_x r + r\partial_z W = 0$.

Using the analogous argument as in [37] Section 2.2.5, we can easily carry out the following estimate:

$$
\begin{aligned}
&\left|\int_0^\tau \int_\Omega \operatorname{div}_x \mathbf{V}\big(p(\rho_\epsilon) - p'(r)(\rho_\epsilon - r) - p(r)\big) dx dz dt\right| \\
&\quad \le C \int_0^\tau h(t)\mathcal{E}(\rho_\epsilon, \mathbf{u}_\epsilon | r, \mathbf{U}) dt
\end{aligned} \tag{33}
$$

According to the periodic boundary condition, it follows that

$$
\int_0^\tau \int_\Omega p'(r)(\rho_\epsilon - r)\partial_z W dx dz dt = \int_0^\tau dt \int_{\mathbb{T}^2} \Big(\int_0^1 \partial_z W dz\Big) p'(r)(\rho_\epsilon - r) dx = 0. \tag{34}
$$

Furthermore, an argument similar to the one used in [29] Section 6.3 shows that

$$
\begin{aligned}
&\int_\Omega \Big(\frac{\rho_\epsilon}{r} - 1\Big)(\mathbf{V} - \mathbf{v}_\epsilon)(\Delta_x \mathbf{V} + \partial_{zz}\mathbf{V}) dx dz \\
&\quad \le C\mathcal{E}(\rho_\epsilon, \mathbf{u}_\epsilon | r, \mathbf{U}) + \delta\|\nabla_x \mathbf{v}_\epsilon - \nabla_x \mathbf{V}\|_{L^2}^2 + \delta\|\partial_z \mathbf{v}_\epsilon - \partial_z \mathbf{V}\|_{L^2}^2.
\end{aligned} \tag{35}
$$

Therefore, putting (4.12)–(4.16) together, we have

$$
\mathcal{E}(\rho_\epsilon, \mathbf{u}_\epsilon | r, \mathbf{U})(\tau) \le C \int_0^\tau h(t)\mathcal{E}(\rho_\epsilon, \mathbf{u}_\epsilon | r, \mathbf{U})(t) dt + o(\epsilon^2). \tag{36}
$$

Then, applying Gronwall's inequality, we finish the proof of Theorem 2.1.

3 Inviscid CPE

In this section, we consider the following compressible Euler system

$$
\begin{cases}
\partial_t \rho + \operatorname{div}(\rho \mathbf{u}) = 0, \\
(\rho \mathbf{u})_t + \operatorname{div}(\rho \mathbf{u} \otimes \mathbf{u}) + \nabla p(\rho) = 0,
\end{cases} \tag{37}
$$

in the thin domain $(0, T) \times \Omega_\epsilon$, where $\Omega_\epsilon = \{(x, z)|x \in \mathbb{T}^2, 0 < z < \epsilon\}$, x denotes the horizontal direction, and z denotes the vertical direction. Here, $\rho, \mathbf{u}, andp$ represent the density, velocity, and pressure, respectively. The velocity $\mathbf{u} = (\mathbf{v}, w)$, where $\mathbf{v}(t, x, z) \in \mathbb{R}^2$ and $w(t, x, z) \in \mathbb{R}$ represent the horizonal velocity and vertical velocity, respectively. Here, we suppose the pressure: $p(\rho) = \rho^\gamma \ (\gamma > 1)$.

The same token as [5], we perform the following rescaling

$$z \to \varepsilon z, w \to \epsilon w$$

while t, x, v, ρ, and p are left unchanged and then get the following rescaled Euler equation in the fixed domain $\Omega := \mathbb{T}^2 \times (0, 1)$:

$$\begin{cases} \partial_t \rho + \mathrm{div}_x(\rho \mathbf{v}) + \partial_z(\rho w) = 0, \\ \rho \partial_t \mathbf{v} + \rho(\mathbf{u} \cdot \nabla)\mathbf{v} + \nabla_x p(\rho) = 0, \\ \epsilon^2\big(\rho \partial_t w_\epsilon + \rho(\mathbf{u} \cdot \nabla)w\big) + \partial_z p(\rho) = 0 \end{cases} \tag{38}$$

We supplement the system with the following boundary and initial conditions:

$$(\rho_\epsilon, \mathbf{u}_\epsilon)|_{t=0} = (\rho_0, \mathbf{u}_0) \tag{39}$$

By setting $\varepsilon \to 0$ in the compressible Euler system (38), we obtain the inviscid compressible PE as the following:

$$\begin{cases} \partial_t \rho + \mathrm{div}_x(\rho \mathbf{v}) + \partial_z(\rho w) = 0, \\ \rho \partial_t \mathbf{v} + \rho(\mathbf{u} \cdot \nabla)\mathbf{v} + \nabla_x p(\rho) = 0, \\ \partial_z p(\rho) = 0 \end{cases} \tag{40}$$

The goal of this part is to investigate the limit process $\epsilon \to 0$ in the system of (38) that converges in a certain sense to the inviscid compressible primitive equations (40).

3.1 Relative Entropy Inequality

Motivated by Brenier [5], Feireisl et al. [17], for any smooth solution $(\rho, \mathbf{u})$, where $\mathbf{u} = (\mathbf{v}, w)$, to the compressible Euler system (38), we introduce the relative entropy functional

$$\begin{aligned} &\mathcal{E}(\rho, \mathbf{u}|r, \mathbf{U}) \\ &= \int_\Omega [\frac{1}{2}\rho|\mathbf{v} - \mathbf{V}|^2 + \frac{\epsilon^2}{2}\rho|w - W|^2 + P(\rho) - P'(r)(\rho - r) - P(r)]dxdz \end{aligned} \tag{41}$$

where $r > 0, \mathbf{U} = (\mathbf{V}, W)$ are smooth "test" functions, $r, \mathbf{U}$ compactly supported in Ω. Here, we have used $rP'(r) - P(r) = p(r)$. For simplicity, we use the notation $\int_\Omega f$ instead of $\int_\Omega f dxdz$.

Lemma 3.1 *Let* $(\rho, \mathbf{u})$ *be a smooth solution to the compressible Euler system* (38), *and let* $(r, \mathbf{U})$ *be smooth solutions to the inviscid compressible PE system* (40). *Then, we have*

$$
\begin{aligned}
&\frac{d}{dt}\mathcal{E}(\rho, \mathbf{u}|r, \mathbf{U}) \\
&\quad = - \int_\Omega \rho w(\mathbf{v} - \mathbf{V})(w - W)\partial_z \mathbf{V} - \int_\Omega (\mathbf{v} - \mathbf{V})(\nabla_x p(\rho) - \frac{\rho}{r}\nabla_x p(r)) \\
&\quad - \int_\Omega \rho(\mathbf{v} - \mathbf{V})^2 \nabla_x \mathbf{V} + \varepsilon^2 \int_\Omega \rho(w - W)(\partial_t w - \partial_t W) \\
&\quad + \varepsilon^2 \int_\Omega \rho \mathbf{v}(w - W)\nabla_x (w - W) + \varepsilon^2 \rho w(w - W)\partial_z (w - W) \\
&\quad - \int_\Omega [P'(\rho) - P'(r)](div_x(\rho \mathbf{v}) + \partial_z(\rho w)) \\
&\quad + \int_\Omega P''(r)(\rho - r)(div_x(r\mathbf{V}) + \partial_z(rW)).
\end{aligned}
\tag{42}
$$

We estimate all the terms on the right side of (41). Firstly, we compute the first term as the following:

$$
\begin{aligned}
&\frac{d}{dt}\int_\Omega \frac{1}{2}\rho|\mathbf{v} - \mathbf{V}|^2 dxdz \\
&= \int_\Omega \frac{1}{2}\rho_t |\mathbf{v} - \mathbf{V}|^2 + \int_\Omega \rho(\mathbf{v} - \mathbf{V})(\partial_t \mathbf{v} - \partial_t \mathbf{V}) \\
&= -\frac{1}{2}\int_\Omega \mathrm{div}_x(\rho \mathbf{v})|\mathbf{v} - \mathbf{V}|^2 dxdz - \frac{1}{2}\int_\Omega \partial_z(\rho w)|\mathbf{v} - \mathbf{V}|^2 \\
&\quad - \int_\Omega (\mathbf{v} - \mathbf{V})(\rho \mathbf{v}\nabla_x \mathbf{v} + \rho w \partial_z \mathbf{v} + \nabla_x p(\rho) - \rho \mathbf{V}\nabla_x \mathbf{V} - \rho W \partial_z \mathbf{V} - \frac{\rho}{r}\nabla_x p(r)) \\
&= \int_\Omega \rho \mathbf{v}(\mathbf{v} - \mathbf{V})(\nabla_x \mathbf{v} - \nabla_x \mathbf{V}) + \int_\Omega \rho w(\mathbf{v} - \mathbf{V})(\partial_z \mathbf{v} - \partial_z \mathbf{V}) \\
&\quad - \int_\Omega (\mathbf{v} - \mathbf{V})[\rho \mathbf{v}\nabla_x(\mathbf{v} - \mathbf{V}) + \rho(\mathbf{v} - \mathbf{V})\nabla_x \mathbf{V} + \rho w \partial_z(\mathbf{v} - \mathbf{V}) \\
&\qquad + \rho(w - W)\partial_z \mathbf{V} + \nabla_x p(\rho) - \frac{\rho}{r}\nabla_x p(r)]
\end{aligned}
$$

$$
\begin{aligned}
&= -\int_\Omega \rho w(\mathbf{v}-\mathbf{V})(w-W)\partial_z \mathbf{V} - \int_\Omega (\mathbf{v}-\mathbf{V})(\nabla_x p(\rho) - \frac{\rho}{r}\nabla_x p(r)) \\
&\quad - \int_\Omega \rho(\mathbf{v}-\mathbf{V})^2 \nabla_x \mathbf{V}, \qquad (43)
\end{aligned}
$$

where we have used the continuity equation $(38)_1$ and momentum equation $(38)_2$.

Similarly, by virtue of the continuity equation, we obtain

$$
\begin{aligned}
&\frac{d}{dt}\int_\Omega \frac{\varepsilon^2}{2}\rho|w-W|^2 \\
&\quad = \frac{\varepsilon^2}{2}\int_\Omega \rho_t|w-W|^2 + \varepsilon^2\int_\Omega \rho(w-W)(\partial_t w - \partial_t W) \\
&\quad = -\frac{\varepsilon^2}{2}\int_\Omega \mathrm{div}_x(\rho\mathbf{v})|w-W|^2 - \frac{\varepsilon^2}{2}\int_\Omega \partial_z(\rho w)|w-W|^2 \\
&\quad + \varepsilon^2\int_\Omega \rho(w-W)(\partial_t w - \partial_t W) \\
&\quad = \varepsilon^2\int_\Omega [\rho\mathbf{v}(w-W)\nabla_x(w-W) + \varepsilon^2\rho w(w-W)\partial_z(w-W)] \\
&\quad + \varepsilon^2\int_\Omega \rho(w-W)(\partial_t w - \partial_t W). \qquad (44)
\end{aligned}
$$

Finally,

$$
\begin{aligned}
&\frac{d}{dt}\int_\Omega (P(\rho) - P'(r)(\rho-r) - P(r)) \\
&\quad = \int_\Omega (P'(\rho)\partial_t\rho - P''(r)\partial_t r(\rho-r) - P'(r)\partial_t\rho) \\
&\quad = \int_\Omega [(P'(\rho)-P'(r))\partial_t\rho - P''(r)\partial_t r(\rho-r)] \\
&\quad = -\int_\Omega [P'(\rho)-P'(r)](\mathrm{div}_x(\rho\mathbf{v}) + \partial_z(\rho w)) \\
&\qquad + \int_\Omega P''(r)(\rho-r)(\mathrm{div}_x(r\mathbf{V}) + \partial_z(rW)). \qquad (45)
\end{aligned}
$$

Getting (43)–(45) together, we complete the proof of Lemma 3.1.

Based on the relative entropy inequality, we can obtain the following lemma from [17]:

Lemma 3.2 *Let $0 < a < b < \infty$. Then, there exists $c = c(a, b) > 0$ such that for all $\rho \in [0, \infty)$ and $r \in [a, b]$ there holds*

$$P(\rho) - P'(r)(\rho - r) - P(r) \geq \begin{cases} C|\rho - r|^2, & \text{when } \frac{r}{2} < \rho < r, \\ C(1 + \rho^\gamma), & \text{otherwise}, \end{cases}$$

where $C = C(a, b)$.

3.2 Main Result

Now, we are ready to state our main result.

Theorem 3.1 *Let $(\rho, \mathbf{u})$ and $(r, \mathbf{U})$ be smooth solutions with $\rho > 0$ and $r > 0$, respectively, the compressible Euler equations* (38) *and the inviscid compressible PE system* (40)*, and T_{max} be the life time of smooth solutions. Supposing that they emerge at the same initial data, that is, $\rho|_{t=0} = r|_{t=0}$, $\mathbf{u}|_{t=0}\mathbf{U}|_{t=0}$, then for all $t \in [0, T_{max}]$, we have*

$$\sup_{t\in(0,T_{max})} \mathcal{E}(\rho, \mathbf{u}|r, \mathbf{U}) \leq C_T \varepsilon^2,$$

where C_T is independent of ε.

Remark 3.1 The compressible Euler system (38) can be written as a positive, symmetric, hyperbolic system, and Kato [28] proved the system possesses a unique, local C^1 solution with $\rho > 0$, provided the initial data are sufficiently regular. We suggest readers refer to [41] for more results and background about the compressible Euler system.

Remark 3.2 Brenier [4], Masmoudi, and Wong [38] obtained the local existence of smooth solutions for inviscid incompressible PE systems. Liu and Titi [35] proved the local existence of strong solutions to the compressible PE model, while the corresponding result for the inviscid compressible PE model is still not known, which is left for future study.

3.3 Proof of Theorem 3.1

Theorem 3.1 can be proved directly by Lemma 3.1. We just need to control the right-hand side of (42). Firstly, we rewrite and control the proceeding term at the right side of (42):

$$
\begin{aligned}
&-\int_\Omega (\mathbf{v}-\mathbf{V})(\nabla_x p(\rho) - \frac{\rho}{r}\nabla_x p(r)) \\
&\quad = -\int_\Omega (\mathbf{v}-\mathbf{V})\frac{(r-\rho)\nabla_x p(\rho) + \rho(\nabla_x p(\rho) - \nabla_x p(r))}{r} \\
&\quad = \int_\Omega (\mathbf{v}-\mathbf{V})(\rho - r)\frac{\nabla_x p(r)}{r} - \int_\Omega (\mathbf{v}-\mathbf{V})\frac{\rho}{r}\big(p'(\rho)\nabla_x\rho - p'(r)\nabla_x r\big) \\
&\quad = I_1 + I_2 \qquad (46)
\end{aligned}
$$

We divide I_1 into two parts and use the Cauchy inequality and then get

$$
\begin{aligned}
I_1 &= \int_{\frac{r}{2}\le\rho\le 2r} (\mathbf{v}-\mathbf{V})(\rho - r)\frac{\nabla_x p(\rho)}{r} + \int_{\rho\le\frac{r}{2},\rho\ge 2r} (\mathbf{v}-\mathbf{V})(\rho - r)\frac{\nabla_x p(\rho)}{r} \\
&= \int_{\frac{r}{2}\le\rho\le 2r} \sqrt{\rho}(\mathbf{v}-\mathbf{V})\cdot(\rho - r)\frac{\nabla_x p(\rho)}{\sqrt{\rho} r} \\
&\quad + \int_{\rho\le\frac{r}{2},\rho\ge 2r} \sqrt{\rho}(\mathbf{v}-\mathbf{V})(\rho - r)\frac{\nabla_x p(\rho)}{\sqrt{\rho} r} \\
&\le \int_\Omega \rho|\mathbf{v}-\mathbf{V}|^2 + C\int_{\frac{r}{2}\le\rho\le 2r}(\rho - r)^2 + C\int_{\rho\le\frac{r}{2},\rho\ge 2r} 1 \le C \\
&\le C\mathcal{E}(\rho,\mathbf{u}|r,\mathbf{U}),
\end{aligned}
$$

where we utilize the fact that $(\rho, \mathbf{u})$ and $(r, \mathbf{U})$ are smooth solution with $\rho > 0$ and $r > 0$, $\int_{\frac{r}{2}\le\rho\le 2r}(\rho - r)^2 \le \mathcal{E}(\rho,\mathbf{u}|r,\mathbf{U})$ and $\int_{\rho\le\frac{r}{2},\rho\ge 2r} 1 \le \mathcal{E}(\rho,\mathbf{u}|r,\mathbf{U})$ from Lemma 2.2.

For the term I_2,

$$
\begin{aligned}
-I_2 &= \int_\Omega (\mathbf{v}-\mathbf{V})\frac{\rho}{r}(p'(\rho) - p'(r))\nabla_x\rho + \int_\Omega (\mathbf{v}-\mathbf{V})\frac{\rho}{r}p'(r)\nabla_x(\rho - r) \\
&= I_{21} + I_{22}.
\end{aligned}
$$

Using the mean value theorem, we see $I_{21} = \int_\Omega (\mathbf{v}-\mathbf{V})\frac{\rho}{r}p''(\xi_1)(\rho - r)\nabla_x\rho$ where $\xi_1 = \theta\rho + (1-\theta)r, \theta \in (0, 1)$. Then, we split the term into two parts as

$$
\begin{aligned}
I_{21} &= \int_{\frac{r}{2}\le\rho\le 2r} (\mathbf{v}-\mathbf{V})\frac{\rho}{r}p''(\xi_1)(\rho - r)\nabla_x\rho \\
&\quad + \int_{\rho\le\frac{r}{2},\rho\ge 2r} (\mathbf{v}-\mathbf{V})\frac{\rho}{r}p''(\xi_1)(\rho - r)\nabla_x\rho
\end{aligned}
$$

$$\leq C\int_{\frac{r}{2}\leq\rho\leq 2r}(\rho-r)^2+C\int_{\Omega}\rho|\mathbf{v}-\mathbf{V}|^2+C\int_{\rho\leq\frac{r}{2},\rho\geq 2r}\rho$$
$$\leq C\mathcal{E}(\rho,\mathbf{u}|r,\mathbf{U}),$$

where we used the Cauchy inequality and Lemma 2.2.

On the other hand, it is similar to I_{22} to get $\int_{\Omega}(\mathbf{v}-\mathbf{V})\frac{\rho}{r}p'(r)\nabla_x(\rho-r)\leq\mathcal{E}(\rho,\mathbf{u}|r,\mathbf{U})$. Moreover, due to the smooth solutions, it is easy to get the following:

$$|\int_{\Omega}\rho(\mathbf{v}-\mathbf{V})^2\nabla_x\mathbf{V}|\leq C\mathcal{E}(\rho,\mathbf{u}|r,\mathbf{U})$$

Next, we turn to estimate the term about P. By the same token, we use Lemma 2.2 to get

$$\int_{\Omega}[P'(\rho)-P'(r)](\mathrm{div}_x(\rho\mathbf{v})+\partial_z(\rho w))$$
$$=\int_{\Omega}P''(\xi_2)(\rho-r)(\mathrm{div}_x(\rho\mathbf{v})+\partial_z(\rho w))\ (\xi_2=\theta\rho+(1-\theta)r,\ \theta\in(0,1))$$
$$=\int_{\frac{r}{2}\leq\rho\leq 2r}P''(\xi_2)(\rho-r)(\mathrm{div}_x(\rho\mathbf{v})+\partial_z(\rho w))$$
$$+\int_{\rho\leq\frac{r}{2},\rho\geq 2r}P''(\xi_2)(\rho-r)(\mathrm{div}_x(\rho\mathbf{v})+\partial_z(\rho w))$$
$$\leq\int_{\frac{r}{2}\leq\rho\leq 2r}|P''(\xi_2)|(\rho-r)^2|\mathrm{div}_x(\rho\mathbf{v})+\partial_z(\rho w)|+C\int_{\rho\leq\frac{r}{2},\rho\geq 2r}1$$
$$\leq C\mathcal{E}(\rho,\mathbf{u}|r,\mathbf{U}),$$

and

$$\int_{\Omega}P''(r)(\rho-r)(\mathrm{div}_x(r\mathbf{V})+\partial_z(rW))$$
$$=\int_{\frac{r}{2}\leq\rho\leq 2r}P''(r)(\rho-r)(\mathrm{div}_x(r\mathbf{V})+\partial_z(rW))$$
$$+\int_{\rho\leq\frac{r}{2},\rho\geq 2r}P''(r)(\rho-r)(\mathrm{div}_x(r\mathbf{V})+\partial_z(rW))$$
$$\leq\int_{\frac{r}{2}\leq\rho\leq 2r}|P''(r)|(\rho-r)^2|\mathrm{div}_x(r\mathbf{V})+\partial_z(rW)|+C\int_{\rho\leq\frac{r}{2},\rho\geq 2r}1$$
$$\leq C\mathcal{E}(\rho,\mathbf{u}|r,\mathbf{U}).$$

Finally, we will control the term about vertical velocity:

$$
\begin{aligned}
\int_\Omega \rho(w-W)(\mathbf{v}-\mathbf{V})\partial_z \mathbf{V} &\le \int_\Omega \sqrt{\rho}(\mathbf{v}-\mathbf{V})\cdot\sqrt{\rho}(w-W)\partial_z\mathbf{V}\\
&\le C\int_\Omega \rho|\mathbf{v}-\mathbf{V}|^2 + C\int_\Omega \rho\\
&\le C\mathcal{E}(\rho,\mathbf{u}|r,\mathbf{U}) + C\int_{\frac{r}{2}\le\rho\le 2r}\rho + C\int_{\rho\le\frac{r}{2},\rho\ge 2r}\rho\\
&\le C\mathcal{E}(\rho,\mathbf{u}|r,\mathbf{U}) + C\int_{\frac{r}{2}\le\rho\le 2r}(\rho-r)^2\\
&\quad + C[\int_{\rho\le\frac{r}{2},\rho\ge 2r}\rho^\gamma + \int_{\rho\le\frac{r}{2},\rho\ge 2r}1]\\
&\le C\mathcal{E}(\rho,\mathbf{u}|r,\mathbf{U})
\end{aligned}
$$

Then, due to the solution being smooth solution, we have

$$
\begin{aligned}
&\varepsilon^2\int_\Omega \rho\mathbf{v}|w-W|\nabla_x(w-W) + \varepsilon^2\rho w|w-W|\partial_z(w-W)\\
&+\varepsilon^2\int_\Omega \rho(w-W)(\partial_t w-\partial_t W)\le O(\varepsilon^2).
\end{aligned}
$$

Therefore, combining all the above estimates together, we have

$$
\frac{d}{dt}\mathcal{E}(\rho,\mathbf{u}|r,\mathbf{U})(t)\le C)\mathcal{E}(\rho,\mathbf{u}|r,\mathbf{U}) + o(\epsilon^2). \tag{47}
$$

Then, applying Gronwall's inequality, we finish the proof of Theorem 3.1.

4 Conclusion and Comment

Compared with the previous results for PE, there are some delicate differences in the process of using the relative energy inequality. Precisely speaking, we take the weak solution of CNS and strong solution of CPE in Sect. 2, while we take both smooth solution of Euler and CPE in Sect. 3. Here, our convergence is between two different systems, and this convergence gives us a system in 3D to "2.5"D. Due to the special structure of CPE, the relative entropy inequalities (13) and (42) are constructed differently from the ones in [20, 37], and we call them *the versatile relative entropy inequalities*.

Acknowledgments The work of Š.N. has been supported by Praemium Academiæ of Š. Nečasová. Finally, the Institute of Mathematics, CAS, is supported by RVO:67985840. T. Tang is partially supported by NSFC No. 12371246,NSF of Jiangsu Province Grant No. BK20221369 and Qing Lan Project of Jiangsu Province.

References

1. A. Abbatiello, E. Feireisl, and A. Novotný, Generalized solutions to models of compressible viscous fluids, *Discrete Contin. Dyn. Syst.* , 41 (2021), 1–28.
2. P. Azérad and F. Guillén, Mathematical justification of the hydrostatic approximation in the primitive equations of geophysical fluid dynamics, *SIAM J. Math. Anal.*, 33 (2001), 847–859.
3. P. Bella, E. Feireisl, A. Novotný, Dimension reduction for compressible viscous fluids, *Acta Appl. Math.*, 134 (2014), 111–121.
4. Y. Brenier, Homogeneous hydrostatic flows with convex velocity profiles, *Nonlinearity*, 12 (1999), 495–512.
5. Y. Brenier, Remarks on the derivation of the hydrostatic Euler equations, *Bull. Sci. Math.*, 127 (2003), 585–595.
6. D. Bresch, F. Guillén-González, N. Masmoudi and M. A. Rodríguez-Bellido, On the uniqueness of weak solutions of the two-dimensional primitive equations, *Differential Integral Equations*, 16 (2003), 77–94.
7. D. Bresch, A. Kazhikhov and J. Lemoine, On the two-dimensional hydrostatic Navier-Stokes equations, *SIAM J. Math. Anal.*, 36 (2004/05), 796–814.
8. D. Bresch and P. E. Jabin, Global existence of weak solutions for compressible Navier-Stokes equations: thermodynamically unstable pressure and anisotropic viscous stress tensor, *Ann. of Math.*, 188 (2018), 577–684.
9. K. Bryan, A numerical method for the study of the circulation of the world ocean, *J. Comp. Phys.*, 4 (1969), 347–376
10. C. S. Cao and E. S. Titi, Global well-posedness of the three-dimensional viscous primitive equations of large scale ocean and atmosphere dynamics, *Ann. of Math.*, 166 (2007), 245–267.
11. C. S. Cao, J. K. Li and E. S. Titi, Global well-posedness of the three-dimensional primitive equations with only horizontal viscosity and diffusion, *Comm. Pure Appl. Math.*, 69 (2016), 1492–1531.
12. C. M. Dafermos, The second law of thermodynamics and stability, *Arch. Rational Mech. Anal.*, 70 (1979) 167–179.
13. D. Donatelli and N. Juhasz, The primitive equations of the polluted atmosphere as a weak and strong limit of the 3D Navier-Stokes equations in downwind-matching coordinates, *Discrete Contin. Dyn. Syst.*, 42 (2022), 2859–2892.
14. B. Ducomet, Š. Nečasová, M. Pokorný and M. A. Rodríguez-Bellido, Derivation of the Navier-Stokes-Poisson system with radiation for an accretion disk, *J. Math. Fluid Mech.,*, 20 (2018), 697–719.
15. M. Ersoy, T. Ngom and M. Sy, Compressible primitive equations: formal derivation and stability of weak solutions, *Nonlinearity*, 24 (2011), 79–96.
16. E. Feireisl, Dynamics of viscous compressible fluids, Oxford Lecture Series in Mathematics and its Applications, Oxford University Press, Oxford, 2004.
17. E. Feireisl, J. B. Jin and A. Novotný, Relative entropies, suitable weak solutions, and weak-strong uniqueness for the compressible Navier-Stokes system, *J. Math. Fluid Mech.*, 14 (2012), 717–730.
18. E. Feireisl and A. Novotný, Singular limits in thermodynamics of viscous fluids, Advances in Mathematical Fluid Mechanics, Birkhäuser, Basel, 2009.
19. K. Furukawa, Y. Giga and T. Kashiwabara, The hydrostatic approximation for the primitive equations by the scaled Navier-Stokes Equations under the no-slip boundary condition, *J. Evol. Equ.*, 21 (2021), 3331–3373.

20. H. J. Gao, Š. Nečasová and T. Tang, On weak-strong uniqueness and singular limit for the compressible Primitive Equations, *Discrete Contin. Dyn. Syst. Ser. A*, 40 (2020), 4287–4305.
21. B. V. Gatapov and A. V. Kazhikhov, Existence of a global solution of a model problem of atmospheric dynamics, *Siberian Math. J.*, 46 (2005), 805–812.
22. P. Germain, Weak-strong uniqueness for the isentropic compressible Navier-Stokes system, *J. Math. Fluid Mech.*, 13 (2011), 137–146.
23. F. Guillén-González, N. Masmoudi and M. A. Rodríguez-Bellido, Anisotropic estimates and strong solutions of the primitive equations, *Differential Integral Equations*, 14 (2001), 1381–1408.
24. E. Grenier, On the derivation of homogeneous hydrostatic equations, *M2AN Math. Model. Numer. Anal.*, 33 (1999), 965–970.
25. B. L. Guo and D. W. Huang, Existence of the universal attractor for the 3-D viscous primitive equations of large-scale moist atmosphere, *J. Differential Equations*, 251 (2011), 457–491.
26. M. Hieber and T. Kashiwabara, Global strong well-posedness of the three dimensional primitive equations in L^p-spaces, *Arch. Ration. Mech. Anal.*, 221 (2016), 1077–1115.
27. M. Hieber, Y. Iida, A. Roy, T. Zöchling, The Lagrangian approach to the compressible primitive equations,arXiv:2502.03630
28. T. Kato, The Cauchy problem for quasi-linear symmetric hyperbolic systems, *Arch. Ration. Mech. Anal.*, 58 (1975), 181–205.
29. O. Kreml, Š. Nečasová and T. Piasecki, Local existence of strong solution and weak-strong uniqueness for the compressible Navier-Stokes system on moving domains, *Proc. Roy. Soc. Edinburgh Sect. A* , 150 (2020), 2255–2300.
30. H. L. Li and C. C. Liang, Global existence and large-time behavior for primitive equations with the free boundary, *Sci. China Math.*, 67 (2024), 2303–2330.
31. J. K. Li, and E. S. Titi, The primitive equations as the small aspect ratio limit of the Navier-Stokes equations: rigorous justification of the hydrostatic approximation, *J. Math. Pures Appl.*, 124 (2019), 30–58.
32. J. L. Lions, R. Temam and S. H. Wang, On the equations of the large-scale ocean, *Nonlinearity*, 5 (1992), 1007–1053.
33. J. L. Lions, R. Temam and S. H. Wang, New formulations of the primitive equations of atmosphere and applications, *Nonlinearity*, 5 (1992), 237–288.
34. J. L. Lions, R. Temam and S. H. Wang, Mathematical theory for the coupled atmosphere-ocean models, (CAO III), *J. Math. Pures Appl.*, 74 (1995), 105–163.
35. X. Liu and E. S. Titi, Local well-posedness of strong solutions to the three-dimensional compressible Primitive Equations, *Arch. Rational Mech. Anal.*, 241 (2021), 729–764.
36. X. Liu and E. S. Titi, Global existence of weak solutions to the compressible Primitive Equations of atmospheric dynamics with degenerate viscosities, *SIAM J. Math. Anal.*, 51 (2019), 1913–1964.
37. D. Maltese and A. Novotný, Compressible Navier-Stokes equations on thin domains, *J. Math. Fluid Mech.*, 16 (2014), 571–594.
38. N. Masmoudi and T. K. Wong, On the H^s theory of hydrostatic Euler equations, *Arch. Ration. Mech. Anal.*, 204 (2012), 231–271.
39. L. Nirenberg, On elliptic partial differential equations, *Ann. Scuola Norm. Sup. Pisa*, 13 (1959), 115–162.
40. W. M. Pedlosky, Geophysical Fluid Dynamics, Springer-Verlag, 1979.
41. D. Serre, Solutions classiques globales des équations d'Euler pour un fluide parfait compressible, Ann. Inst. Fourier Grenoble, 47 (1997), 139–159.
42. T. Tang and H. J. Gao, On the stability of weak solution for compressible primitive equations, *Acta Appl. Math.*, 140 (2015), 133–145.
43. R. Temam and M. Ziane, *Some mathematical problems in geophysical fluid dynamics*, Handbook of Mathematical Fluid Dynamics, 2004.
44. F. C. Wang, C. S. Dou and Q. S. Jiu, Global weak solutions to 3D compressible primitive equations with density-dependent viscosity, J. Math. Phys. 61 (2020), no. 2, 021507, 33 pp.

MIX
Papier aus verantwortungsvollen Quellen
Paper from responsible sources
FSC® C105338

If you have any concerns about our products,
you can contact us on
ProductSafety@springernature.com

In case Publisher is established outside the EU,
the EU authorized representative is:
Springer Nature Customer Service Center GmbH
Europaplatz 3, 69115 Heidelberg, Germany

Printed by Libri Plureos GmbH
in Hamburg, Germany